D0363745

The Living Land

Agriculture, Food and Community Regeneration in Rural Europe

Jules Pretty

Earthscan Publications Ltd, London

For

Gill, Freya and Theo

and

John and Susan

11690658
UNIVERSITY OF GLAMORGAN
PRIFYSGOL MORGANNWG
Learning Resources Centre

First published in the UK in 1998 by
Earthscan Publications Ltd

Copyright © Jules Pretty, 1998

All rights reserved

A catalogue record for this book is available from the British Library

ISBN: 1 85383 516 1

Typesetting and page design by PCS Mapping & DTP, Newcastle upon Tyne
Printed and bound by Biddles Ltd, Guildford and King's Lynn
Cover design by The John Burke Graphic Design Consultancy
Cover photo © Bryan and Cherry Alexander/Still Pictures

For a full list of publications please contact:

Earthscan Publications Limited
120 Pentonville Road
London N1 9JN
Tel: (0171) 278 0433
Fax: (0171) 278 1142
Email: earthinfo@earthscan.co.uk
http://www.earthscan.co.uk

Earthscan is an editorially independent subsidiary of Kogan Page Limited and publishes in association with WWF-UK and the International Institute for Environment and Development.

'This book is a helpful antidote to some of the guff that is written about the countryside, well argued and based on hard fact rather than sentiment. The essential message is one of hope. We can live better. We can eat well and eat safely. But not if we carry on the way we have been going for the past two generations. We cannot continue to exploit the environment as if there were no tomorrow. If we do not heed the message of this book, there may very well not be – not, at least, for a way of life most of us treasure'

John Humphrys, *Today*, BBC Radio Four

'This is an important book which combines careful analysis with a positive vision of the future – if only the rest of us stop sleepwalking into it'

Jonathan Dimbleby, journalist and broadcaster

'Proves beyond doubt that sustainable agriculture can work to conserve the soil, the environment and rural life itself... a splendid book'

Derek Cooper, *The Food Programme*, BBC Radio Four

'If you don't know what sustainable agriculture is now, you will by the time you finish The Living Land. *Above all, it demonstrates that those still making the case for yet more earth-bashing intensive agriculture, as the only solution to world hunger in the future, are entirely wrong. And hundreds of thousands of farmers around the world are already making a success of the sustainable alternatives'*

Jonathon Porritt, Director, Forum for the Future

'A tour de force of careful research and lateral thinking. Spanning regeneration activity in both developed and developing countries and both urban and rural areas, this book convincingly makes the universal link between the use of participatory planning methods and environmental sustainability. A milestone for participation methodology and essential reading for anyone involved in rural development'

Nick Wates, senior research associate,
The Prince of Wales's Institute of Architecture, London

'A very useful and thought-provoking book'

Professor Gordon Conway, President of the Rockefeller
Foundation, New York and former Vice-Chancellor, University of Sussex

'This is an excellent book; it does that rare thing, giving in equal measure a summary of key problems in our current food system and practical suggestions for significant improvement'

Professor Tim Lang, Centre for Food Policy, Thames Valley University

'This book is very important in the current discussions on agricultural policy reform across Europe. It shows how to internalise the environmental costs without punishing farmers and consumers so hard that farmers go out of business and consumers can't afford the products'

Inger Källander, farmer and President of
Ecological Farmers' Association, Floda, Sweden

'This book is a valuable contribution to the debate on sustainable agriculture and is important reading for anyone wanting to understand the complexity of issues involved'

Marie Skinner, Norfolk farmer and journalist

'The Living Land *is a story of the facts and mishaps that have affected our fragile earth. It points to real alternatives and sustainable ways that can still be put to work. It is fascinating reading; a complex subject but written in a simple language that everyone can understand'*

Peter Eden, farmer, Elvas, Portugal

'A very impressive book – it shows the value of a thriving and sustainable agriculture for the future of the countryside, its communities and its wildlife'

Simon Lyster, Director General, The Wildlife Trusts, London

'A unique and up-to-date book covering all the main aspects of sustainable agriculture'

Professor Adel El Titi, agricultural scientist,
Landesanstalt für Pflanzenschutz, Stuttgart, Germany

'A very interesting book with relevant information for researchers and farmers across Europe'

Philippe Viaux, Institut Technique des Céréales
at des Fourages, Boingeville, France

'A strong political pamphlet in defence of sustainable development'

Gérald Assouline, Theys, France

'A timely and very useful contribution to the current policy debate. It provides an enthusiasm that is lacking in most policy analysis'

Natacha Yellachich, World Wide Fund For Nature (WWF) Belgium

'A valuable contribution to our understanding of rural people and their communities... a fluent read'

James Garo Derounian, Cheltenham and Gloucester
College of Higher Education, Cheltenham

'Stakeholding, local participation, social capital and communitarianism – it's all here'

'Slowly, little by little we are beginning to see the flaws in our systems of living: rampant individualism, apathy, rootlessness and consumerism. A way must be found which helps people to re-engage, where their voice matters and where they are central to behaviour, systems and processes that will in turn lead to stronger, more equitable and sensitive community living. The Living Land *opens that debate'*

Geoff Fagan, University of Strathclyde

'Good stuff! This will be a great help for teaching many courses'

Hugh Ward, Department of Government, University of Essex

'A timely, well-written and thorough publication on the value of localising food systems'

Harriet Festing, Wye College, University of London

'Shows how communities can take back control of the food system with projects and policies that have wider transformative potential'

Rod MacRae, Toronto Food Policy Council, Canada

'For those of us who refuse to be sanguine about the sustainability of our food supply and system, The Living Land *is essential reading'*

'The oligopolistc forces of multinational agribusiness would have us believe that all's well in the foodshed. Jules Pretty tells a different story that should be listened to by anyone concerned with the sustainability of our food system'

Mark Winne, Director, The Hartford Food System, Connecticut, USA

'This book fills a huge void in the literature regarding real-life examples – some large scale, many just glimmers of hope – of how alternative agriculture is challenging what seemed an inevitability: the decline, demoralisation and homogenisation of farms and rural communities'

Bill Vorley, Director, Environment and Agriculture Program,
Institute for Agriculture and Trade Policy, Minnesota, USA

'This book draws on many examples of recent changes in practice in food and farming systems, and points to how these could transform our whole economies'

Andy Fisher, Community Food Security Coalition, Venice, California, USA

Contents

Acronyms and Abbreviations

ACORA	The Archbishops' Commission on Rural Areas
ACRE	Action with Communities in Rural England
ADAS	executive agency formed from the former Agricultural Development and Advisory Service of MAFF
ARG	Agricultural Reform Group
BMA	British Medical Association
BSE	bovine spongiform encephalopathy ('mad cow disease')
Bt	*Bacillus thuringiensis*
BTO	British Trust for Ornithology
bu	bushell
CADISPA	conservation and development in sparsely populated areas
CAP	Common Agricultural Policy
CC	Countryside Commission
CCW	Countryside Council for Wales
CES	Centre for Environment and Society, University of Essex
CJD	Creutzfeld–Jakob Disease
CLA	Country Landowners Association
CPRE	Council for the Protection of Rural England
CS	Countryside Stewardship
CSA	community-supported agriculture
CSO	Central Statistical Office
DAC	Development Assistance Committee of the OECD
Danida	Danish International Development Agency
DBCP	dibromochloropropane
DETR	Department for Environment, Transport and the Regions (formerly the DoE)
DFID	Department for International Development (formerly the British ODA)
DM	Deutschmark
DoE	Department of the Environment (to 1997)
DoW	Duke of Westminster
DTI	Department of Trade and Industry
EA	Environment Agency
EC	European Commission
E coli	*Escherichia coli*
ECU	European Currency Unit (approx £0.70 and $1.40 in 1997)

EFA	Ecological Farmers' Association, Sweden
EPA	US Environmental Protection Agency
ESA	Environmentally Sensitive Area
EU	European Union
FABBL	farm assured British beef and lamb
FABPIG	farm assured British pigs
FAO	Food and Agriculture Organisation of the United Nations
FF	French franc
FOE	Friends of the Earth
FPC	Toronto Food Policy Council
FTE	full time job equivalent
FWAG	Farming and Wildlife Advisory Group
G8	Group of Eight (leading industrialised nations)
GATT	General Agreement on Tariffs and Trade
GDP	gross domestic product
GFB	Good Food Box, Toronto
GJ	giga joule (a measure of energy; 1 joule = 4.2 calories)
GMO	genetically modified organism
GNP	gross national product
GP	General Practitioner doctor
GPS	global positioning system
GT	GreenThumb, New York
ha	hectare (equivalent to 2.47 acres)
HC	House of Commons
HFS	Hartford Food System
HSE	Health and Safety Executive
IACS	Integrated Administrative and Control System
IAFS	Integrated Arable Farming Systems
IBA	important bird area
ICM	integrated crop management
IEEP	Institute for European Environmental Policy
IGER	Institute of Grassland and Environmental Research
IMF	International Monetary fund
IIED	International Institute for Environment and Development
IPM	integrated pest management
IPPR	Institute for Public Policy Research
IRRI	International Rice Research Institute
ISO	International Standards Organisation
ITCF	Institut des Céreales et des Fourages
K	potassium
kg	kilogramme
LEAF	linking environment and farming

LETS	Local Exchange Trading Systems
LIFE	less intensive farming for the environment
Mg	magnesium
MJ	mega joule (a measure of energy: 1 joule = 4.2 calories)
MAFF	Ministry of Agriculture, Fisheries and Food
MOD	Ministry of Defence
mt	million tonnes
N	nitrogen
NACAB	National Association of Citizens Advice Bureaux
NAF	North-West Area Foundation, US
NEPP	national environmental policy plan
NFA	National Food Alliance
NFFO	non-fossil fuel obligation
NGA	American National Gardeners Association
NGO	non-governmental organisation
NHS	National Health Service
Nimby	not in my backyard
Nodam	no development after mine
NPA	National Park Authority
NSA	nitrate sensitive area
NVZ	nitrate vulnerable zone
NYM	North Yorks Moors
ODA	Overseas Development Administration (to 1997: now DFID)
OECD	Organisation for Economic Cooperation and Development
OM	organic matter
OPs	organophosphate pesticides
OPIN	Organophosphate Information Network
OTA	US Office of Technology Assessment
P	phosphorous
PA	Participatory Appraisal
PFI	Practical Farmers of Iowa
PGR	plant growth hormone
PLA	Participatory Learning and Action
PRA	Participatory Rural Appraisal
PSE	producer subsidy equivalent
PVCP	Pang Valley Countryside Project
PYO	pick-your-own operation
RAP	recommended action point
RCC	Rural Community Council
RCEP	Royal Commission on Environmental Pollution
RDC	Rural Development Commission
RSPB	Royal Society for the Protection of Birds

SAFE	Sustainable Agriculture, Food and Environment Alliance
Sida	Swedish International Development Co-operation Agency
SME	small- and medium-sized enterprises
SNH	Scottish Natural Heritage
SPEC	species of European conservation concern
SRB	single regeneration budget
SRC	short rotation coppice
SSB	South Shore Bank
SSSI	Site of Special Scientific Interest
t	tonne (1000 kg)
t/ha	tonnes per hectare
TEC	Training and Enterprise Council
TIBRE	targeted inputs for a better rural environment
UCIRI	Union of Indian Communities in the Isthmus Region
UKROFS	UK Register of Organic Food Standards
USDA	United States Department of Agriculture
VA	village appraisal
VAT	value-added tax
WCED	World Commission on Environment and Development
WHO	World Health Organisation
WI	Women's Institutes
WIC	women, infants and children
WRI	World Resources Institute
WTO	World Trade Organisation
WWF	Worldwide Fund For Nature

A Few Words on Terminology

There is no satisfactory single set of terms for describing broad categories of countries. Where the distinctions once were clearer, now the boundaries are blurred. In this book, the poorer countries of the world are described as 'developing countries' or 'countries of the South'. These terms are not perfect. They encompass a great variety of economies and cultures, ranging from some with per capita GNP of below US$250 to some with over $2000. The richer countries are called 'industrialised countries', 'developed' or 'countries of the North', which are broadly defined as those belonging to the OECD.

There is much overlap between categories. Few would call Singapore a 'developing' country now that its rates of economic growth exceed those of most European countries. North and South work well for many people, but not generally for those in Australasia. Some recent members of the OECD, such as Mexico, belong in many people's minds to 'developing' or Southern country categories. And what was once called a 'developed' country, such as Britain, now has, in some rural districts, three to four out of ten households under the poverty line, whereas a 'developing' country such as India has some rural areas vibrant with community groups and economic activity.

The distinctions between categories will become less clear over time. In this book, I move between these terms fairly interchangeably. It is not a perfect solution. Indeed it is somewhat of a fudge. Generally, it should be clear from the context when reference is made to a Southern, developing country. Where it is not clear, then further notes are made. The same goes for Northern, developed, industrialised, or OECD countries.

Acknowledgements

I am very grateful to many colleagues and friends who kindly gave critical comment and constructive suggestions on chapters of this book or earlier related material. They are Melinda Appleby, Gérald Assouline, Gordon Conway, Peter Eden, Adel El Titi, Geoff Fagan, Tom and Giana Ferguson, Harriet Festing, Andy Fisher, James Garo Derounian, Tara Garnett, Andy Inglis, Inger Källandar, Tim Lang, Simon Lyster, Rod MacRae, Duncan McLaren, James Morison, Jonathon Porritt, Marie Skinner, Philippe Viaux, Bill Vorley, Perry Walker, Hugh Ward, Nick Wates, Mark Winne, and Natacha Yellachich.

Many others, including researchers, farmers, community development specialists and environmentalists, have contributed by providing great insights into their current work. These include David Atte, Ted Benton, Wes Beery, Eric Bignal, Helen Browning, Roland Bunch, Ed Cooper, Caroline Cranbrook, John Devavaram, Caroline Drummond, David Dubois, Eva Tjelke Eckborn, Peter Edling, Thierry de l'Escaille, Wyatt Fraas, Philippe Girardin, Hal Hamilton, Vicki Hird, Vic Jordan, Hal Hamilton, Gary Huber, Ian Hutchcroft, Kiki Kalbutji, James Kelly, J K Kiara, Andrew Kinnear, Andy Langford, Patrick Madden, Chris MacKenzie-Davey, Jimmy Mascarenhas, Miguel Naveso, Matt Phillips, Michel Pimbert, Bruno de Ponteves, Ralph Raistrick, Matt Rayment, Gaston Remmers, Frank Rennie, Nick Robins, Niels Röling, Walter and Dorothy Schwarz, Charles Secrett, Parmesh Shah, Liz Shephard, G W Stokes, Mary Taylor, Stephen Thake, Camilla Toulmin, Ed Tyler, Pieter Vereijken, Tim and Susie Wall, Drennan Watson, Albert Weale, Jane Weissman, the Duke of Westminster, John Wibberley, Frank Wijnands and Michael Wildenhayn. I am grateful to them all. To all those who have inadvertently slipped off the list, I also give my thanks and hope they will forgive the omission.

I have particularly benefited from membership of the Agricultural Reform Group, whose active and vigorous members have, since its foundation in 1993, included Richard Aylard, Ewan Cameron, Charles Clover, Oliver Doubleday, Simon Gourlay, Patrick Holden, Simon Lyster, Jonathon Porritt, Hugh Raven, Marie Skinner, Oliver Walston, the Duke of Westminster and Roger Young, with continued support from HRH The Prince of Wales. In a similar vein, I have also benefited greatly from fellow members of the Neighbourhood Think Tank, convened by Tony

Gibson and including, amongst others, Mary Barnes, Pat Conaty, Brian Davey, Ed Mayo, Mike Roberts, Stephen Thake, John Turner, and Perry Walker. The insights and experience of friends from both these groups is beyond measure.

Research time for this book came from several sources. This was mainly in the form of a direct grant from the Prince of Wales Trust via the International Institute for Environment and Development (IIED), who also gave support and time during my sabbatical at the University of Essex, where the Department of Biological Sciences kindly provided space and support. I am particularly grateful to Richard Sandbrook and John Thompson of IIED and Neil Baker, Steve Long and Dave Nedwell of the University of Essex for making it all possible. Duncan McLaren and Tim Lang played a vital role by asking me to write a review of sustainable agriculture for Friends of the Earth. Helpful administrative support came from both Marie Chan and Hilary Workman.

Throughout the whole process, Jonathan Sinclair Wilson and colleagues at Earthscan have, as ever, been supportive and given valuable advice and insight on many occasions.

It is also important to state the normal disclaimer on behalf of all those mentioned here: any omissions or errors in this book are, of course, solely my own responsibility.

Jules Pretty
University of Essex
January 1998

All comments to the author at: Centre for Environment and Society, University of Essex, Wivenhoe Park, Colchester, CO4 3SQ, UK. Tel: 01206 873 323. Fax: 01206 873 416. Email: jpretty@essex.ac.uk.

Chapter 1

A Living Land for Rural Europe

What are those blue remember'd hills
What spires, what farms are those?
That is the land of lost content,
I see it shining plain,
The happy highways where I went
And cannot come again

A E Housman (1859–1936), from *England*

Recreating a Living Land

The Challenge Ahead

This book is about getting back something we have lost. It is also about creating something new we never had. We value our countryside, our rural landscapes, our wildlife. We value our rural communities and their many idyllic settings. Yet we are still losing many valued features of our natural environments, such as meadows, wetlands, woodlands, birds and other wildlife. Our rural communities are suffering too. There are fewer rural livelihood opportunities and fewer basic services. Hardship and poverty are common.

This book's message is about getting back some of these natural and social aspects of our countryside and rural economies that we value. It is also about getting more from less by using fewer resources. We can live better in more connected communities, we can protect our natural environment, we can eat well and safely. These are simple ideas, but difficult to put into practice. According to some measures, rural communities and farmers throughout Europe are very successful. Farms are more efficient, and food is cheaper and more abundant. But this 'success' has come at some cost. The state of both natural resources and rural societies is vital for our welfare and economic growth. However, as soils become depleted or erode, water is polluted, trees, hedges and other habitats lost, and wildlife threatened; and as trust falls, social institutions are rendered ineffective, and reciprocity and exchange mechanisms lost, so it is increasingly difficult to sustain vibrant farming and rural communities. As these stocks of natural and social capital diminish, it becomes more difficult to make a living from what remains.

Fortunately, it is not all doom and gloom. Throughout Europe and North America, there are initiatives and experiments underway that are not only repairing the damage, but also showing that alternatives are economically viable. Sustainable agriculture works for farmers and consumers. It is also good for wildlife and other natural resources. Food can be produced in adequate amounts for all and at a quality that is both nutritious and safe. Rural communities can take a major role in their own social and economic development.

These initiatives show that there is potential for a large sustainability dividend. Using less resources and less fossil fuel, it is possible to create more wealth. Instead of depleting natural and social capital, these can be regenerated to provide everyone with enriched and varied livelihood opportunities. At the same time as birds are protected, jobs can be created. While soils are regenerated, so rural communities can become more cohesive and pleasant places to live. As less pesticide is used, so food quality improves. And as farming becomes increasingly sustainable, so a greater involvement of different groups in development processes can regenerate local democracy. This alternative vision is also about spreading the benefits from our countryside and its economies more evenly. There are many different groups who have a stake in our rural, countryside and food systems, but returns to these stakeholders differ. Some do very well, others poorly. Large farmers do proportionally much better than small family farmers, even though smaller ones may protect the environment better. Agrochemical companies, food manufacturers, processors and retailers capture much more of the value in the food system than they used to do. As a result, much less of the food pound, franc, mark, or dollar gets back to farmers and rural communities. This unevenness undermines the whole system.

The interesting fact about technologies, processes and policies that produce a more even spread of proceeds is that the whole system benefits. A multiple stakeholder approach produces a bigger and better pie. A community-based approach to development which builds on existing social and natural resources actually produces more jobs and services in rural areas than an externally driven, exogenous approach relying, in the main, on distant technologies and mobile capital. What is now clear is that sustainable agriculture can yield as much or more food than conventional systems, but does so without damaging the natural environment. It also produces more jobs and business opportunities. Community food systems capture more value for local people. Rural partnerships bring together different actors in new networks that develop mutual trust and new opportunities for exchange and reciprocity.

But none of this will happen without a helping hand. Most national and international policies do not, as yet, support a sustainability-led approach to rural development. Many say they do, since sustainability is now in fashion. But, in reality, governments are yet to create the necessary enabling conditions. Nonetheless, some have taken small steps that have already delivered a dividend – a community food security act in the US, a national food and health policy in Norway, green taxes on pesticides and fertilisers in various European countries, pesticide reduction policies in The Netherlands, and the gradual switching of payments under the Common Agricultural Policy (CAP) from food production to environmental goods and services. However, no industrialised country has a national policy framework that properly integrates these various challenges and needs. In addition, some of these advances are coming under pressure as countries have to make their internal policies compatible with the World Trade Organisation (WTO) and General Agreement on Tariffs and Trade (GATT) agreements.

Rural Europe and the Rest of the World

This book takes Europe as its primary focus and draws mainly upon recent experience throughout the industrialised countries of Europe and North America. It is intended to complement an earlier book of mine entitled *Regenerating Agriculture: Policies and Practice for Sustainability and Self-Reliance*, which almost entirely focused on sustainable agriculture and rural development in Africa, Asia and Latin America.

At first glance, it may appear that there is little commonality between India, Kenya or Honduras with Britain, Sweden or Spain. But the similarities of challenge and opportunity are remarkable. Of course, their national economies, political systems and social institutions are structured in different ways. But in local communities, the needs, aspirations and hopes of individuals and their families are parallel. In many ways, the improvements brought about by community-based activities in developing countries are far more advanced than those in the North. There is much to learn from this experience.

In Bangladesh and elsewhere in South Asia, the Grameen Bank system of small-scale credit has brought dignity and better welfare to hundreds of thousands of poor women and their families. Soil conservation and organic farming programmes in Kenya, working in a participatory fashion with local people, have improved environments, farm yields and diversity of produce for some 200,000 farming families. In the south Brazilian states of Paraná and Santa Catarina, there have been massive increases in agricultural productivity with locally adapted and sustainable technologies for some 300,000 farming families. In Central America, 45,000 farmers in Honduras and Guatemala have tripled agricultural yields by adopting collective approaches to sustainable technology development.

In South and South-East Asia, farmer-field schools have helped some 800,000 farmers to learn systematically about the ecology of their rice fields and farms, leading to new technology development and substantial cuts in pesticide use. Despite the cuts in inputs, rice yields have increased by a small amount, perhaps 8 to 12 per cent, but the great benefit has been the extra fish, frogs and vegetables that can now be raised and grown with the rice. In Bangladesh, rice farmers now say 'our fields are singing again' – for 30 years they have produced only rice; now productivity has grown, but on a sustainable basis.

These are only a small selection of some recent remarkable achievements. Clearly, no single country has all the answers, whether in technology, knowledge, processes or policies. But knowing that agriculture can be both sustainable and productive at one location encourages us to identify ways to help it emerge elsewhere.

What Went Wrong in Rural Europe?

From the middle of the 20th century, most governments in Europe began to adopt policies that encouraged their farmers to produce much more food. Visions of hunger and concerns about low self-sufficiency pushed through a wide range of changes. Farmers responded by massively increasing output. Between the 1940s and 1990s, the amount of food produced on European farms grew at an unprecedented rate. Compared with the 1940s, we now produce three times the amount of wheat and barley per hectare and more than twice as much potato and sugar beet; cows now produce twice the milk per lactation.

These improvements were both extraordinary and historic. Never had such rapid growth occurred – perhaps it never will again in industrialised countries. Unfortunately, they also came at considerable cost. In the process of increasing output with greater use of non-renewable inputs, we have lost hedgerows, ponds, woodlands and other natural habitats; soils have been depleted; water polluted with pesticides and fertilisers; human health damaged by pesticides; and birds and other wildlife have been lost. In the process of making agriculture more efficient, commonly measured by the amount produced per unit input of labour or land, farms have become larger and fewer in number. Jobs have gone, and farming has become increasingly isolated both within rural communities and from its clients – the food consumers. And in the process of producing ever more food with ever increasing 'efficiency', we have managed also to produce some foods and food products that are harmful to our health. Yet in rural areas, natural and social environments are bound closely together. We cannot have one without the other. It is farming that has given almost every landscape, community and environment throughout Europe its distinctiveness. It is also modern farming that has taken it away.

Over the past ten years, we have seen a series of major public health and countryside issues come to dominate the food and agriculture industry. We now have the bovine spongiform encephalopathy (BSE) scare in beef and probable links to Creutzfeld–Jakob Disease (CJD); the effect of organophosphate pesticides on sheep farmers and other agricultural workers; the cloning of sheep; conflicts between landowners and the public over access to the countryside; poisonous strains of bacteria, such as *Salmonella* in eggs, *Listeria* in soft cheeses, and *E coli* in cooked meats; animal welfare concerns, particularly over young cows in veal crates and the live transport of sheep and cattle; residues of pesticides in our foods and drinking water; and genetically modified organisms entering the food chain. All of these have steadily undermined the confidence which the public once had in the farming industry.

Biotechnology and the genetic modification of organisms is an emerging area of concern for many. Biotechnology involves making molecular changes to living or almost-living things such as proteins; genetic engineering is the horizontal transfer of DNA from species to species. In agriculture, the first generation changes have been to incorporate genes into crops that confer resistance to a particular herbicide, or that encourage plant cells to produce materials that harm insects. In 1994, there were no genetically modified crops grown anywhere in the world. By 1997, more than 12 million hectares of genetically modified crops were cultivated, mainly in the US, and involving a quarter of all cotton, 14 per cent of soya and 10 per cent of maize.

Despite the many alleged biotechnology goods, such as the potential for reduced pesticide and fertiliser use, and the many human health benefits from 'molecule pharming' – the production of proteins such as insulin and blood-clotting factor in the milk of genetically-modified sheep and pigs – there are many concerns over the potential biotechnology bads (see Chapter 4). The greatest concerns are over genetic pollution, where genetically modified crops cross with wild relatives, transferring the genes into the environment. No one knows what will happen in the long term. Already resistance genes in oil seed rape are found in France to have become incorporated into wild radish. In Scotland, potatoes containing snowdrop genes that confer resistance to peach potato aphid have been found to reduce the lifespan of female ladybirds by half. Other concerns are over the potential spread of antibiotic resistance to humans from the antibiotic markers used in these crops.

Does It Matter?

What we have gained with one hand has been taken away with the other. Should we be concerned? Surely cheap and abundant food is more important than a few songbirds, or a few lost trees and ponds? Rural people have enough of the good life anyway. Why should they receive special treatment?

Remarkably, a third to a half of urban people in Britain want to move to the countryside. Most hold dear some notion of a rural idyll. They would be surprised to learn that it rarely exists. Wildlife habitats have disappeared. Birds are fewer in number. Freshwater is polluted. And while some families live in thatched cottages, with glorious flower and vegetable gardens, a quarter of rural households are under the official poverty line. A third or more parishes have no shop, pub or village hall; one half have no school; only a third still have a public transport link, even though one in eight households have no car at all. Rural communities are under growing threat across Britain. Elsewhere in Europe, particularly in southern countries, it is even more severe, with land being abandoned, families emigrating, and a growing loss of the distinctive character of farmed habitats and their valued wildlife.

Farming in Britain currently consumes more than £3 billion (some 4.3 billion ECUs) of direct subsidies every year. The Common Agricultural Policy (CAP) cost the whole EU more than 40 billion ECUs per year in the mid 1990s. Yet, at the same time, there are 18 million unemployed in Europe – few of whom have assets or the means to make a living, unlike most farmers. Farm subsidies are also inequitably spread. Big farmers receive the largest slice of the pie. Just 12 per cent of all farms produce 60 per cent of all agricultural output and, in the EU as a whole, 1 per cent of farms rear 40 per cent of all animals. Do taxpayers get good value for this money, or might they wish for something different?

Farming is at a crossroads. The simplest path to take is 'business as usual', with farming becoming more and more 'efficient' but less and less linked to rural communities, the environment and the urban consumers of food. If farming were to lose its public support, then it would have to make a straightforward transition. Let us have a relatively or even completely unsubsidised farming industry which competes at world market prices, with a few dominant companies farming huge fields and running giant livestock operations. In another direction, however, there lies something quite different. This future requires some imagination, even leaps of faith. It implies a future with dynamic rural economies, with sustainable systems of food production, and with cohesive rural communities. Here are places where stocks of natural capital and social capital are high, and where people like to live. As we shall see, it does not cost more to go down this route. Far from it. There is even a substantial dividend to be had. But it will be difficult. History constrains us all, and we may be unable to break free. This book sets out some of the methods for grasping the opportunities before us.

The Value of Natural and Social Capital

Aspects of Capital

We are all familiar with the term capital. It refers to the stock of materials or information that exists at a given point in time. Most commonly, we think of it as the amount of money we have, either tied up in buildings, land, cars or jewellery, or saved in a bank or pension fund. Each form of capital stock generates a flow of valuable services that may be used to transform materials or the way they interact to enhance human welfare. The wellbeing of people in a rural society depends, therefore, on the value of services flowing from the total capital stock existing in their local economy.

Although capital stock and assets take many forms, we have tended to undervalue two critical types. These are natural capital and social capital. Some kinds of capital are easy to value. We know how much a house or a car is worth. Others, though, are more difficult to assess. How much is a hedgerow worth? A cohesive rural community? Organic matter in the soil? It is difficult to answer these questions, even if we ask them in the first place.

Natural capital refers to the stocks of plants and animals and the ecosystems they make up: minerals, atmosphere and water. These stocks of capital create 'services' that comprise flows of material, energy and information which we can combine with manufactured and human capital to produce welfare. Natural capital is vital: it is difficult to imagine generating welfare without it. Although it is impossible to give an absolute value to some capital stocks – the atmosphere, for example, has infinite value to us – it is instructive to see how much the services that come from the capital are worth. Robert Costanza of the University of Maryland and 12 colleagues' study (Constanza et al, 1997) of the value of the world's ecosystem goods and services – which include water regulation and supply, climate regulation, nutrient cycling, soil formation, waste treatment, wild food production, biological control of pests, and recreation – estimate it to be in the range US$16–54 trillion per year. The best guess of $33 trillion is almost twice the global gross national product of US$18 trillion.

There are three features of natural capital that are important: stocks, ecological relationships and diversity. Levels of capital stocks are important, since they determine the amount of ecosystem services we can derive. Ecological relationships and processes, such as predation, herbivory, competition, nitrogen fixation and water absorption by soils, are important as they define the exchanges between the different elements of natural capital. Diversity is vital since it is a measure of horizontal and vertical connectedness in ecosystems.

Social capital is an equally fundamental basis for economic growth. It lowers the costs of working together (the transaction cost) and so facilitates cooperation between people. One of the first people to use the term social capital, James Coleman (1990), has described it as 'the structure of relations between actors and among actors' that encourages productive activities. There are aspects of social structure and organisation that act as resources that people can use to realise their personal interests. Robert Putnam (1993) and colleagues' study of Italian society has suggested that the greater economic success of the north compared with the south of the country has come about because of the greater social capital:

> *...norms of reciprocity and networks of civic engagement have been embodied in tower societies, guilds, mutual aid societies, co-operatives, unions and even soccer clubs and literary societies. These horizontal civic bonds have undergirded economic and institutional performance generally much higher than in the south, where social and political relations have been vertically structured.*

Drawing mainly on the work of Coleman, Putnam, Elinor Ostrom and Michael Taylor, it is possible to identify four central aspects of social capital (Putnam, 1993; Coleman, 1990; Ostrom, 1990; Taylor, 1982). These have close echoes in the aspects of natural capital, and are: trust; rules and sanctions; reciprocity; and connectedness. Trust is important as it lubricates cooperation and reduces transaction costs. There are two types of trust: the trust we have in individuals whom we know; and the trust we have in those we do not know, but which arises because of our confidence in a known social structure. It takes time to build but is easily broken. Rules and sanctions are the mutually agreed or received norms of behaviour that place group interests above those of individuals. They give individuals the confidence to invest in collective or group activities, knowing that others will also do so. Individuals can take responsibility and ensure that their rights are not infringed. Mutually agreed sanctions ensure that those who break the rules know they will be punished.

Reciprocity and exchanges increase trust. There are two types of reciprocity. Specific reciprocity refers to simultaneous exchanges of items of equal value; diffuse reciprocity refers to a continuous relationship of exchange that at any given time is unrequited, but over time is repaid and balanced. Connectedness, networks, and civic engagement of all types are vital for social capital formation and maintenance. This may be of many types – from guilds and mutual aid societies, to soccer clubs and credit groups, to forest, fishery or pest management groups, to literary societies and mother and toddler groups. Connectedness can be horizontal or vertical, though horizontal bonds and networks are the most vital for effective institutional performance. New platforms for horizontal collaboration and cooperation are commonly needed for the sustainable management of natural resources.

Common Principles of Natural and Social Capital

These two types of capital share important common features. Both provide the basis for economic growth and enhanced human welfare, supplying vital services that people can use. Unlike conventional capital, both tend to be public goods, and so rarely have a market value. Both social and natural capital are actually more often like club goods, in that they are indivisible, and outsiders are often excluded from the benefits. Like all public goods, they are often undervalued and undersupplied by private individuals. When they decline, it is difficult to say who is at fault. They are both diminished by externalities which arise from the activities of individuals or institutions, such as factories or farming. These costs do not accrue to the producers of the costs, but are borne by whole societies and ecosystems. Both, however, are also special kinds of capital in that they can increase with use. Under certain circumstances, the more they are used, the more they regenerate. Natural capital is augmented if regenerative technologies are used that give a return while improving the capital stock. Social capital is also self-reinforcing when exchanges and reciprocity increase connectedness between people, leading to greater trust and confidence.

Equally, though, both types can be rapidly diminished with the wrong kind of social and economic development approach. Whales and trees, for example, reproduce themselves so slowly that the most profitable strategy for harvesting them is to convert them all into money which can be put in the bank. Natural capital may be lost, but financial capital is not. Modern agriculture can diminish natural capital stocks of buffering insects, soil organic matter, water quality, hedgerows, while continuing to enhance financial capital in the short term. The exertion of individual rights without regard for others can reduce social cohesion in communities.

According to some economists, critical capital does not have any substitutes. This implies that no depletion of, or damage to, these stocks can be compensated for by increases in other capital stocks. In many places, they have been lost or eroded away, even to the point where there is no hope of recovery. But an important message of this book is that natural and social capital can be regenerated, recreated and reinforced. This does not justify development processes that degrade natural or social capital, just because recovery is possible. But if it has happened, as it has in every country worldwide, then some sort of restoration must occur.

How do we rebuild critical natural and social capital? There are three entry points: making farming more sustainable (Part I); improving the sustainability of food systems and adding local value for communities (Part II); and improving the participation of rural people in their own development, thereby increasing connectedness between different groups, agencies and organisations (Part III). Each of these entry points can be multifunctional. Improving the sustainability of farming would have a clear impact on natural capital, but it would also contribute to social capital since it embodies greater cooperation amongst farmers and between farmers and local communities. If these farmers also spend more in local communities on goods and services, then this helps to create local financial capital by ensuring that money recycles within communities several times before leaking out. Improving the connectedness between consumers and farmers leads to greater biodiversity as farmers respond more easily to consumer demand for diverse foods.

Who Are the Stakeholders and Who Gets the Value?

An issue as important as the state of our natural and social capital is who gets the value from these assets. We all, to a certain extent, derive benefit from our farming and rural systems. But could we, as a whole, get more value?

The food and drinks system is one of the largest industrial and commercial sectors in the world, with global production amounting to some US$1500 billion per year. Half a century ago, at least half of the pound, franc, mark or dollar spent on food found its way back to the farmer and the rural community (see Chapter 5). The rest was spread amongst suppliers of various inputs (feeds, pesticides, fertilisers, seeds, machinery, labour and so on) and manufacturers, processors and retailers. Since then, the balance of power has shifted increasingly away from the middle, with value concentrated on the input side by agrochemical, feed and seed companies, and on the output side by those who move, transform and sell the food. Food consumers have benefited as the real cost of food has fallen. There is now greater choice, and processed foods increasingly reduce preparation and cooking time – though, ironically, as we spend more time travelling to shops, the total time spent on food has grown (Raven and Lang, 1995). But as far as farmers and rural communities are concerned, the effect has been largely negative. Farmers get a smaller share, typically only 10 to 20 per cent, and they also pass on less. They spend less in rural communities and they employ fewer local people. Tens of millions of jobs have been lost in farming throughout western Europe and North America in the past 50 years. And farms continue to be abandoned and farm labour laid off.

There are eight basic types of individuals and groups who have a stake in the farming and food system. These stakeholders are:

- input suppliers delivering physical inputs of pesticides, fertilisers, feedstuffs, seeds and machinery; financial inputs of capital; energy inputs of electricity and fuel; and knowledge inputs from research and extension services;
- farmers engaged in the business of food production and management of natural resources;
- rural people and their communities who live in the farmed landscape and who may work in the farming and food system, but who increasingly have no direct contact with farming, save for experiencing the amenity value of the countryside;
- manufacturing, processing, trading and retailing companies and their shareholders, who take farm produce from the farm gate, and move it, transform it and sell it;
- food consumers, both urban and rural, who increasingly buy food from large retailers;
- countryside visitors and tourists who also value the countryside for landscape, wildlife and other services;
- environmental organisations who act on behalf of the natural environment and wildlife;
- governments, both local and national, which act on behalf of the public at large and set policies that affect all stakeholders.

These stakeholders are connected by webs of mutual rights and responsibilities, and shared understanding, values and exchange systems. But many of these relationships have been weakened during the modernisation project. Value has been captured by fewer stakeholders in the system, leading to diminished contact between and within each group of stakeholders. Declining reciprocity leads to a fall in trust and confidence, further strengthening the vicious circle. As connections are weakened, understanding quickly gives way to distrust and suspicion. Some farmers no longer trust environmentalists because they feel they are trying to limit their freedoms and their right to farm as they wish. Some urban consumers no longer trust farmers to produce wholesome and safe food.

What Happens when the Value is Captured by the Few

The positive spiral of reciprocity, a feature of communities with abundant social capital, has been identified in the business sector as a result of a more balanced stakeholder approach (Hampden-Turner, 1996). Wealth is created when employees, shareholders, customers, communities and governments work together. Satisfying all these agents does not mean that one group has to suffer unduly. However, the past two decades in particular, coinciding with the individualistic dogma that has dominated Europe and North America, have seen huge imbalances emerge. In the 1980s, governments sought to crush trade unions and put power into the hands of companies and their shareholders. But now shareholders have become too strong, and financial institutions with great power have been transferring the rewards of industry away from customers, employees and governments towards themselves.

Shareholders get more from three sources. Companies provide poorer services, in effect now competing with customers. There are takeovers and acquisitions. And there is 'downsizing' – a jargon term for making staff redundant. Downsizing is usually followed by a leap in share price, which pleases shareholders. However, evidence from the American Management Association shows that nearly 60 per cent of 547 downsizers failed to improve profits, had to rehire staff within a year, had more complaints from customers, and had more of their remaining staff suffer stress-related illness (Hampden-Turner, 1996).

In Britain, when one utility company cut 25,000 jobs after privatisation, there was a 150 per cent increase in complaints from customers. When a conglomerate acquired a famous battery company in the early 1980s, it closed the research and development department, shedding 900 jobs; when it sold the company to a US rival in 1993, its market share had fallen from 80 to 30 per cent (Trapp, 1995). By contrast, the successes during the early 1990s of the 'Asian tigers' appear to have been founded on corporations' willingness to invest in internal social capital, with their staff repaying them in ever-improving performance. The reciprocal benefits escalate and mutually reinforce each stakeholder group.

From this and much more evidence a realisation is growing that the apparently lowest-cost solution is not necessarily the best: hidden costs eventually do appear. The agriculture industry has sought the lowest cost solution in the name of increasing efficiency. Farming has cut jobs and investment in natural capital; it has externalised as many costs as possible. It has been very successful according to narrow food production measures, but it has simply shifted these costs onto others in the system. If we could properly measure the value of social and natural capital, balance sheets would look quite different.

This is also the basic argument made by Will Hutton in his book *The State We're In* (1995). His central argument is that the weakness of the British economy originates in the financial system: 'the targets for profit are too high and the time horizons too short'. Everything is geared to quick returns at all costs, and the result is a lack of investment in human capacity, in innovations and in long term benefits. This has resulted in a whole series of exclusions that are at the core of societal disorders – exclusions of people from a decent standard of living; exclusions of groups from political power; and exclusions of individuals and groups from the major benefits, unless you are a company shareholder. If we are to see sustainable development, then value will have to be spread more evenly amongst different stakeholders.

Individual Rights and Collective Responsibilities

Is there any likelihood of this value being spread more evenly amongst stakeholders? Have things gone so far that we no longer care? Some would say that if our farming and food system is a consequence of greater individual freedoms, and the result of governments removing the constraints from individuals and companies, then so be it. We each have the opportunity as individuals to capture some of the economic benefits in our societies. If we fail so to do, then that is our own fault.

Amitai Etzioni (1995), one of the founders of the communitarian movement, called the 1980s the 'I' decade, since it was characterised by increased individual and institutional freedoms, freedom from state intervention, privatisation and deregulation, and 'unbridled self-interest and greed'. The modernist project has undermined social capital by disrupting the balance between rights and responsibilities, created new external dependencies, weakened communities and undermined the role they play in healthy society, and allocated value to a limited number of stakeholders. This is the age, as another American academic, Francis Fukuyama (1992), put it, of 'perfect rights and defective duties'. We are now all too ready to claim our rights, but less and less willing to grant that the universal claiming of all rights may reduce the capacity of someone else to get their rights. Collective responsibility has gone. Individual rights are in. This has brought security to some, but has deprived many others.

The former British Prime Minister Margaret Thatcher declared in the 1980s that there was 'no such thing as society'. The market would deliver all that every individual would need, and if it did not, then we were urged to 'get on our bikes' to find that opportunity, as another minister famously put it. Yet the market does not deliver everything. It fails to value natural capital, and so allows development to pollute or destroy wildlife, habitats and groundwater. It fails to value social capital, and so allows our communities to fracture and fall apart, leading to growing disillusionment, stress, vandalism and crime. But many are fearful of moving away from the apparent freedoms of individualism. Community and comanagement mean working with others and giving up a little freedom for the greater good. Have we lost this capacity? The ultimate goal of totalitarianism, as Fukuyama indicates, is not to deprive someone of freedom, but more subtly to make people fear freedom in favour of security and the status quo. They would rather stay coerced than take responsibility to break out onto a new path. When something does not happen – the road outside our home is unrepaired, or the local pub or shop closes – we all too easily blame others, or wait for them to do something. We wait for the local council or national government to act. And when they do not come, we blame them. We feel rightfully indignant – we pay our taxes and so expect them to do their job properly. But we are stuck in a dependency deadlock. External institutions have neither the capacity nor the resources to do everything for us, yet we have almost entirely become dependent upon them.

Nevertheless, there has to come a time when we see through this fog. We can do things differently if only we can find ways of working with others in the same predicament. The country with one of the most vibrant community development sectors in the world, in my opinion, is Nigeria. Every rural community has groups working together on delivering health, water, education and farming services. Villagers who have migrated to urban centres, or even to other countries, stay networked and send back money to finance development in their village. One reason why this has happened is precisely because people know that government has failed them. And if the government will not do it, then people will have to do it themselves. It may not be fair, but it is logical.

This is not to suggest that governments should be removed as key stakeholders. It simply illustrates that where the state is no longer capable, through centralisation, corruption, decadence or overburdening bureaucracy, to deliver services, then people simply have to do it themselves.

A Broader Responsibility in Agriculture?

Where does that leave us now? In Europe we have evolved a strong sense of entitlement coupled with a weak sense of obligation. People want the freedom and licence to do as they please, but tend not to participate with or serve others. Etzioni has argued that there is an urgent need to rebuild a sense of both personal and social responsibility. But this call for increased social responsibility is not a call for curbing rights since 'strong rights presume strong responsibility'. This right to engage in individual enterprise at all costs has meant that many have benefited – particularly the better-off. But many have lost out too. We see less and less civic participation throughout Europe and North America and a growing social decay. We see more economic growth, and yet 18 million people are still unemployed in the European Union.

Asserting one's rights without considering those of others is morally wrong. But as Etzioni put it: 'soon "I can do what I want as long as I do not hurt others" becomes "I can do what I want, because I have a right to do it"'. In the context of farming, this translates as 'I can apply pesticides because I have a right to do so'; or 'I can grub up hedgerows because they reduce the efficiency of my machinery operations'; or 'I can feed ruminant animals the remains of other animals because it is cheap'; or 'I can employ as few people on the farm as I want, because I have a right to do so'. All of these rights, though, affect others. They are rights asserted without due regard for other stakeholders.

Let us take the example of pesticides. These are rigorously tested by companies and only given a licence by the UK Government's Advisory Committee on Pesticides when the company concerned can persuade a group of eminent scientists that the products are safe and harmless when used at specified doses. This regulatory system has, however, broken down on many occasions – often because the possible harmful effects of a particular product are simply unknown or not predicted. Examples of this breakdown include organophosphate sheep dips in the 1980s to 1990s or organochlorines in the 1950s and 1960s. Morbidity and mortality from pesticides is still very high, despite all the increased scientific knowledge and statements of safety (see Chapter 2). But the question a user should ask is, although I have a right to use product X (it has been passed by the government, after all), does this right affect anyone or anything else? If so, are there alternative practices that would mean the rights of others (people as well as wildlife) are not damaged?

Fortunately, there are now alternatives. It could be argued that during the height of the modernisation project, farmers did not have alternatives or did not know of them. Organic agriculture was seen as a fringe activity. But there are no excuses today. Farmers have many more sustainable options open to them. Organic farming is well proven, and there are now some 50,000 organic farms in Europe. There are also intermediate alternatives that use far fewer pesticides and other purchased inputs per hectare than conventional farming at its modernist peak. In developing countries, at least two million farmers have sustainably increased their food production by incorporating a huge diversity of resource-conserving technologies.

With this type of stakeholder approach, farmers can correctly expect other stakeholders to act in a responsible way that does not damage farming. Ramblers and walkers should be permitted access to footpaths, but not to abuse those rights by, for example, permitting dogs to scare animals or run in crops. In turn, farmers should accept the responsibility to produce the best possible food and the best possible natural environment.

Towards Sustainable Agriculture, Food Systems and Rural Development

A central theme of this book is that changes can be made to farming, food systems and rural economic development that will deliver a substantial 'sustainability dividend' to the system as a whole, and to certain stakeholder groups in particular. The following sections of this chapter contain summaries of Chapters 2 to 3 (on sustainable agriculture), of Chapter 4 to 5 (on food systems), and of Chapters 6 to 7 (on rural development and participatory processes).

Sustainable Agriculture

Agricultural and rural development policies have in the latter part of the 20th century successfully emphasised external inputs as the means to increase food production. This has produced remarkable growth in global consumption of pesticides, inorganic fertilisers, animal feedstuffs, and tractors and other machinery. These external inputs have, however, substituted for natural control processes and resources, rendering them more vulnerable. Pesticides have replaced biological, cultural and mechanical methods for controlling pests, weeds and diseases; inorganic fertilisers have substituted for livestock manures, composts, nitrogen-fixing crops and fertile soils; information for management decisions comes from input suppliers, researchers and extensionists rather than from a combination of local and external sources; and fossil fuels have substituted for locally generated energy sources. What were once valued local resources often now have become waste products.

The basic challenge for sustainable agriculture is to make better use of available natural and social resources. This can be done by minimising the use of external inputs, by utilising and regenerating internal resources more effectively, or by a combination of both. This ensures the efficient and effective use of what is available, and ensures that any dependencies on external systems are kept to a reasonable minimum. Sustainable agriculture offers many opportunities to integrate a wide range of economic, social and environmental concerns in the countryside. Farming does not have to produce its food by damaging or destroying the environment: resource-conserving technologies allow farmers to be productive and earn a living while protecting the landscape and its natural resources for future generations. Farming does not have to be dislocated from local rural communities: sustainable agriculture, with its need for increased knowledge, management skills and labour, offers new upstream and downstream work opportunities for businesses and people in rural areas. Farming does not have to be a 'closed-doors' industry, with the public and food consumers peering in from the outside and wondering, suspiciously, what is going on; sustainable agriculture seeks to bring consumers and farmers closer together, both to ensure the quality of the food and to increase understanding, trust and support.

Chapters 2 and 3 show how resource-conserving technologies can be hugely beneficial for both farmers and rural environments. Such technologies make a positive contribution to natural capital, either returning it to predegraded levels or improving it yet more. At the same time, direct costs are reduced. There is more from less. More natural capital from fewer external inputs. More social capital by involving farmers and rural communities in change. More food output from fewer inputs. The best example of reforming farming systems to incorporate sustainability goals comes from a wide range of countries in Africa, Asia and Latin America (Pretty, 1995a; Hinchcliffe et al, 1996). Here a major concern is to increase food production in the areas where farming has been largely untouched by the modern packages of externally supplied technologies. In these lands, farming communities which adopt regenerative technologies have substantially improved agricultural yields, often using few or no external inputs. In the higher-input systems where the Green Revolution of the 1960s and 1970s had such an impact on food output, particularly of rice, the dividend comes from a reduction in pesticide use while slightly increasing yields. Most of these successes are community-based activities that have involved a complete redesign of farming and other economic activities at local level with the full participation of farmers and local people in the process (see Chapter 7).

A wide range of more sustainable forms of agriculture are now emerging and spreading in Europe and North America. Many different terms are used, including, in no particular order, organic, sustainable, alternative, integrated, regenerative, low-external input, balanced input, precision farming, targeted inputs, wise use of inputs, resource conserving, biological, natural, eco-agriculture, agro-ecological, biodynamic, and permaculture. Some of these have precisely defined standards; more do not. Some make substantial contributions to natural capital; others very little. What is clear, however, is that these are all forms of *more* sustainable agriculture. But it is difficult to say how much more sustainable one is than another unless we know about the specific local conditions and technologies in use.

Some important misconceptions persist about sustainable and regenerative agriculture (Pretty, 1995a). The most common characterisation is that sustainable agriculture represents a return to some form of low-technology, backward or traditional agricultural practices. This is manifestly untrue. Sustainable agriculture involves the incorporation of recent innovations that may originate with scientists, with farmers, or both. Some are very high technology; some are practices proven over thousands of years. It is also commonly stated that any farming using low or lower amounts of external inputs can only produce low levels of output. This is not true for two reasons. Firstly, many sustainable agriculture farmers show that their crop yields can be better than or equal to those of their more conventional neighbours. Most show that their costs can be reduced. In developing countries, this offers new opportunities for economic growth for communities that do not have access to, or cannot afford, external resources. Secondly, sustainable agriculture produces more than just food – it significantly contributes to natural and social capital, both of which are sometimes difficult to measure and cost.

Despite recent changes to modern agriculture, there are still millions of hectares of land farmed in environmentally sensitive and low-intensive ways throughout Europe. Their extent dwarfs recent transitions to sustainably intensified agriculture. These systems contribute to local communities and economies in important social ways. Most are also relatively unintensive, unchanged or unimproved. More than half of Europe's highly valued habitats for wildlife occur in low-intensity farmland (Bignall and McCracken, 1996). In a recent study of low-intensity agriculture in nine countries, Eric Bignall and Davy McCracken indicate that there are some 56 million hectares of farmland in this category, representing some 38 per cent of all utilisable agricultural land (see Table 3.2). Many are under threat from modern farming methods and from rural abandonment (see Chapters 2 and 6).

Organic farming is one form of sustainable agriculture in which maximum reliance is put on self-regulating agro-ecosystems, locally or farm-derived renewable resources, and the management of ecological and biological processes. The use of external inputs, whether inorganic or organic, is reduced as far as possible. In some countries, organic agriculture is known as ecological or biological agriculture. In Western Europe, there have been dramatic increases in organic agriculture in recent years. The extent has increased tenfold from about 120,000 hectares in 1985 to 1.2 million hectares in 1996 (Lampkin, 1996). Nearly 50,000 farmers are now engaged in certified organic farming. In the UK, there were about

820 organic farms in 1997, with an area of 47,900 hectares.

Years of research across Europe into integrated farming systems have seen integrated farms emerge as another type of more environmentally friendly agriculture. Once again, the emphasis is upon integrating technologies to produce site-specific management systems for whole farms, incorporating a higher input of management and information for planning, setting targets and monitoring progress. There are important historical, financial and policy reasons why still relatively few farmers have taken the leap from 'modern' high-input farming to organic agriculture. But it is possible for anyone to take a small step which can, in theory, be followed by another step. Integrated farming in its various guises represents a step, or several steps, towards sustainability.

It is difficult to generalise about what happens to outputs. It used to be thought that more sustainable agriculture, whether organic or integrated, would mean reductions in crop and livestock yields (Pretty and Howes, 1993). However, this generalisation no longer stands. It appears that farmers can make some cuts in input use (at least 10 to 20 per cent) without negatively affecting gross margins. By adopting better targeting and precision methods, there is less wastage and so the environment benefits. Yields may fall initially but will rise over time. Farmers can then make greater cuts in input use (20 to 50 per cent) once they substitute some regenerative technologies for external inputs, such as legumes for inorganic fertilisers or predators for pesticides. And, finally, they can replace some or all external inputs entirely over time once they have adopted another type of farming characterised by new goals and technologies. Sustainable agriculture farmers get better over time.

If all these good things are already happening across Europe and North America, what can be done to encourage a further transition to sustainability? One way of reconciling some of the differences is to regard sustainable agriculture as a series of steps. The transition towards sustainable agriculture is often seen as requiring a sudden shift in both practices and values. But not all farmers are able or willing to take such a leap. However, everyone can take one small step. And small steps, added together, can bring about big change in the end. Drawing on the work of Rod MacRae, Stuart Hill and colleagues in Canada, four steps have been conceived of in this process, three of which are on the path to sustainability (adapted from MacRae et al, 1993). Step 0 is, of course, conventional modern farming. The other three incorporate changes in economic and environmental efficiency (step 1); changes in the integration of regenerative technologies (step 2); and wholesale redesign with communities (step 3). The number of farmers in each of these steps is best expressed as a pyramid. The base is step 0; the apex is step 3. Most farmers worldwide are still part of step 0. The challenge is to find a wide range of ways to help invert the pyramid – putting the great majority of farmers and rural communities at the redesign and regenerative end of the spectrum.

How can more farmers be encouraged to begin the transition through the three steps of sustainability? The first thing to note is that sustainable agriculture should not prescribe a concretely defined set of technologies, practices or policies. This would only serve to restrict the future options of farmers. As economic, ecological, climatic and social conditions change, so farmers and communities must be encouraged and allowed to modify their practices. Sustainable agriculture is, therefore, not a simple model or package to be widely applied or fixed with time. It is, rather, a process for social learning with targets to measure progress. Farmers must therefore invest in learning. Lack of information and management skills is a major barrier to adopting sustainable agriculture. We know much less about these resource-conserving technologies than we do about the use of external inputs in modernised systems. In addition, much less research on resource-conserving technologies is conducted by research institutions. It is clear that the process by which farmers learn about technology alternatives is crucial. If they are enforced or coerced, then they may adopt for a while, but will only continue if their incomes are dependent on the technology. Cross-compliance and green conditionality of this type may only buy grudging support, or indeed none at all. But if the process is participatory and enhances farmers' capacity to learn about their farms and their resources, then it appears that the foundation for redesign is laid (Pretty, 1995b).

Policies, though, are still slowing the transition, and there is a need for fundamental reform to increase the incentives for more sustainable agriculture and to reduce the distortions that still give an unfair advantage to modernist farming. Reforms in the food system will help farming become more sustainable.

Sustainable Food Systems

The goods and services produced by the natural and social capital in rural areas are extremely valuable. We all, to a certain extent, derive benefit from our farming and rural systems. But can we, as a whole, get more? In the middle of the 20th century, about half of the money spent on food found its way back to the farmer and rural community. The rest accrued to suppliers of various inputs (feeds, pesticides, fertilisers, seeds, machinery, labour and so on) and manufacturers, processors and retailers. Since then, the middle has steadily lost out, with value increasingly captured on the input side by agrochemical, feed and seed companies, and on the output side by those who move, transform and sell the food. Food consumers appear to have benefited, as the real cost of food has fallen. There is now greater choice, and more processed foods reduce preparation and cooking time.

In order to create a more sustainable food system, more value needs to be concentrated with rural communities and farmers. The Kansas Rural Center and the Center for Rural Affairs in Nebraska call this 'taking back the middle'. There are several ways to take back the middle, spreading the benefits more evenly amongst stakeholders. The first is for farmers to find ways of selling their produce directly to consumers. There are a variety of proven mechanisms, including farm shops and direct mailing, farmers' and produce markets, community-supported agriculture and box schemes. Farm shops have long been one way for farmers to sell directly to consumers. When successful, they can make a very significant contribution to individual farm income. However, they can be costly to establish and run, and usually rely on consumers driving to farms to make purchases. Direct mailing, however, can be successful. Produce markets, in which growers sell directly to customers, are another option for taking back more of the food pound. Produce markets also offer the opportunity for consumers to buy organic or sustainably produced food without paying a premium.

There are thought to be some 2400 farmers' markets in the US. Each market is unique, though all usually offer farm-fresh and organic vegetables, fruits and herbs, as well as flowers, cheese, baked goods and sometimes seafood. Each week, about one million people visit these farmers' markets, 90 per cent of whom live within 11 kilometres of the market. The annual national turnover is about US$1 billion. The benefits these farmers' markets bring are substantial. They improve access to local food; they improve returns to farmers; they also contribute to community life and social capital, bringing large numbers of people together on a regular basis (Festing, 1995; Fisher, 1996; see also Chapter 5). Farmers' markets are central to the growing food security movement in the US, now supported by the passing of the Community Food Security Act as part of the 1996 Farm Bill. In Britain, although the first farmers' markets were only held in Bath in 1997, the 538 Women's Institutes (WI) cooperative markets have long operated on similar principles. They comprise weekly markets for the direct sale of home-grown and homemade produce from gardens and kitchens to the public. These markets have a social as well as an economic function since they help local people develop skills, market their produce by working together, and provide a friendly meeting place.

Community-supported agriculture (CSA) is a partnership arrangement between producers and consumers designed, again, to take back more value and also to provide a guaranteed quality of food. The basic model is simple: consumers provide support for growers by agreeing to pay for a share of the total produce, and growers provide a weekly share of food of a guaranteed quality and quantity. CSAs help to reconnect people to farming, and farmers to their customers. Members know where their food has come from, and farmers receive payment at the beginning of the season rather than when the harvest is in. In this way, a community shares the risks and responsibilities of farming.

Organic farmers and growers in the UK have recently made great strides in alternative marketing and sales of their produce directly to consumers. In addition to farm shops, subscription systems or box schemes have become very popular. Here a grower prepares a budget and projected yields for the season and offers shares in advance; the sharers then receive a proportion of the season's produce in the form of weekly boxes of vegetables and fruit. Over the past four years, the growth in box schemes for the marketing of vegetables has been spectacular. By 1997, about one third of British organic farmers were involved in marketing directly to some 25,000 households (Booth, 1996). Over time, many of the supplying farms or smallholdings have become highly diverse in response to customer wishes – some growing 50 to 60 different crop varieties.

The second theme for taking back the middle is to enhance links with urban communities through community cooperatives and community gardens and farms. This can enable communities to take back more control of the food system from currently dominant institutions. Poverty is usually associated with ill-health and poor diets. When money is tight, the cheapest way to get sufficient calories is to purchase sugary, fatty and processed foods. But these usually lack many vital ingredients for the diet, such as proteins, vitamins and minerals. Direct links between consumers and farmers have had spectacular success in Japan, with the rapid growth of consumer cooperatives, *sanchoku* groups (direct from the place of production) and *teikei* schemes (tie-up or mutual compromise between consumers and producers). This extraordinary movement has been driven by consumers rather than farmers, and mainly by women. There are now some 800 to 1000 groups in Japan, with a total membership of 11 million people and an annual turnover of more than US$15 billion. These consumer–producer groups are based on relations of trust and put a high value on face-to-face contact. Some of these have had a remarkable effect on farming, as well as on other environmental matters (see Festing, 1997).

One option for those on low incomes is to find ways to grow their own food. In southern countries it is common for very large numbers of urban dwellers to be directly engaged in food production. In some Latin American and African cities, up to 30 per cent of vegetable demand is met by urban production. But in industrialised countries, far fewer people grow their own food. Although allotments are still widespread in cities and towns, their numbers and areas have been falling as land has been claimed for other purposes. In the US, the National Gardening Association estimates that some 35 million people are engaged in growing their own food in back gardens and allotments. Their contribution to the informal economy is huge. The Gross National Home Garden Product is estimated to be some US$12 to 14 billion per year (National Gardening Association, 1986).

Community gardens have a great impact on social capital. In the UK, there are now several hundred city farms or community gardens which cultivate formerly derelict land. These are important for a variety of reasons. They provide food, especially vegetables and fruit for poorer urban groups, and other natural products – wood, flowers and herbs. They add some local value to produce before sale. They result in the transformation of derelict or vacant land into desirable areas for local people to visit and enjoy – and create quiet, tranquil places for the community. They help to reduce vandalism through the involvement of schoolchildren. They provide local children with an educational opportunity to learn about farming and animals. They allow mental health patients to engage in work that builds self-esteem and confidence (Garnett, 1996). Once again, though, the scale is much greater in North America (Cook and Rodgers, 1996). Poverty, unemployment and malnourishment are mounting, and there is now an emerging policy commitment to local food production through the US Community Food Security Act of 1996. New York alone has some 700 community gardens involving about 20,000 households in local regeneration. GreenThumb, coordinated by Jane Weissman of the New York City Department, has been working since 1978. Community gardens have tended to create employment, foster community pride, produce fresh, nutritious food, and act as a meeting place for local youngsters. Some cities have now adopted citywide food policy councils to integrate better food, poverty, hunger and farming issues.

Another way that farmers can create extra local value is to work together in groups, cooperatives and alliances in order to create social capital by learning and working together. Examples include participatory research and experimenting groups, machinery rings, comarketing groups and community food cooperatives. Sustainable agriculture is knowledge and management intensive; it needs timely and relevant information to produce value. Farmers have to be engaged in a continuous learning process to develop new technologies and practices. The normal mode of agricultural research has been to conduct experiments under controlled conditions on research stations; the results are then passed on to farmers.

In this process, farmers have no control over experimentation and technology adaptation. Farmers' organisations can, however, help research institutions to become more responsive to the diversity of local needs if scientists are willing to relinquish some of their control over the research process. But this implies new roles for both farmers and scientists, and it takes a deliberate effort to create the necessary conditions for such research-oriented local groups. Nonetheless, there have been successes in both industrialised and developing countries.

At first, all farmers lack the knowledge and courage to engage in sustainable agriculture, but their skills do develop over time. By experimenting, farmers increase their own awareness of what does and does not work. But not all farmers can engage in the new methods alone. Collective efforts are vital for sharing ideas and findings. Farmers appreciate the atmosphere of these study groups. The challenge for extension is to shift from being solely technical experts to being experts in facilitating others' learning. In this way, farmers can add value to their activities by increasing their inputs of knowledge and management skills.

The fourth option is to enhance labelling and traceability of foodstuffs in order to increase consumer confidence about both the source and quality of foods. This can be done through ecolabelling and other assurance schemes, organic standards, and fair trade schemes. The question consumers are increasingly asking is: can we trust the food we see on the shelves? Labelling is important, as it tells consumers something about the way that the food was produced. Although organic products have long been clearly labelled for consumers, it is only recently that there has been an expansion in other labelling schemes. Nonetheless, labelling practice and policy are still full of contradictions: there is full disclosure for additives, but not for pesticides.

Various assurance schemes have been launched in recent years. Traceability will be an increasingly important part of the industry – the intention is that any food product can be traced back to the individual farmer. Not only will this provide a link when food scares do happen, it is also hoped that consumer confidence and awareness will rise as the information with which they are provided also increases. 'Fair trade' is now emerging as another type of labelling scheme. The basic assumption is that the conditions of poverty in which people live are due, at least in part, to the manner in which industrialised countries trade with them. Fair trade schemes seek to guarantee a better deal for Southern producers as well as to ensure that consumers receive good-quality produce. The opportunities are significantly greater in Europe. The EU imports three times as much from the least developed countries than does the US, and twice as much as the other G8 members put together.

The agricultural products most commonly subject to fair trade production and marketing are chocolate, coffee and tea. Producers are paid a higher than world price and are guaranteed a market price for at least three years ahead. There are considerable social as well as economic benefits for local people. It has been estimated that there are some 100,000 Mexican farmers now producing fair trade coffee with organic methods (Geier, 1996).

The emergence of community food security and the idea of the foodshed as integrating concepts help communities to take back more of the middle. The foodshed, by bringing consumers and producers literally and figuratively closer, regenerates and reinvigorates natural and social capital. Few say that autarkic systems with no external linkages are best. Rather, it is a question of making the most of local capacities and resources before turning to externally sourced products. Community food security has now become the way to implement this notion of the foodshed. System-wide changes are possible, as experiences of the Toronto Food Policy Council in Canada and the Hartford Food System in Connecticut show just how. Fundamental changes to the agriculture and food systems are two parts of the living land triumvirate. The third is sustainable rural development and the enhancing of rural social capital.

Sustainable Rural Development

The 20th century's period of remarkably successful agricultural growth brought great social change in rural areas throughout Europe. In the quest for greater food production, landscapes, rural livelihoods and farming systems have all been progressively simplified. Where there were diverse and integrated farms employing local people, there are now operations specialising in one or two enterprises that largely rely on farm or contractor labour only. Where processing operations were local, now they are centralised and remote from many rural communities.

In every European country, farms have become both fewer in number and larger during this century. When the CAP was established in 1957, the six member countries had 22 million farmers. Today that number has fallen to about seven million. Many argue that progressively larger farms are an economic necessity. They permit economies of scale to be made. Such individuals mean that more efficient producers can take over or absorb the operations of smaller and more inefficient producers. But small farms provide other benefits to society. They employ more people than larger farms, probably because access to labour is easier than to capital. In the UK, farms under 40 hectares provide five times the per-hectare employment than those over 200 hectares. Small farms contribute to both rural social capital and natural capital. Their greater on-farm diversity maintains both plant and animal biodiversity; they are more efficient users of energy; they are better at preserving and enhancing landscape and wildlife; and they tend to have a better record with animal welfare (Raven and Brownbridge, 1996; Rawles and Holland, 1996).

Changing farm size and abandonment have also brought a dramatic decline in the numbers of people employed in agriculture across Europe. Most countries now have unemployment rates higher in rural than in urban regions. For the OECD as a whole, there were some 25 million people officially unemployed in 1995, a doubling over the previous two decades (Bollman and Bryden, 1997). Fewer farms, fewer jobs and larger-scale farming operations have also played a role in the rise of rural poverty and lack of services. A range of UK national inquiries conducted in the 1990s has shown that about a quarter of rural households are living on or below the official poverty line. The decline in rural services has also been marked. The poorest and most deprived live in areas where key services, such as schools, doctors, pharmacies and shops are concentrated in the larger towns, while people in outlying areas have to travel for these services. Four out of ten parishes in rural England have no shop or post office; six out of ten no primary school; and three-quarters no bus service or GP practice. The result of this poverty, deprivation and declining services is the gradual unravelling of communities. Although there has been an increase in material affluence, this is not linked to social, cultural and spiritual strength. Throughout the world, external agencies have routinely undermined social capital in order to encourage economic 'development'. It may not be intentional, but the effect of doing things for people, rather than encouraging or motivating them to act as much as possible for themselves, is substantial.

All of this has led to a decreased capacity amongst local people to cope with environmental and economic change. As once thriving communities are plunged into dependency, so people have increasingly lost a sense of belonging to a particular place. This spiral of decline in local communities has diminished social capital. Another result of the decline of social capital arising from changes in farming is a marked increase in mental distress amongst farmers. In Britain, farmers and farmworkers are two and a half times more likely to commit suicide than the rest of the population; suicide is also the second most common form of death for farmers and agricultural workers aged 15 to 44 years. In France, suicide rates amongst both men and women are highest in rural communities and fall steadily with increasing size of village and then town. This is the exact opposite of the reported situation at the turn of the century – with the highest rates recorded in the cities (Philippe, 1974).

Despite all these economic and social problems in rural areas, large numbers of people want to live there. Surveys in the UK indicate that nearly half of those who live in cities prefer living in a village or country town; and about a quarter expect to move out in the next decade. Most of these people will be looking for a traditional country life in some form or other. Incomers, once they are there, are more likely to object to new developments intended to supply housing or to contribute to jobs. They show not only Nimby attitudes (not in my back yard), but also Nodam (no development after mine). They are also less likely to understand farming and its positive contribution to landscape and natural capital (Derounian, 1998).

There are two competing schools of thought about what is the best way to promote development in rural areas. One seeks to build on locally available natural and social resources – so called 'endogenous' development. The other seeks external (exogenous) investments and capital to help 'modernise' the countryside. The latter exogenous model currently predominates. It says that we, in industrialised countries, must look to other sectors of the economy to provide new jobs and wealth creation. Farming is seen inevitably to lose labour as countries get richer. It also contributes proportionately less and less over time to gross domestic product (GDP). Farming has become more mechanised and modern at the same time as countries have become richer, and so the way for them to get richer is to encourage the same process and trends. This school of thought implies that agriculture is best left to its own devices. Agriculture is for producing food. We are very good at it now – but it is unlikely to have a major role in rural economic development. Where inward investment is to be encouraged, it is best targeted at alternatives to agriculture and natural resource-based activities. Industrial relocation, golf courses, high-technology industry, craft-based rural industry and tele-cottaging can all bring jobs; but few have much to do with agriculture.

The alternative school of thought focuses on endogenous patterns of development – that is, growing or originating from within. Here the priority is to look first at what resources are available in rural areas, primarily agriculture, people, natural resources and wildlife. Then ask: can anything be done differently that results in better use of these already available resources? Can it be done without incurring environmental and social costs? This will be difficult since it means operating in a very different way than the past 50 years. The current patterns of institutions, policies, funding and intellectual thought all discriminate against endogenous development. When asked what they need, most people's reflex is to turn to external solutions: 'we are waiting for the government to solve our problems'; 'we need a change in exchange or interest rates to give us more money'; 'we need more tractors and fertilisers'; 'we need the council to fix our pavement, gutter or potholes'; 'we need an industry to relocate to bring us new jobs'. These are all legitimate needs. But they cannot happen for everyone all the time.

These attitudes foster the dependency deadlock. Local people become entirely dependent upon external agencies and actors to provide solutions to local problems. Again, this may appear reasonable. If you pay your taxes, why should you not expect the local council to fix the road or supply other services? But there are always things that local people can do better – they know the local conditions. They will be living there long after external bodies have departed. They just may not realise their own capabilities. For endogenous or community-based economic development to occur, new partnerships and connectedness between different actors are needed. As social capital has been lost in our rural areas, with inevitable increases in deprivation, stress and unhappiness, so it needs to be recreated with the new approaches to policy and practice. New participatory processes are required to bring together different stakeholders in the renewal of the countryside. History tells us that coercion does not work. We may have technologies and practices that are productive and sustainable, but if they are imposed on people, they do not work in the long term. These processes and technologies must be locally grounded to produce different solutions for different places.

Fortunately, we do have somewhere to turn. There has been a revolution in the past ten years in methodologies for creating social capital. Emerging from a range of different traditions and disciplines, participatory methods have expanded in use and efficacy during the 1980s and 1990s. The greatest expansion has occurred in the developing world context, where participatory approaches are now being used in almost every country. Recent years have also seen an expansion in their use in Europe, including village or parish appraisals, participatory appraisal, future search, community audits, parish maps, action planning, planning for real, and citizens' juries (see cases in Chapter 7). The terms people's participation and popular participation are now part of the normal language of many development agencies, including non-governmental organisations (NGOs), government departments and banks. It is so fashionable that almost everyone says that participation is part of their work. This has created many paradoxes. The term participation has been used to justify the extension of state control as well as the build-up of local capacity and self-reliance; it has been used to justify external decisions as well as to devolve power and decision-making away from external agencies. The many ways that organisations interpret and use the term participation can be resolved into seven distinct types. These range from manipulative and passive participation, where people are told what is to happen and act out predetermined roles, to self-mobilisation, where people take initiatives largely independently of external institutions. What this typology implies is that the term participation should not be accepted without careful clarification.

A major challenge is to institutionalise these participatory approaches and structures that encourage learning. Most organisations have mechanisms for identifying departures from normal operating procedures. This is single-loop learning. But most institutions are very resistant to double-loop learning, since this involves the questioning of, and possible changes to, the wider values and procedures under which they operate. For organisations to become learning organisations, they must ensure that people become aware of the way they learn, both from mistakes and from successes.

Recent years have seen rapid innovation in new participatory methods and approaches for stakeholder learning and interaction in the context of community development. There are now more than 50 different terms for these systems of learning and action, some more widely used than others. This diversity is a good thing. It is a sign that many different groups are adapting participatory methods to their own needs. These groups have common principles. There is a defined and organised methodology for cumulative learning by all actors. The inquiry and learning processes are user-friendly, with a particular emphasis on simple visual and dialogue methods. Diversity and inclusion are emphasised throughout in order to reveal multiple perspectives. External actors facilitate learning and are concerned with transformations that people regard as improvements. These self-assessments lead to new visions for the future. And the analysis and debate about change lead to an increased motivation to act.

There is a need for new rural European partnerships. Sustainable agriculture and rural development are not just about developing resource-conserving and regenerative technologies on individual farms. They are about the complex business of creating institutional linkages that encourage coordinated action. This comanagement is important for pest and predator management, nutrient management, groundwater protection, maintaining landscape value, conserving soils, and sustaining access to the countryside. There are good social reasons for working together in partnerships. Regular exchanges and reciprocity increase trust and confidence and lubricate cooperation. There are great assets already within communities. It is just that local people often do not realise they have them. The existing assets are primarily in social capital: everyone knows each other, and they tend to help each other when in need. They bring good expertise; they experience problems at first hand; they are aware of local networks; and they are much more likely to sustain initiatives if they feel involved.

In industrialised countries, agriculture is no longer a significant contributor to jobs and rural livelihoods. However, sustainable agriculture's need for more management skills, knowledge and labour represents a huge and emerging opportunity. What is becoming increasingly clear is that sustainable agriculture farmers do three things. They spend more money locally on knowledge-based goods and services; they employ slightly more people in the business of farming; and they add value where they can through processing and direct marketing. The jobs dividend could be very significant – in the range of 0.4 to one job per farm.

We know that community partnerships work. They can be good for social capital. They can be good for the environment. They can mean more local jobs. But we also know that they are given only patchy coverage and small amounts of resources – only a very small number of people are currently benefiting. There is a pressing need for the right policy environment to nurture these local processes and to help them spread on a wider scale.

The Sustainability Dividend

The Opportunity

This book argues that there is a net benefit to the whole farming, countryside and food system with a shift towards a more sustainable and community-led approach to rural economic development. But what shape would such a dividend take? Is it just a matter of more songbirds in the countryside or of a more appealing view? Or would it make a difference to jobs and services in rural areas too? With some 18 million people unemployed in the EU in mid 1997, and a massive 25 million throughout the whole OECD, it is job opportunities and economic growth that concern people and politicians more than the state of the environment. But is there a win–win opportunity here? Is it possible to attain a clean environment, vibrant rural communities, good food, and economic growth to produce new jobs and business opportunities? There are two ways to create this sustainability dividend. The first is to reduce the external costs imposed on the environment. The second is to transfer productive activities from one group of stakeholders to another.

Resource Flows in Rural Areas of the UK

Before we can assess the potential for this sustainability dividend, it is important, first, to characterise the flows of resources to and from rural communities (Figure 1.1). Situated at the centre of the diagram are

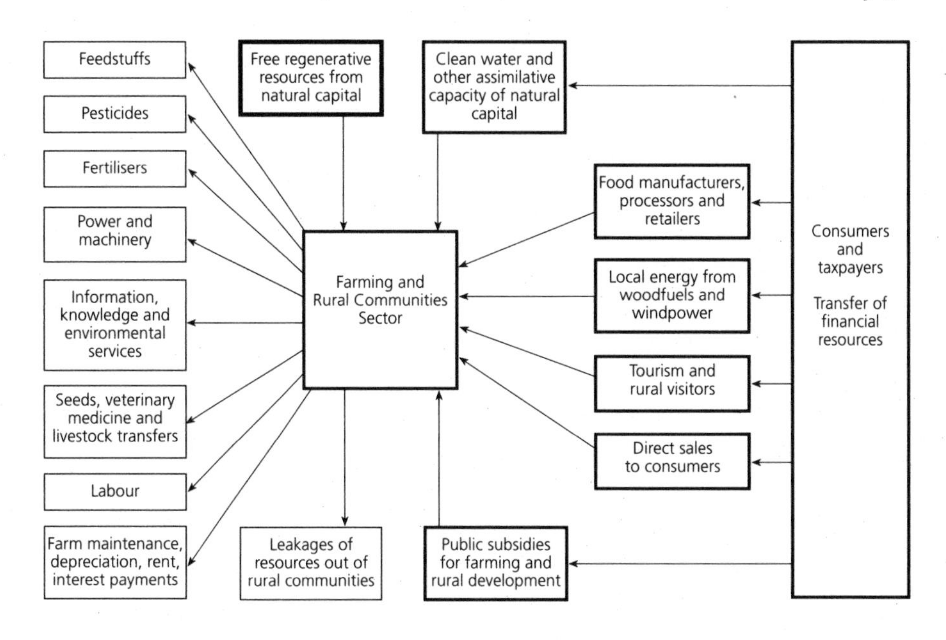

Figure 1.1 *Resource Flows in Rural Areas*

farms, surrounded by their local rural communities. The arrows indicate flows of money.

On the left side can be seen the major expenditure items – farmers spend money on feedstuffs; on pesticides and fertilisers; on power and machinery; on knowledge, information and environmental services; on seeds, veterinary services and livestock; on labour; and on farm maintenance, depreciation, interest payments, rents and insurance. On the right side can be seen the major sources of money for farmers and rural communities. They sell food to manufacturers, processors and retailers; they sell directly to consumers through farm shops, direct markets and box schemes; they receive income from tourists and visitors; they produce energy through wind and wood power; and they receive money from the public purse in the form of direct subsidies, both for farming via the CAP and for rural development via EU cohesion policy or national institutions.

Farming also receives some free resources. These include 'clean' natural resources that act as a sink for farm-produced pollution, especially surface and groundwater, but also the atmosphere. It includes the free environmental services derived from natural capital on-farm, such as nitrogen fixed from the air or insect predators which consume pests. In order to assess the potential impact on farming and rural communities of changed patterns in this model, it is important, also, to investigate two further factors. These are the amounts of expenditure made in rural communities, and the amount of income that immediately leaks away (the leakage rate). It makes a difference, for example, whether farmers buy services and goods from local businesses or from distant companies. It also makes a difference whether tourists spend their money on goods and services that are locally sourced, adding value locally, or whether they simply buy 'imported' goods.

Combining the whole model, it becomes clear that transfers make a difference too. A total of £720 million is spent on fertilisers by farmers in Britain annually; yet the nitrogen in these fertilisers is available to farmers free of cost. Some £200 million is spent on electrical power annually; however, with some land devoted to arable coppicing or wind power, much of this expenditure could be saved. Some £490 million is spent on pesticides each year; yet sustainable agricultural systems are more self-regulating, relying on free environmental services for natural control; they also rely more on information and knowledge. In this way farmers who switch to lower-input and precision farming shift payments from pesticide companies to local economies. At least £5 billion is spent each year by urban visitors to the countryside during day visits, but little of this value is captured by farmers and businesses in rural communities.

The potential dividend is estimated for three scales of change. The first assumes that all farmers and rural communities have switched to sustainable practices. This is called the SD_{100}. The second represents a more realistic future, and assumes that half have made the switch. This may still be ambitious, but at least we can imagine this kind of change happening over a five- to ten-year period. This is termed the SD_{50}. The third is the sustainability dividend that has already accrued. Some farmers are already farming organically, or have adopted integrated technologies; box schemes and direct markets already have shifted resources into the direct sales box in Figure 1.1; tourists are visiting the countryside and spending their money. These levels allow us to set realistic targets for all the stakeholders in the system, with particular focus on policy-makers who need to know both what is possible and how to get there. Realistic targets for sustainable agriculture are crucial so that progress towards sustainability can be measured and assessed.

Over a five-year period, we ought reasonably to expect that the following changes will have occurred:

- half of all farmers are farming sustainably;
- half of all rural communities are engaged in community group action for local development;
- a 50 per cent reduction in pesticide use has occurred, with at least half of farmers using integrated pest management (IPM) technologies and practices, with a consequent 50 per cent reduction in surface and groundwater pollution;

- a 50 per cent reduction in soil erosion and the external costs caused by soil erosion;
- a 50 per cent improvement in the energy balance in agriculture (with both reductions in the energy used, and increases in the energy produced);
- a 50 per cent reduction in farmers relying solely on fertilisers, with increased use of regenerative and resource-conserving technologies;
- 50 per cent of tourists to rural areas spending half as much again, but all of this on locally derived goods and services.

Dividend from Reducing External Costs in the UK

It is almost impossible to assess fully all the environmental costs of industrial agriculture. The direct and indirect impacts are extensive, affecting different systems, such as natural capital (soil erosion, wildlife and habitat loss, water pollution, genetic loss); or people (pesticide poisoning); or social capital (loss of local jobs, farm abandonment). Chapters 2 and 6 detail this extraordinary legacy of modern farming. But monetary values can only be attributed to some of these consequences. The loss of a traditional breed of livestock cannot be costed, yet some 700 breeds have become extinct in Europe during the 20th century.

Nevertheless, there are some costs that can be measured and, although these do not represent the full costs, they do illustrate the kinds of savings that would release a sustainability dividend. Some of these costs are not actually the true cost of lost natural capital, but are what we have to pay to clean up the damage to an agreed or acceptable level. For example, the costs of pesticide pollution of ground and surface water are impossible to assess; but we do know that it has cost water companies and their customers £1000 million of investment in machinery plus a further £121 million annually for removal running costs. In the UK, farmers use 29.2 million kilogrammes of pesticide active ingredient annually. This means that for every kilogramme used by farmers, it costs £4.14 to repair the pollution of water, rising to £7.57 if the investment costs are added and spread over ten years. The cost per hectare of farmland on which pesticides are applied is between £19.39 and £22.10. This is unlikely to be acceptable to domestic and industrial water users for long.

Other known external costs include £24 to 50 million per year for soil erosion; £24 million per year to remove nitrates from drinking water, on top of water treatment measures since 1986 costing £275 million; and £5 million per year on farm pollution incidents arising from livestock wastes (see Chapter 2). There is clearly much room for savings to be made. In the US, annual sales of pesticides and fertilisers are about

US$15 billion (approximately £9.4 billion), while annual costs for monitoring pesticides and nitrates in groundwater are $1.8 billion per year, equivalent to some 12 per cent of sales. Who pays for this? It is neither the agrochemical industry nor farmers. It is the taxpayers and the public.

None of these costs are currently paid by farmers. They represent, in effect, a transfer of resources or cross-subsidy from the public directly to farming. A reduction in water pollution, and consequent reduction in costs, means a transfer of benefits to water companies and their customers. Indeed, some water companies have already concluded that it is cheaper and more efficient to work with farmers to reduce pesticide pollution before it occurs. In Germany, some are paying farmers to convert to organic husbandry. A recent internal UK government report also concluded that catchment management to reduce pollution would be much cheaper than paying to clean up pollution after it had occurred.

A 50 per cent reduction in pesticide pollution to water, combined with a halving in soil erosion and nitrate contamination, would yield a staggering annual dividend (SD_{50}) of about £100 million at current prices. If we were able fully to measure the other external costs (such as to hedgerows, wildlife, landscape and amenity value, as well as pesticide poisonings), then this figure would be far exceeded.

Dividend from Transferring Productive Farm Activities to Sustainability Sectors in the UK

The second dividend comes from the transfer of productive activities to sustainability sectors. As will be described in later chapters, the size of the present sustainability dividend is much larger than most people would expect. Again, it is impossible to cost this exactly. What we do know is that in the UK there are 820 organic farms (and some 50,000 in the whole of the EU); 150,000 hectares of conventional farms that have been precision mapped, leading to reductions of input use and savings of £3.75 to £7.5 million per year; 538 county markets which attract 40,000 weekly visitors; 25 to 30,000 families who are members of organic box schemes and direct marketing; and 660 million day-visitors to the countryside each year, who spend some £5.28 billion.

In the US, the numbers are much greater and include some 600 CSA farms, with some 100,000 members and an annual turnover of US$10 to 20 million per year; 2000 farmers' markets that attract three million visitors weekly and turn over at least $900 million per year; and 750 community gardens in New York City alone, with 20,000 people involved and producing $1 million worth of vegetables and fruit per year for home consumption. All of these activities deliver a substantial dividend to some of the stakeholder groups in the food and farming system. By looking closely at the current patterns of resource use in the various boxes in Figure 1.1, it is possible to assess where both transfers and savings can be made on the input and marketing sides (see Table 1.1).

In 1995, British farming spent some £13.42 billion on all inputs of agrochemicals, labour, feedstuffs, seeds, power and maintenance. The net sustainability dividend with a 50 per cent shift to sustainable agriculture is £630 to 810 million, rising to £1.17 to 1.53 billion for a complete shift by all farmers. This counts extra labour as a loss to farming. But for the whole rural economy, expenditure on increased labour demand can be seen as a positive outcome. The transfer from pesticides, fertilis-

Table 1.1 *Summary of projected dividends available for a sustainable UK farm sector from savings and transfers in input sectors (£ billion, 1995 prices)*

Sector	*Total expenditure by British farmers (£ billion)*	*Dividend* SD_{50}	*Dividend* SD_{100}
1 Pesticides	0.49	+ 0.1 – 0.25	+ 0.25 – 0.39
2 Fertilisers	0.72	+ 0.11 – 0.22	+ 0.36 – 0.58
3 Labour			
• hired	1.71	– 0.34	– 0.68
• family and partners	1.02	– 0.10	– 0.15
4 Feedstuffs	3.28	+ 0.82	+ 1.31
5 Seeds, veterinary medicines and livestock	0.52	0	0
6 Power and machinery (electricity)	2.60	+ 0.04	+ 0.08
7 Maintenance, depreciation, interest and rent	3.07	0	0
Total (net) (£ bn)	13.42	+ 0.63 – 0.89	+ 1.17 – 1.53
Total (with labour converted to a positive contribution to local communities) (£ bn)		+ 1.51 – 1.77	+ 2.83 – 3.19

Notes: the assumptions for SD_{50} and SD_{100} projections are drawn from empirical data presented elsewhere in this book. They are as follows:
1 Pesticide use: SD_{50} down 20–50%; SD_{100} down 50–80%
2 Fertiliser use: SD_{50} down 15–30%; SD_{100} down 50–80%
3 Labour hired: SD_{50} up 20%; SD_{100} up 40%
Labour family and partners: SD_{50} up 10%; SD_{100} up 15%
4 Feedstuffs: SD_{50} down 25%; SD_{100} down 40%
5 Seeds, veterinary and livestock: no change
Power and machinery: no change except for electricity – SD_{50} down 20%; SD_{100} down 40%
7 Maintenance, depreciation, interest and rent: no change

ers and feedstuffs to labour is £440 to £830 million. The total sustainability dividend thus rises to a massive £1.5 to 1.8 billion with half of farmers shifting to sustainable agriculture, and rising to £2.8 to 3.2 billion if all do so.

Both pesticide and fertiliser use fall with a shift to more sustainable agriculture. Based on empirical evidence from farms and research sites (see Chapter 3), it is easier to cut pesticide use than fertilisers in the early stages of transition, but 50 to 80 per cent savings from both can be expected when all farmers have made the transition. Purchase of feedstuff also falls, with farmers increasingly using legume-based systems to supply feed for animals. The SD_{50} and SD_{100} assumptions are more conservative than for crop-based systems. It is assumed for these projections that there is no change to seeds, veterinary medicine and livestock purchases. Similarly, no change is assumed to occur to current maintenance, depreciation, interest and rent payments. There is little change to power and machines, except for 20 to 40 per cent savings made to electricity bills with the substitution of locally produced energy from arable coppicing and wind power (see Chapter 7). Labour use, however, is assumed to increase, since knowledge and management skills substitute for external inputs. Spending on labour will increase for both hired and family–partners labour, up by 20 to 40 per cent for hired labour and 10 to 15 per cent for family members.

The Total Dividend

There is also a substantial jobs dividend from a transition to sustainable agriculture. Assuming one job per £12 to 15,000 of expenditure on labour, the jobs dividend would be 30 to 70,000 new jobs for the SD_{100}. This accords reasonably well with the estimates derived from the amount of extra labour needed on sustainable agriculture farms (see Chapters 3 and 8). These data indicate an increased labour demand for all Britain's farms of 80 to 150,000 new jobs. Some of these data, however, refer to the jobs dividend coming from value captured in the marketing and processing operations. They do not account, however, for job losses in distant agrochemical factories.

Substantial value can also be added by changing operations on the output side too. The total farm receipts in 1995 were £17.44 billion, of which £4.9 billion were for cereal and root crops, £2.1 billion for horticulture, £6.2 billion for livestock, and £4 billion for milk and eggs. If farmers engage much more in direct selling through farm shops, farmers' markets, box schemes and the like, then more value can be added to some of these products. In the UK, some 5 per cent of farmers already sell directly to consumers, and this is said to account for about 9 per cent of fresh produce. In Germany and the US, a sixth of farmers sell direct; in France and Japan this rises to a quarter (Festing, 1997). It is unlikely, however, that direct sales to customers will affect cereals, sugar beet and oil seeds (total of £3.62 billion). But there are clearly opportunities for direct marketing of products in other sectors, such as vegetables, potatoes, fruit, meat products and milk products.

If we assume that there is an increase of direct marketing to 15 per cent at the half-way transition (SD_{50}) and to 25 per cent for the full transition (SD_{100}), then the total dividend would amount to between £2.1 billion and £3.5 billion. This is just under one half of what is released from transfers in the input side to regenerative and management-intensive technologies and practices. The final source of dividend comes from tourism, which brings large amounts of resources into rural communities. There are two types of visitors – day visitors and those taking longer holiday breaks. According to the Countryside Commission, there are some 660 million day visits to the UK countryside each year. According to a range of recent studies in Scotland, Wales, Dorset and Surrey (Cuff and Rayment, 1997), each visitor spends between £4 to £12 (at 1996 prices). Assuming a mid range figure of £8 per person, this means that £5.28 billion is spent locally by these visitors. But only expenditure on local goods and services (such as foods, handicrafts and accommodation) adds local value, and much leaks out again.

Assuming a 70 per cent leakage rate based on tourist industry estimates, this means that £1.58 billion currently accrues to rural communities. In addition to day visitors, some 32 million Britons took holidays of four nights or more in Britain in 1995. A conservative estimate suggests that these 32 million visitors spend a further £6.4 billion, of which 70 per cent leaks away, leaving £1.92 billion. This puts the net current contribution after leakage by British visitors to their countryside and rural communities at £3.5 billion.* A vision for a more sustainable and vibrant rural economy would have more of the total spend by visitors, both domestic and overseas, captured by rural and land-based businesses. Assuming that the leakage rate is cut to 50 per cent at the

* The Scottish tourism estimate is that of each £1000 spent, some £300 of local income is generated. See the *Economist Data Book* 1997. Altogether, 26 million Britons took holidays abroad, which compares with 30 million at home and five million abroad in 1965. Britain has the fourth highest number of foreign visitors in Europe, with 23.8 million per year who spend £475 per person. Spain, France and Italy top the league, each with more than 50 million visitors per year. Here, it is assumed that none of this income comes to rural communities, as the majority of visitors spend their holidays at the big attractions – castles, heritage sites and cities. It is possible that ways could be found to capture some of this value.

SD_{50} stage and 20 per cent for SD_{100}, this gives a net benefit to rural communities of between £5.85 and £9.36 billion. Once again, this could have a substantial impact on jobs. In the tourist industry, a spend of £1500 is assumed to generate 0.04 full-time equivalent (FTE) jobs. (The Scottish multiplier assumes that £1500 spent generates 0.04 FTE jobs; that is £37,500 per job (Rayment, 1997).) This puts the total number of jobs created by tourists spending money that is held for longer in rural communities at between 234,000 and 374,000. The total sustainability dividend for Britain's rural communities is therefore between £9.5 and £16 billion in cash terms and 320–590,000 jobs (Table 1.2).

Table 1.2 *Total sustainability dividend for rural Britain*

Dividend sector	*Total sustainability dividend (£ billion)*[1]		*Total sustainability dividend (thousands of jobs created)*	
	SD_{50}	SD_{100}	SD_{50}	SD_{100}
Reduced external costs, mainly pesticides and nitrates in drinking water and soil erosion[2]	0.075	0.100	neg	neg
Input transfers to regenerative resources and to labour[3]	1.51–1.77	2.83–3.19	15–35	30–70
Direct marketing and sales[4]	2.1	3.5	70–87	116–145
Tourism[5]	5.85	9.36	234	374
Total (£ billion or thousand jobs)	9.5–9.8	15.8–16.2	319–356	520–589

Notes: these are likely to be low estimates because:
1 SD_{50} assumes that half of British farmers have adopted sustainable agriculture methods leading to a substantial reduction in use of non-renewable inputs. The estimates for SD_{50} and SD_{100} are based upon limited data sets. As more farmers engage in sustainable agriculture, it is likely that the learning and innovation will be much greater, increasing returns and reducing input use;
2 Reduced external costs only account for costs we can measure. All those that cannot be measured will also be reduced;
3 Sustainability dividends are conservatively calculated (see Table 1.1). The benefits delivered by an entirely organic agriculture would be substantially higher than here. While unlikely in the immediate future, more and more people are entertaining the possibility that whole farming industries could become entirely organic;
4 Assuming a conservative half, the increased returns to farmers from direct marketing and sales are turned into local jobs;
5 Tourism dividend does not account for overseas visitors, but it does assume that the leakage rate of resources out of rural communities is cut.

neg = negative.

Winners and Losers

The sustainability dividend brings net benefits to rural communities and farming. It does not, however, represent the full benefit to the economy at large. Not all stakeholders in the British food and farming system will be winners. Some will be losers. What is important is that the losers are the groups already doing well; it can be argued on equity grounds that they can afford to capture less of the value from the whole system.

The major loser is clearly the input companies, currently supplying fertilisers, pesticides, feedstuffs and seeds. They lose out in the shift towards sustainable agriculture and localised food systems unless they change the nature of their business. Losers too, but not significantly, are the food manufacturers, processors and retailers. Since greater value from food is captured by rural communities, so some market share is lost by these larger players unless they too change the nature of their business.

Farmers as a whole are winners. They are better off as gross margins improve with sustainable agriculture. Their environments are better and more healthy. They benefit from increased business opportunities through jobs and spending in rural communities. They also benefit through their repaired relationships in society as stewards of the countryside. Rural communities are major winners. There are more jobs and wealth created in rural communities. There is more social capital developed through better interchanges and participation in development. There is more natural capital delivered by a more sustainable farming industry. However, some groups of farmers may benefit more than others. It will be difficult to reverse the loss of family and mixed farms without greater attention focused on the extra social and environmental benefits they can bring to rural areas.

Food consumers are also winners. Better quality food is available for all. Closer links to farming encourage people to eat more varied and better diets, which reduces ill-health. Wildlife and other natural resources are winners, as these are valued and protected through sustainable systems of land use. Tourists stay longer and spend more in landscapes and environments that give them pleasure. Governments, both local and national, are also winners. Fewer taxes are spent on cleaning up environmental damage caused by modern farming. The health costs brought about by poor food and diets are reduced. A more robust and resilient natural capital delivers goods and services of benefit to many sectors of the economy. And rural communities with reinforced social capital are more self-regulating and pleasant places in which to live.

Financial Support and New Policies for a Living Land

Priorities for Change

This book sets out the case for a more sustainable food and farming system that will deliver substantial natural, social and economic advantages to local people as well as to nations. But the fact that there exists an overall social dividend does not guarantee that changes will emerge. Well-resourced vested interests have much to lose and will state their case clearly to governments and the public at large. However, national governments can do much to help deliver this dividend. The problem is that public policies at national, European and international levels largely discriminate against sustainable agriculture and community-based rural development. Chapters 7 and 8 contain many cases of successful schemes and initiatives that bring components of the sustainability dividend to rural communities across Europe and North America. Yet many of these occur despite existing policy frameworks. Fundamental reforms will be needed.

This will not be easy since vested interests in maintaining the status quo will clearly resist any change. Why should fertiliser companies support a transition to legume-based farming when this would cost them £350 to £500 million in a year in Britain? Why should a pesticide company be balanced in its presentation of different types of farming when it knows some types of sustainable agriculture mean little or none of its products will be used? Each year, some £3 billion of public money in the UK and 45 billion ECU in the European Union are spent on farming. But only a small proportion ends up helping local economies. It is inefficiently used; the value is captured by only a few stakeholders; and much ends up helping to degrade natural and social capital.

What is needed are fundamental shifts in public policy principles and practice. Some will ask, why do we need public support at all? We now produce enough food from a modern and highly efficient agricultural industry. If people no longer want to live in rural areas, then so be it. If farms are abandoned, then these are likely to be the inefficient ones which cannot compete with their larger counterparts. These are just a few of the prices we have to pay for our modernisation and advancement in society. Rural areas will still be there for urban dwellers to visit.

If we have completely 'free' markets with the so-called level playing field, then the most financially efficient farmers will survive and the more marginal will fail. But this will be hugely counterproductive. As land is abandoned, so valued habitats and landscapes disappear. People are essential. Without them the character and value of many rural environments and economies diminish. Much depends, again, on how we define efficiency. If it is on the basis of food production alone, then we end up with one type of farming. If it is to include natural and social capital, then our decisions will be quite different. If we now say that we need more explicit support for public environmental and social goods, then we will need a very different kind of policy context compared with today's.

Others may say, if there are so many benefits from sustainable systems, why is there not an immediate shift? But history constrains us all. Dependencies are fully internalised, and it is very difficult to begin new processes with nothing to build upon and no help. The modernisation project has been so successful that few believe that alternatives exist or can succeed.

Towards National Frameworks

No industrialised country has a national framework for sustainable agriculture and rural regeneration. This is the first step. A national policy for sustainable agriculture sets out a vision and embodies different values. It reveals to all stakeholders what is expected to occur, and how we are to get there. It clarifies policies and policy processes that will support the transition. A national policy for sustainable agriculture would contain a mix of approaches and instruments that would either penalise polluters or reward resource conservers. Penalties on polluters would include a balance of ecotaxes and levies to discourage polluting practices. Taxes on pesticides and fertilisers are proving successful in raising revenue from farming for training, research and extension. Other important instruments include regulations on farming activities in certain areas, such as above aquifers, which are prone to contamination.

The alternative to penalising farmers is to encourage them to adopt alternative low- or non-polluting or degrading technologies by acting on subsidies, grants, credit or low-interest loans. These could be in the form of direct subsidies for low-input systems or the removal of subsidies and other interventions that currently work against alternative systems. Acting on either would have the effect of removing distortions and making the sustainable options more attractive. Although some resource-conserving technologies and practices are currently being used, the proportion of European farmers employing them is still small. This is because adopting these technologies is not a costless process for farmers. They cannot simply cut their existing use of fertiliser or pesticides and hope to maintain outputs, thereby making their operations more profitable. They will need to substitute something in return. They cannot simply introduce a new productive element into their farming systems and hope it succeeds. They will need to invest labour, management skills and knowledge. But these costs do not necessarily go on forever, and much can be done to support this transition. Key components of policy reform would include:

- reform of the Common Agricultural Policy to switch all payments away from production to environmental and social goods;
- national schemes for rural partnerships and rural development support;
- national strategies for integrated pest management, integrated nutrient conservation and soil regeneration;
- a national scheme to guarantee the quality of foods through approved standards and certification schemes;
- support for education in agricultural colleges, universities and training organisations which focuses on sustainable agriculture and rural development.

A number of countries have taken significant steps to help shift agriculture towards sustainability. In Germany, some state governments have adopted pro-sustainability policies. In Switzerland, a new tiered system of support encourages farmers to take steps towards habitat protection, 'greening' of the crop habitat, and eventually the adoption of organic methods. And in Austria, subsidies that favour organic farming have helped to turn 10 per cent of agricultural land over to organic production.

In Europe, the greatest change would occur from a reformed CAP. Instead of paying money simply for food production, this could support farming methods that deliver environmental and social goods. A reformed CAP would support sustainable agriculture, localised food systems and community regeneration. Payments would be coupled to what people want and need, rather than to productive processes that diminish natural and social capital (see indicators for sustainability, in Chapter 3).

But perhaps the biggest challenge will be to ensure that sustainability policies arise through public participation rather than through green conditionality. Throughout the world, environmental policy has tended to take the view that rural people are mismanagers of natural resources. The history of soil and water conservation, rangeland management, national parks, and modern crop dissemination shows a common pattern: technical prescriptions are applied widely with little or no regard for diverse local needs and conditions. These diverse needs often make the technologies unworkable and unacceptable. When they are rejected locally, policies shift to manipulating social, economic and ecological environments, eventually using outright enforcement.

For sustainable agriculture to succeed, policy formulation must not repeat these mistakes. Policies must arise in a new way. They must be enabling, creating the conditions for sustainable development which is based more on locally available resources and local skills and knowledge. Achieving this will be difficult. In practice, policy is the net result of the actions of different interest groups. It is not just the normative expression of governments. Effective policy will have to recognise this and seek to bring together a range of actors and institutions for creative interaction.

Outline of the Book

Part I of this book considers the transition towards sustainable agriculture by assessing first the environmental legacy of modern agriculture (Chapter 2), and then the status of sustainable agriculture in Europe (Chapter 3). New evidence is presented for sustainable farming in many different systems.

Part II takes a look at food systems. Chapter 4 assesses the current problems with a global food system that produces so much food and so much hunger. Chapter 5 presents ways in which farming and rural communities can add value to the food they produce, helping to 'take back more of the middle'.

Part III makes the case for more sustainable rural communities by assessing the social costs of countryside modernisation (Chapter 6). Participation and partnerships for community economic and social regeneration are discussed and presented in detail in Chapter 7.

Part IV describes how we can make a difference. Sustainable agriculture, localised food systems and community-based rural development could make substantial improvements to rural economies as well as to natural and social capital. The sustainability dividend could be enormous. Chapter 8 concludes by assessing the sources of financial support and the changes in policy that are necessary to bring about a sustainable and living land.

Part 1

Towards a Sustainable Agriculture

Chapter 2

The Dying Land: Modern Agriculture's Legacy

There are monstrous changes taking place in the world, forces shaping a future whose face we do not know. Some of these forces seem evil to us, perhaps not in themselves but because their tendency is to eliminate other things we hold good... When our food and clothing and housing all are born in the complication of mass production, mass method is bound to get into our thinking and so eliminate all other thinking.

John Steinbeck (1952) *East of Eden*

The Hidden Environmental Costs

Improvements and Losses

During the last hundred years, and particularly since the 1940s, farmers across Europe have been spectacularly successful at increasing food production. They have intensified the use of non-farm resources to produce much more from the same amount of land. They now get three times the amount of wheat, barley and other grains, potatoes and sugar beet from the same area of land, while milk yields per cow have more than doubled. And these are just averages – increases on many individual farms are far greater.*

*European wheat yields in the 1940s were of the order of 1.5 to 2 tonnes per hectare, and potatoes 13 to 18 tonnes per hectare (t/ha). By the late 1990s, they were three to four times higher. In the UK, average wheat yields rose from 2.1 to 7.5 t/ha; barley yields from 2.1 to 5.9 t/ha; potato yields from 17 to 41 t/ha; sugar beet from 27 to 44 t/ha; and milk yields from about 6.3 to over 14.5 litres per day per cow. Of the food that can be produced in the UK, some three-quarters is grown here compared with just 60 per cent in 1970. The UK now exports some grain. From MAFF Statistics, at www.maff.gov.uk

But these remarkable achievements have been costly, bringing new environmental and social problems. Mixed farms produce few external impacts: crop residues are fed to livestock or incorporated in the soil; manure is returned to the land; legumes fix nitrogen; trees and hedges bind the soil and provide valuable fodder, fuelwood and habitats for wildlife and predators of pests. These highly integrated systems are now largely a thing of the past. Farms have become more specialised and crop and livestock enterprises are separated. Livestock are often reared indoors on farms whose arable land is too small to absorb the livestock wastes. Livestock enterprises have become removed from the centres of arable cropping.

Agriculture has increased its use of external inputs as it has intensified. Between 1945 and 1995, inorganic nutrient use on winter wheat increased in England and Wales from 50 to 325 kilogrammes per hectare (RCEP, 1996). But inputs of nutrients and pesticides are never used entirely efficiently by the receiving crops or livestock. As a result, some are lost to the environment. Some 30 to 80 per cent of nitrogen in fertilisers and small but significant quantities of pesticides are lost directly to the environment to contaminate water, food and fodder and the atmosphere. It is impossible to generalise about how much of applied pesticides are lost to the environment. Some are very mobile (such as herbicides) and get into water very easily; others rapidly adsorb onto soil particles and remain bound for long periods. What is clear is that sufficient amounts of pesticides are lost from agricultural systems to cause considerable external damage (Conway and Pretty, 1991).

Pollution problems arising from intensive agriculture have been widely documented over recent decades. Nitrate in water has in the past given rise to blue-baby syndrome in infants. Pesticides contaminating water harm wildlife and now regularly exceed strict drinking water standards; nitrates and phosphates from fertilisers, livestock wastes, and silage effluents all contribute to algal blooms in surface waters, causing deoxygenation, fish deaths and general nuisance to leisure users. Water abstracted for irrigation threatens aquifers and rivers when rates exceed replenishment. (In this book the term pesticide is used generically, incorporating insecticides, herbicides, fungicides, acaricides, nematicides and miticides.)

Soil erosion is another problem: when lost from fields it disrupts watercourses, and runoff from eroded land causes flooding and damage to housing and natural resources. Farmers and farmworkers are harmed by some pesticides, and consumers are exposed to residues in foods. Birds, fish and other wildlife are harmed and also suffer through the loss of habitats and food sources. The atmospheric environment is contaminated by methane, nitrous oxide and ammonia derived from livestock, their manures and fertilisers. Despite considerable research, advocacy and policy change over the years, these costs to national economies and environments are still growing.

Calculating the Costs

Several attempts have been made to put a cost on polluting activities associated with agriculture (Pimentel et al, 1992; Steiner et al, 1995; Pearce and Tinch, 1998). In theory, we should be able to calculate the benefits of, say, pesticide use, the financial costs of their use, and the external costs. The benefit–cost ratio would then be calculated as benefits divided by the financial costs plus external costs. For several reasons, however, it is impossible to make this simple calculation.

The first problem is the difficulty of putting a price on non-marketable goods. How do we value, for example, skylarks singing on a summer's day, or a landscape with hedgerows and trees, or clean water, or even a human life? The second problem is that the data that do exist often represent the cost of cleaning up a problem, such as groundwater contamination or replacing honeybee colonies killed by pesticides. They do not reflect an individual's willingness to pay to avoid the problem occurring in the first place. These difficulties with methodology make the whole business of calculating a total cost somewhat arbitrary.

Two studies of the costs of pesticide use in US agriculture by David Pimentel and colleagues at Cornell University and R Steiner and others at the Rockefeller Foundation put the total external cost at between US$1.3 and $8 billion per year. These costs were for public health damage; domestic animal deaths; loss of natural enemies; pesticide resistance; honeybee losses and reduced pollination of crops; crop and fisheries losses; bird losses; groundwater contamination; and government regulations to prevent damage. But the cost of bird deaths (a massive $2 billion) is arrived at by multiplying 67 million losses by a nominal $30 a bird. The groundwater contamination ($1.8 billion) is based on the cost of cleaning up groundwater, and fishery losses are estimated partly on the numbers of fish kills and known costs of monitoring pesticide residues in wildlife.

Equally contentious is the 'value' of a statistical life – a figure needed if total costs are to be estimated. In the UK and US, $1 to $2 million is commonly used as the value of a statistical life. As some 6000 US children may develop cancer as a result of exposure to pesticides in fruit and vegetables, then the total cost would be $12 billion (Pearce and Tinch, 1998). Where these figures get even more difficult is when a lower value is put on a human life in developing countries – typically ten to 20 times less than in the North. Despite the economic rationale, economies, wages and opportunities do differ from country to country: few can surely be at all comfortable with the moral and ethical issues that these kinds of calculations raise.

As a result of these methodological problems, it is impossible to arrive at a reliable figure for the environmental costs of modern agriculture. What is clear, however, is that as these figures give only a partial cost, the real costs are almost certainly higher. Many believe that they are getting worse, though, again, there are no 'hard' data to show this to be true. It is, however, possible to assess accurately some of the costs of the polluting activities. In the EU, pesticides in drinking water have to be removed if they exceed 0.1 microgrammes per litre (μg/l) for one product or 0.5 μg/l in total. The cost of removal for the UK water industry is £121 million annually, after the one-off investment of £1000 million for detoxification machinery (Soil Association, 1997; Tye, 1997). As some 29.2 million kilogrammes of active ingredient are used each year in the UK, this represents an annual cost of £4.14 per kilogramme of active ingredient used in agriculture. If the investment costs are spread over ten years, this rises to £7.57 per kilogramme of active ingredient. This is a cost that water consumers pay, and so represents an indirect transfer from the public to farmers. The cost per hectare of farmland on which pesticides are applied is between £19.39 and £22.10.

Interestingly, the costs per kilogramme of pesticide active ingredient are similar in the US. Each year, some 500 million kilogrammes of pesticide active ingredient are applied to crops. If the high 'Pimentel' figure of US$8 billion for the costs of damage is used, this puts the cost at $16 or £10 per kilogramme of active ingredient. If the more recent but lower figure of $1.8 billion that is spent just on groundwater monitoring is used, the cost is $3.6 or £2.25 per kilogramme of active ingredient used by farmers (Center for Science in the Public Interest, 1995). The cost would escalate massively if carbon filtration were to be used, as some estimates put the clean-up costs at $25 million per well (Curtis et al, 1993).

Nitrates in drinking water are also costly. These can come from both inorganic fertilisers and organic wastes and cost £24 million per year to remove, on top of £275 million invested between 1989 and 1996. This puts the annual cost to water consumers at £0.03 per kilogramme of nitrogen fertiliser used, assuming three-quarters of the nitrates come from nitrate fertiliser and one quarter from organic manures. But the cost per hectare of farmland rises to £3.70, owing to the large amounts of fertilisers still applied to crops and grassland. Other known costs in the UK include those caused by soil erosion and water runoff, which run at some £24 to 50 million per year, and the costs of farm wastes polluting water, which cause some £5 million of damage per year.

However, not all agree that these environmental and health effects are significant. Various studies in the US put the total benefits of pesticide use, derived by calculating the increased output and 'better' quality food arising from pesticide use, at some four times greater than the US$4 billion annual financial cost (Pearce and Tinch, 1998; Zilberman et al, 1991). To some this would mean that the external pesticide costs would have to exceed $12 billion (the benefits are said to equal $16 billion, less costs of $4 billion) before any controls should be exerted on pesticide use. Some make other comparisons. Dennis Avery, author of the book *Saving the Planet with Pesticides and Plastic*, has stated that 'agro-chemicals are not a documented threat to the survival of a single species... the health and ecological consequences of chemically based and intensive systems are overstated' (Avery, pers comm 1996). Others still say that without pesticides, the whole success of agriculture in terms of food production will be threatened, making the poor and hungry suffer yet more (Borlaug, 1992; 1994a,b). Some of these claims are investigated in detail in Chapter 3, where recent evidence for the emergence of more sustainable agriculture systems is investigated in detail. First, though, this chapter assesses some of the environmental problems to which 'real' costs are difficult to attribute.

Pesticides and Human Health

Pesticides in the Environment

In the past 50 years, the use of pesticides in agriculture has increased dramatically. The value of the global market is now about US$30 billion (Dinham, 1996). Herbicides account for nearly half, insecticides a quarter, and fungicides less than a fifth. The largest individual consumer is the US, now followed by countries of the Far East (see Table 2.1). Japan is the most intensive user per hectare of cultivated land, closely followed by The Netherlands. The highest growth rates for the market occurred in the 1960s, reaching some 12 per cent per year; they then fell back to 2 per cent during the 1980s and have declined further during the 1990s. But this hides some huge variation: pesticide use is rapidly increasing in developing countries, but falling in Europe as a result of recent policy reforms and the trend towards more concentrated products and the adoption of more sustainable technologies One industry estimate suggests that developing countries will buy 35 per cent of all pesticides in the year 2000, up from 19 per cent in the mid-1990s (Dinham, 1996).

When a pesticide is applied to crops, most is either taken up by plants and animals or degraded by microbial and chemical pathways. But a proportion is dispersed to the environment: some is vapourised, eventually to be deposited in rainfall, some remains in the soil, while some reaches surface and groundwater by runoff or leaching.

Table 2.1 *Global consumption of pesticides, 1996*

Area	*% of global market of US$30 billion*
North America (mainly US)	29.5%
East Asia	25.1%
West Europe	25.8%
South America	10.8%
East Europe	3.4%
Rest of the World	5.4%

Source: British Agrochemicals Association

Pesticides have been detected in rainfall. Most deposited are of local origin, though lindane found in a remote Japanese lake with no inflows of surface or groundwater probably travelled 1500 kilometres from China or Korea (Anderson, 1986). Organophosphates have been found at concentrations of 10 to 50 microgrammes per litre (μg/l) in the US, well above the maximum 'acceptable' level of 0.1 μg/l for drinking water (Glotfelty et al, 1987). But even at very low concentrations, the total loading on natural environments can be huge: 0.005 μg/l of DDT in rainfall over Canada in the 1970s put an annual loading on Lake Ontario of 80 kilogrammes from rainfall alone (Conway and Pretty, 1991).

Pesticides in groundwater, surface waters and drinking water are now the most serious environmental problem associated with pesticide use. Pesticides reach water by leaching, runoff, transport on soil particles, and rapid flow though cracked soils and field drains. Most pesticides found in the environment come from surface runoff or leaching. The proportion lost is normally low – usually of the order of 0.5 per cent but sometimes rising to 5 per cent (Conway and Pretty, 1991). The early generations of pesticides, the organochlorines, arsenicals and paraquat, are strongly adsorbed to soil particles and tend only to be lost when soil itself is eroded. This can be a significant source of pollution. Soil erosion is responsible for the reappearance of aldrin and dieldrin, formerly used on bulb fields and long since banned, in watercourses in Cornwall in the 1990s (RCEP, 1996). Nonetheless, the concentration of some of these herbicides has increased over time, with particular concerns in The Netherlands over paraquat and diquat in sandy soils, and in Northern Ireland over paraquat. Modern herbicides, however, are much more mobile and readily transfer from soil to water. Isoproturon, mecoprop, atrazine and chlorotoluron are frequently associated with breaches in drinking water standards. They are used on winter cereals and are often applied when the soils are wet and when lack of soil cover contributes to leaching and runoff.

Numerous ground water supplies now exceed the EC maximum admissible concentration of 0.1 μg/l for any individual compound, and 0.5 μg/l for total pesticides. Groundwater samples with residues above 0.1 μg/l range from about 5 per cent in Denmark to 50 per cent in Italy, Spain and The Netherlands (*Agrow*, 1996). In the US, some 9900 wells out of 68,800 tested between 1971 and 1991 had residues exceeding Environmental Protection Agency (EPA) standards for drinking water. Some compounds have been found long after their supposed cessation of agricultural use. In the late 1980s, the hazardous DBCP (dibromochloropropane) was detected in some 2000 wells in California over an area of 1.8 million hectares, even though its registration had been cancelled a decade earlier (Conway and Pretty, 1991). In England, the greatest contamination is under farmland on chalk, though it is important to note that farming is not the only source of contamination. Industry of various types has been implicated in point–source pollution of very high concentrations at several locations.

Resistance to Pesticides

Another cost of pesticide overuse is induced resistance in pests and weeds. Resistance can develop in a pest population if some individuals possess genes which give them a behavioural, biochemical or physiological resistance mechanism to one or more pesticides. These individuals survive applications of the pesticide, passing these genes to their offspring so that with repeated applications the whole population soon becomes resistant.

Resistance has now developed to all insecticide groups, and some 500 species of insects, mites or ticks have been recorded as resistant to one or more compounds. Resistance has also developed in weeds and pathogens. Before 1970, few weeds were resistant to herbicides but now at least 180 withstand one or more products. Some 150 fungi and bacteria are also resistant (Georghiou, 1986; WRI, 1994). Unfortunately, natural enemies evolve resistance to pesticides more slowly than herbivores, mainly because of the smaller size of the natural enemy populations relative to pests and their different evolutionary history. The coevolution of many herbivores with host plants that contain toxic secondary compounds means they have metabolic pathways easily adjusted to produce resistance.

New problems continue to emerge in the UK, where resistant pests and weeds are causing great problems for farmers. Blackgrass resistant to one or more herbicides has now been found on 750 farms in 30 counties – about 3.7 per cent of the total 20,000 arable farms in the UK. In early 1997 there were 117 farms in Lincolnshire where blackgrass resistance was recorded; 103 in Essex; 92 in Oxfordshire and 75 in Cambridgeshire. Two particularly important new problems are resistance to organophosphate pesticides (OP) in peach potato aphid and to fungicides in cereal eyespot. In Canada, resistant wild oats now infect 1.2 million hectares of Manitoba cropland, and in Australia more than 3000 large wheat farms covering one million hectares have weed biotypes resistant to virtually all herbicides (Vorley and Keeney, 1998). There are also increasing concerns that *Bacillus thuringiensis* (*Bt* for short) resistance in pests may be promoted by genetically modified organisms (GMOs) containing the *Bt* gene. In 1996, transgenic varieties of maize, cotton and potatoes carrying a *Bt* endotoxin gene were marketed and planted on a wide scale for the first time. The selection pressure for *Bt* resistance has now increased several fold.

Human Morbidity and Mortality

Pesticides are intended to be hazardous – their value lies in their ability to kill unwanted organisms. But they are rarely selective. Most act by interfering with fundamental biochemical and physiological processes that are common to a wide range of organisms – not only to pests and weeds but to humans too. Pesticides affect workers engaged in their manufacture, transport and disposal; operators who apply them in the field; the general public; and wildlife in the environment. In the first case, the hazard is common to all industries dealing with toxic chemicals. Of greater concern is the hazard to the field operator because of the high variability in field conditions and frequent lack of protective clothing. But most at risk are the general public and the wider environment. They are difficult to protect directly.

At very high dosages many pesticides are lethal both to laboratory animals and people and can cause severe illness at sublethal levels. But the risks differ from pesticide to pesticide. Some are known to be carcinogenic, others suspected but unproven; some are acutely toxic but produce no long-term effects, others of long-term concern. But much of the information on the health effects of pesticides is still highly contested, and there is no clear picture of how much harm they do to farmworkers and consumers of food. As the British Medical Association put it in 1992: 'there is an incompleteness of existing knowledge concerning the effects of pesticide exposure on human health' (Baldock and Bishop, 1996). This is still true.

Fatalities from pesticides at work in Europe and North America are rare – one a decade in the UK and eight a decade in California – and there are many other more common causes of death on the farm. One problem is that the systems for recording pesticide poisoning vary within and between countries and are difficult to compare. In the UK, a variety of institutions collect mortality and morbidity data, indicating that there are some 40 to 80 confirmed cases each year amongst farmers, farmworkers and the public (Conway and Pretty, 1991). In 1995–96, none of the 169 incidents investigated by the HSE was 'confirmed', with only eight regarded as 'likely' (HSE, 1996). But in California, where there is the most comprehensive system of reporting in the world, official records show that some 1200 to 2000 farmers, farmworkers and the general public are poisoned each year (California Department of Food and Agriculture, 1972–current). Unfortunately, there is also great risk from pesticides in the home and garden where children are most likely to suffer. In California alone, 6000 to 8000 children of less than six years of age are treated for pesticide poisoning each year. In Britain, 600 to 1000 people need hospital treatment each year from home poisoning (Conway and Pretty, 1991). Here the problem is that pesticides are rarely perceived as being in the same category as medicines and so are less carefully guarded.

The most recent concern, and possibly one of the most significant yet to emerge in Europe, is over the health effects of organophosphates and carbamates commonly used in sheep dips. In the UK, it has long been compulsory for farmers to dip their sheep annually to control sheep scab. Organophosphates (or OPs) and carbamates affect insects and humans by reacting with the enzyme acetyl cholinesterase, which plays a key role in terminating the transmission of nerve impulses. As OPs inhibit this enzyme, they result in continuous nerve stimulation, leading to such symptoms as headaches, giddiness, nausea, blurred vision and rapid heart action. In Latin America, some 10 to 30 per cent of tested farmworkers show inhibition of cholinesterase, and a 1996 study put the annual level of poisoning in the US due to OPs alone at 3000 to 5000 people (Steenland, 1996). But, until recently, little has been known of the long-term effect of OPs. During the 1990s, sheep farmers started to report long-term ill health, which they put down to the OPs used in the dips. Official bodies and industry continued to claim that the products were both safe and efficient, and indicated that there was no cause for their removal. However, it is now clear that OPs have caused great harm. In 1993, Ministry of Agriculture, Fisheries and Food (MAFF) finally accepted that there were 529 cases of human illness related to sheep dips. And the numbers continue to rise.

Even so, the government and its Veterinary Products Committee continued to argue until early 1997 that OPs were safe if used in accordance with manufacturers' instructions. Others, though, do not think so. Department of Health officials have reported that low levels of exposure to OPs can be dangerous; and the OP Information Network has argued for several years that nerve damage does occur. Doctors from the Southern General and Guys Hospitals who have conducted studies on the effects of OPs on sheep farmers have found higher than supposed levels of poisoning. But their evidence was sceptically received by representatives of the Advisory Committee on Pesticides, the Medical Research Council and the Ministry of Defence.

The emerging evidence is still contested, with some saying that nothing should be done until more research is completed. Meanwhile, many sheep farmers clearly suffer ill-health, symptoms that are not repeated in other livestock sectors of the farming industry. Paul Tyler, the Lib Dem spokesman for rural affairs, has indicated that as many as 6000 sheep farmers may be suffering ill-health through the use of OPs; there are 90,000 sheep farmers in Britain in all (*Farmers' Weekly*, 1996). The Countess of Mar is a farmer who has suffered ill-health from sheep dips. In a 1997 interview with David Buffin of the Pesticides Trust, she said:

> *...we had our first flock of sheep in 1986 and I used to help with the sheep dipping. I would actually dunk the sheep in the dip, and only in retrospect did I realise that I had what is known as 'dipping flu' each time we dipped. This meant having a runny nose, a tight chest and cough, but it used to disappear in 24 to 48 hours, and I did not think any more of it... In 1989, I accidentally spilt some dip in my boot... I went on until I finished dipping... Again the ill effects went after 48 hours. About three weeks later I got up in the morning to find I was too tired to go out and help around the farm. I then spent nearly a year just going between the bed and settee. I could not concentrate, was not able to read, nor absorb what I had read and developed strange muscle aches.*

Since then, she has asked several hundred questions in the House of Lords and has led calls for restrictions to be put on the use of OPs until much better information is available.

The first Minister for Environmental Protection of the 1997 government, Michael Meacher, agreed in late 1996: 'It is now clear that OP products, particularly sheep dips, can be extremely dangerous. Public knowledge about health risks is still far too low, and OPs are still used without proper safety precautions when safer alternatives exist.' Some believe that OPs may also play a role in encouraging people to take their own lives. Suicides and stress-related illnesses are already common in rural areas (see Chapter 6). According to Robert Davies, a consultant psychiatrist from Somerset, farmers using OPs are four times more likely to commit suicide (OPIN, 1996). They are also more likely to develop signs of mental illness, ranging from personality changes, impaired understanding and loss of short-term memory and language disorders (*Farmers' Weekly*, 1997a).

Pesticides and Health in Developing Countries

In developing and particularly tropical countries, mortality and illness due to pesticide exposure are much more common relative to the amount of pesticide used. Lack of legislation, widespread misunderstanding of the hazards involved, poor labelling and the discomfort of wearing protective clothing in hot climates greatly increase the hazard both to agricultural workers and to the general public. Moreover, many pesticides known to be highly hazardous and either banned or severely restricted in industrialised countries, such as DDT, parathion, mevinphos and endrin, are still widely available.

According to the latest estimates by the World Health Organisation (WHO), at least three million and perhaps as many as 25 million agricultural workers are poisoned each year. Of these, some 20,000 die. At least 3000 people die in China alone each year. The data paint a bleak and worsening picture (Pretty, 1995a; Dinham, 1993; Repetto and Baliga, 1996; Singh, 1996; *Pesticides News*, 1997):

- In Malaysia and Sri Lanka, 7 to 15 per cent of all farmers experience poisoning at least once in their lives.
- In Thailand, surveys of government hospitals and health centres indicate that there are some 40,000 people poisoned by pesticides each year.
- In the Philippines, 50 per cent of rice farmers have suffered from sickness due to pesticide use.
- In Latin America, 10 to 30 per cent of agricultural workers show inhibition of the blood enzyme cholinesterase, which is a sign of poisoning by organophosphate pesticides.
- In India (Himachal Pradesh), farmers in 1996 reported that 46 per cent had suffered respiratory difficulties and generalised body pain, 26 to 40 per cent had eye problems and 94 per cent had headaches after pesticide applications.
- In Venezuela, 10,300 cases of poisoning with 576 deaths occurred between 1980 and 1990.
- In Brazil, 28 per cent of farmers in Santa Catarina say they have been poisoned at least once, and in Paraná, some 7800 people were poisoned between 1982 and 1992.
- In Egypt, more than 50 per cent of cotton workers in the 1990s suffer symptoms of chronic pesticide poisoning, including neurological and vision disorders.
- In China, some 13 to 15,000 new cases of pesticide poisoning in agriculture were reported annually between 1993 and 1995, with some 3200 fatalities each year – many were said to be victims of

> home-made cocktails marketed illegally; some 30 per cent of products were unlicensed by authorities, and 90 per cent of the incidents were caused by OPs.

One of the best studies to quantify the social costs of pesticide use was conducted at the International Rice Research Institute (IRRI) in the Philippines. Researchers investigated the health status of Filipino rice farmers exposed to pesticides and found statistically significant increased eye, skin, lung and neurological disorders. Two-thirds of farmers suffered from severe irritation of the conjunctivae; and about half had eczema, nail pitting, and various respiratory problems. Other studies in Taiwan show that at least 30 to 50 per cent of exposed farmers are affected by skin disorders; in Egypt 50 to 60 per cent of farmers and workers suffer optical damage, dizziness and permanent sensory loss; and in Central Java, Indonesia, 69 symptoms of ill-health associated with pesticide application have been documented (Rola and Pingali, 1993; Pingali and Roger, 1995; Guo et al, 1996; Jishi and Hirschern, 1995; Amr, 1995).

Agnes Rola and Prabhu Pingali calculated the health costs of these pesticide problems, taking into account impact on exposed farmers and the costs of restoring individuals to normal health (see Table 2.2). They then compared the economics of various pest control strategies. The so-called 'complete protection' strategy, with nine pesticide sprays per season, returned less per hectare than the other two control strategies. It also cost the most in terms of ill-health (7500 pesos per hectare). The 'natural control' strategy resulted in no ill-health amongst farmers. As Rola and Pingali put it: 'the value of crops lost to pests is invariably lower than the cost of treating pesticide-related illness and the associated loss in farmer productivity. When health costs are factored in, the natural control option is the most profitable pest management strategy.' Any expected positive production benefits of applying pesticides are clearly overwhelmed by the health costs.

It would appear that, in rice systems at least, both farmers and national economies would be better off by cutting pesticide use and adopting sustainable farming practices. It is the recognition of these benefits that partly accounts for the massive increase in uptake of sustainable agriculture in irrigated rice systems in South-East Asia, with at least 790,000 farmers now using little or no pesticides, yet still increasing their yields by some 8 to 12 per cent, and further benefiting from the harvest of fish, frogs and snails they now get from the unpolluted rice fields (Pretty et al, 1996; Pretty, 1995a).

Table 2.2 *Benefits and health costs of pest management strategies in lowland irrigated rice, the Philippines*

Pest management strategy	*Agricultural returns, excluding health costs (pesos/ha)*	*Health costs (pesos/ha)*	*Net benefit (pesos/ha)*
'Complete' protection: nine applications of pesticide per season	11,846	7500	4346
Economic threshold: treatment only when threshold passed, usually no more than two applications used	12,797	1188	11,609
Natural control: pest control emphasises predator preservation and habitat management, alternative hosts and resistant varieties	14,009	0	14,009

Source: Rola A and Pingali P. 1993. *Pesticides, Rice Productivity and Farmers – An Economic Assessment*. IRRI, Manila and WRI, Washington

Countryside, Wildlife and Genetic Diversity

Habitats under Threat

The countryside is one of our most precious resources. It provides food, timber, wildlife and habitats, jobs, landscape, and opportunities for recreation. In most parts of Europe, these goods and services have been maintained by traditional farming systems. In Britain, the patchwork quilt lowland fields, the moorlands and hill pastures, the blanket bogs and sandy coastal pastures, the acid heaths and the woodlands – all of these have been created and maintained by farming. On the continent, habitats are equally varied – the wood pasture *dehesas* and *montados* of Spain and Portugal; the dry *maquis* and *garrigue* grasslands of France; the integrated crop–livestock systems of *tanya* in Hungary, *minifundia* in Portugal, *coltural promiscura* in Italy, and lowland mixed farms of Ireland; the pastoral systems in northern Sweden and Finland; and the permanent vine, olive and other tree crop systems of Mediterranean countries (Bignall and McCracken, 1996; Viera and Eden, 1995; IEEP and WWF, 1994).

The Portuguese *montados* are mixed systems of cork and holm oak trees, with pigs, cattle, and sheep and various cereals (Viera and Eden, 1995). Few external inputs are used, and though the outputs of the individual components are low, these systems have great value. They provide diverse local livelihoods, produce 70 per cent of the world's cork, have aided the survival of several rare breeds of cattle, sheep, goats and pigs, and are home to many rare animals and birds, including wolf, lynx, black vulture, wild boar and many others. There are thought to be some 2.5 million hectares of *montados*. But there has been much recent loss, as trees have been grubbed up to be replaced by intensive cereals farmed with heavy machinery (Yellachich, 1993).

In Italy, a million hectares of olive groves are mostly still traditionally managed. But this is another system in decline. Older trees are larger and less productive, but they provide good habitats for wildlife. Newer intensive groves are much more productive, and so, increasingly, traditional ones are being abandoned. It is thought that roughly 110,000 tonnes of olives are left unharvested each year in traditional groves, equivalent to about 22 million kilogrammes of olive oil. In Germany, the Black Forest mid-altitude mountainous region has seen a dramatic decline in agriculture over the past 30 to 40 years (Luick, 1996). Yet agriculture has long been crucial for the Black Forest itself. It has preserved areas of high ecological importance and landscapes with a deep cultural value. It has produced high quality food, maintained traditional breeds of livestock, and provided recreational areas for urban populations. Declining agriculture is threatening all these values. In the Frieburg district, for example, the number of farms has fallen from 52,000 to 29,000 in the past two decades. Over the same period, the number of traditional Black Forest cattle, a small indigenous and hardy breed, has fallen by 80,000. As a result, the area of woodland has doubled, leading to the loss of many rich biotypes, particularly the diverse grasslands. Prospects for the future are very poor, as no one wants to enter farming. In the 1950s, Frieburg was annually training 750 to 800 young people for farming; in 1994, there were only 26 trainees.

In Spain, there are many areas suffering from decline and abandonment. In the Contraviesa, part of the Alpujarra Mountains of southern Granada, agricultural systems are very diverse. Farmers intercrop trees with cereals and legumes, growing up to ten varieties of almonds, five of figs, and 25 of vines, as well as raise sheep, goats and pigs (Remmers, 1996). The area is noted for its strong self-dependence, diverse agriculture and deep mistrust of external authorities. Yet it is being threatened by the emigration of younger people to cities and unstable prices for some primary products. The *dehesas* are highly integrated systems contributing significantly to both natural and social capital. But these

Box 2.1 The value of Spanish *dehesas* to natural and social capital

The term *dehesa* refers to a mix of woodland pastures and open grassland, dominated by holm oak (*Quercus ilex*) and cork oak (*Q suber*), with cereal crops and livestock – sheep, cattle, pigs and goats – feeding on grass and acorns. These are highly integrated systems, with the trees providing charcoal, firewood, shade, acorns and cork; the cereals providing grain and fodder; and the animals providing livestock products (meat and milk). There are usually four to 20 year rotations for arable practices, so the whole landscape is a mosaic of mixed habitats. They are very rich with wildlife: up to 60 plant species per square metre, butterflies, birds and animals, including threatened species such as the Spanish imperial eagle, Egyptian vulture, black stork and Iberian lynx. *Dehesas* are also important sources of employment for local people when managed properly. For example, the publicly owned 7000 hectare estate in Jérez de la Frontera, Andalucía, produces cork, timber, firewood, livestock (sheep, deer and cattle) and wild plants (herbs and fungi). The arable and grassland systems are low input. There is high labour use, with 13 full-time jobs per 1000 hectares per year – much higher than in neighbouring estates not managed in an integrated fashion.

Sources: Diaz M, Campos P and Pulido F J. 1997. 'The Spanish dehesas: a diversity in land use and wildlife' in: Pain D J and Pienkowski M W (eds) *Farming and Birds in Europe*. Academic Press, London. Cuff J and Rayment M. 1997. *Working with Nature: Economies, Employment and Conservation in Europe*. RSPB and Birdlife International, Sandy, Beds.

areas are generally in decline, with old trees removed for more intensive and large-scale agriculture or land given over to reafforestation with pines and eucalyptus (see Box 2.1).

In the British countryside, every habitat is in decline. Hedgerows and drystone walls are cherished features of the landscape. In Britain's 450,000 kilometres of hedgerows, there are some 600 plant, 1500 insect, 65 bird and 20 mammal species. A few are relics of ancient woodland that covered Britain until about 3000 years ago. But these are fast disappearing. In 50 years, we have lost over 40 per cent of our hedgerows. The losses continue, despite increasing public concern, and may even be increasing. Some 13 to 16,000 kilometres are lost each year, of which two-thirds are uprooted and one third destroyed through neglect. There was anecdotal evidence that some farmers were actually accelerating the removal of hedgerows in late 1996 and early 1997 prior to the passing of a law to protect them. Much of the recent loss, though, is due to neglect rather than to deliberate removal.

Drystone walls are attractive landscape features of many regions.

There are 112,000 kilometres of drystone walls in England, but a 1996 Countryside Commission survey found that half of these are now derelict or not stockproof, and a staggering 96 per cent are in need of some form of restoration. In the past 40 years, about 7000 kilometres have been lost altogether (CPRE, 1997). Over the same period, the number of ponds and lakes in Britain fell from 470,000 to 330,000 (DoE, 1996).

Apple orchards used to be a distinctive feature of the rural landscape. In 1970, there were 62,200 hectares but, by 1994, only 31,800 hectares remained. Again, the loss is continuing. MAFF pays farmers £4700 per hectare, via an EU scheme, to grub out their 'uneconomic' orchards. And many are taking up the offer. The EC proposed in 1997 to spend £77 million on grubbing a further 20,000 hectares across the EU. But orchards should not be judged just on food-production efficiency grounds alone. They are habitats for wildlife, such as orchids, cowslips, bees, hares and birds, and they are also storehouses of biological diversity. There used to be some 2000 varieties of apples in Britain's traditional orchards. In shops, consumers will now very rarely find more than ten local varieties (Common Ground, 1991).

Special features of the British landscape are given protected status as Sites of Special Scientific Interest (SSSI). These are considered to be the most valuable jewels in the countryside crown. Yet there is also evidence of their diminishing status and protection. In the year ending to March 1996, English Nature recorded 163 cases of damage on 121 SSSIs – a 20 per cent rise on the previous year in which some 1110 hectares of SSSIs were damaged, and 111 SSSIs so damaged that they had lost their special nature (English Nature, quoted in HC, 1996). Overgrazing and ploughing of grasslands were the main problems.

Vanishing Wildlife

Modern farming has had a severe impact on wildlife. It has been estimated that 170 native species in the UK have become extinct this century, including one in 14 of our dragonflies, one in 20 of our butterflies and one in 50 of our fish and mammals (DoE, 1996; Fuller et al, 1994). Since 1945, we have lost 95 per cent of our wildflower-rich meadows; 30 to 50 per cent of ancient lowland woods; 50 per cent of our heathland; 50 per cent of lowland fens, valley and basin mires; and 40 per cent of hedgerows. Despite increasing public concern, rates of loss are still increasing.

Species diversity is also declining in the farmed habitat itself. There was a 30 per cent fall in the number of plant species in arable fields from the late 1970s to 1990 alone (Barr et al, 1993). Overgrazing of upland grasslands and moorlands has reduced species diversity. Draining and fertilisers have replaced floristically rich meadows with ryegrass monocultures. There are also fewer butterflies and, most noticeably, fewer bird species. Farmland birds appear to have particularly suffered. The first recorded incidents occurred in 1804, when wheat seed dressed with arsenic to control smut caused the deaths of pheasants and partridges in Norfolk (Young, 1804). And it was direct poisoning of this kind in the 1950s that first alerted the public and scientific community to the hazards caused by pesticides to wildlife, prompting Rachel Carson to write her seminal book *Silent Spring* in the early 1960s.

Wild birds are killed in three ways – by direct poisoning, by indirect effects on their reproductive systems, and by destruction of their habitats (Conway and Pretty, 1991). Direct poisoning was common in the 1950s and 1960s, but less so now that most pesticides are less directly toxic to birds. Indirect effects on reproduction have also been important. Eggshell thinning was recognised during the late 1960s as the main cause of the collapse of predatory bird populations. With the withdrawal of the organochlorines, most populations of peregrines and sparrowhawks have returned to their pre-1940s levels. Destruction of habitats and loss of food sources are now the causes of population decline amongst birds. Many studies show that there have been rapid declines in the numbers of many threatened birds dependent on farmed habitats throughout Europe – from the songbirds of England to the bustards and birds of prey of the Spanish steppes (Tucker and Heath, 1994; Donázar et al, 1997; Suárez et al, 1997).

Several recent reports commissioned by the then DoE and produced by the Joint Nature Conservation Committee, English Nature, the Royal Society for the Protection of Birds (RSPB) and the British Trust for Ornithology (BTO) have recorded heavy losses of farmland birds over the past 25 years (Table 2.3) (Campbell and Cooke, 1995; DoE, 1996). The familiar and best-loved birds are being lost at an alarming rate; the corncrake, snipe, yellow wagtail and corn bunting are now rarities. The numbers of skylarks, a symbol of summer for so many, have fallen by nearly 60 per cent. Pesticides have affected these birds by reducing the abundance of invertebrate food sources during the breeding season; herbicides reduce the number of host plants, affecting the invertebrates that depend on them; and herbicides reduce the abundance of weeds and seed important as food for birds in winter. Several studies show that there is more bird and other wildlife, particularly butterflies, on organic farms than on conventional farms (Wilson, 1993; Feber, 1996; Chamberlain et al, 1996).

On the plus side, scavengers such as magpies, carrion crow and jackdaw are thriving – probably because more birds are being killed by

Table 2.3 *Changes in populations of farmland birds in rural Britain, 1970–1995*

Birds in decline		*Birds on the increase*	
Tree sparrow	–89%	Mallard	+36%
Grey partridge	–82%	Cuckoo	+38%
Turtle dove	–77%	Chaffinch	+40%
Bullfinch	–76%	Great tit	+42%
Song thrush	–73%	Pheasant	+57%
Lapwing	–67%	Coal tit	+82%
Reed bunting	–61%	Magpie	+138%
Skylark	–58%	Carrion crow	+151%
Linnet	–52%	Woodpigeon	+154%
Swallow	–43%	Blackcap	+170%
Blackbird	–42%	Jackdaw	+195%
Mistle thrush	–39%	Stock dove	+246%
Moorhen	–37%	Great spotted woodpecker	+303%
Sedge warbler	–35%		
Starling	–23%		

Source: Campbell L H and Cooke A S (eds). 1997. *The Indirect Effects of Pesticides on Birds*. Joint Nature Conservation Committee, Peterborough

road traffic. Other species, such as the coal and great tits and woodpeckers, are doing better because farmers are managing their woodlands more sensitively than they were a generation ago.

In Spain, the richly biodiverse 'pseudo-steppes' and other low-intensity farmland have long been home to many bird species. The mosaic of habitats, characterised by cereals, dry legumes, winter fallows and three- to five-year leys, and situated on many types of soils, is closely linked to the numbers of species and populations inherent to each. These systems are vast, covering four million hectares, one half of the total cereal growing area. More than half of this area is denoted as IBAs – important bird areas. Miguel Naveso and colleagues (Suárez et al, 1997; Donázar et al, 1997) have documented how these areas are under threat from fundamental changes in farming practices. The area under irrigated land is now some 3.21 million hectares – up 77 per cent since 1960, and fertilisers and pesticides are increasingly used. Land holdings are also being amalgamated – since the 1960s some five million hectares have been affected, with resulting larger units more intensive and more simplified. And where farmland has been abandoned, the reversion to scrub has taken away important habitats. These changes have affected many birds.

A measure of conservation concern is given by the Species of European Conservation Concern (SPEC) status (Tucker and Heath,

1994). For Europe as a whole, 38 per cent of bird species are SPECs. For the Spanish pseudo-steppes, this increases to more than 80 per cent. Thirty species are in decline and only nine are increasing. The great bustard, lesser kestrel and little bustard are globally threatened species; locally endangered include the Egyptian vulture and black wheatear. Skylarks, quails, stone curlews and hen harrier are all now vulnerable. These areas will continue to be under threat. As Francisco Suárez and colleagues put it in 1997:

> *...from an economic perspective, the pseudosteppes are fairly marginal. The fact that dry cereal farms, low-intensity livestock farms and mixed farms all have low-profit margins means that traditional and current agricultural users depend on income obtained from subsidies.*

According to national forecasts, plans exist to irrigate a further 600,000 hectares of *dehesas* and pseudo-steppes in the next 20 years. But if this and other agricultural intensification happens, much of the natural value will be lost forever.

> *The viability of pseudosteppes in the medium term depends on the implementation of an aid system that fully incorporates agricultural practices which contribute to different aspects of production – such as the conservation of natural values or the landscape.*

Much the same is true of diverse habitats in Scotland. Long-term research by Eric Bignall and Davy McCracken (1997) on the island of Islay in the Inner Hebrides has revealed again the crucial role of a complex and diverse landscape for bird life. Islay has a number of important bird species: barnacle goose, chough, corncrake, golden eagle, golden plover, hen harrier, merlin, peregrine falcon and white-footed goose. By dividing the island's semi-natural and agricultural systems into eight types, and classifying all 687 one-kilometre squares, they found that different species use different land types at different times of the year. It was the mosaic that was crucial: 'one of the most important features [is] ... how all of the land types are selected at some time in the year. Consequently we concluded that it is the diversity of land types that supports such a large number of bird species.' It is clearly not a question of maintaining one or two habitats or remnants amongst intensive farmland. Whole landscapes need to be protected through mixed and sustainable farming practices.

Disappearing Countryside

Britain's countryside is also being lost. Some 7 per cent of Britain, and 10 per cent of England, is now urban or suburban. A further 2.5 per cent is taken up by roads and railways outside urban areas. The CPRE says that a sixth of England's land area has already been lost to urban areas at a current rate of 11,000 hectares per year; while the former DoE said that the rate is only about half this, with a total of 12 per cent expected to be lost by 2016 (up from 10 per cent in 1980) (CPRE, 1992; 1993; DoE, 1993; 1996; RCEP, 1996). Both agree that the greatest pressures are in the south and east of the country, and forecast that a significant increase in urbanisation will occur during the late 1990s and early 21st century. By the middle of the next century, a fifth of England could be urbanised if rates continue.

Most people associate tranquillity with rural areas. But tranquillity in rural areas has fallen substantially. A 1995 study for CPRE and the Countryside Commission has documented the changes in tranquil places since the 1960s (CPRE, 1995). These are defined as being sufficiently far away from the visual or noise intrusion of development or traffic to be considered unspoilt by urban influences. These are areas that urban people would like to live in or visit when they think of the 'rural idyll'. Yet the area of tranquillity has fallen by a quarter over the past 30 years (Table 2.4). The trend is driven mainly by increasing road and air traffic and will continue to get worse for the foreseeable future.

Something similar has happened in Norway, where 22 per cent of the land is forested, 3 per cent under agriculture, and almost all the rest comprised of mountains, glaciers and lakes. But developed areas are expanding. In 1900, nearly half of the country was located more than five kilometres from encroachment by roads, hydroelectric schemes, powerline corridors, industrial projects and housing. By 1994, this had fallen to just an eighth of the country (Skauge, 1996).

The increased numbers of cars is an important reason for reduced tranquillity. CPRE also reported that in 1996 the character of as many as 5000 small country lanes and roads is being ruined by heavy traffic and the speed at which it travels. People are discouraged from cycling, walking and horse-riding and the roads themselves are damaged by the increased volume of cars and lorries. Many are being used as short cuts or 'rat-runs' to escape other congestion (CPRE, 1996).

Table 2.4 *Changes in the area of tranquillity in rural England, 1960s–1990s*

Area	*1960s (area of tranquillity: million hectares)*	*1990s (area of tranquillity: million hectares)*	*% loss (1960s–1990s)*
East Anglia	0.90	0.81	–11
East Midlands	1.09	0.87	–20
North East	0.65	0.59	–9
North West	0.95	0.77	–18
South East	1.58	1.03	–35
South West	1.97	1.58	–20
West Midlands	0.91	0.72	–21
Yorks and Humberside	1.14	0.93	–18
Total for England	9.19	7.30	–21%

Note: tranquil areas lie 4 km from the largest power stations; 3 km from highly trafficked roads and major industry; 2 km from most motorways and major trunk roads; 1 km from medium-disturbance roads and mainline railways. A tranquil area also lies beyond military and civil airfield noise and beyond open cast mining. Tranquil areas have a minimum radius of 1 km.
Source: Council for the Protection of Rural England and Countryside Commission. 1995. *Tranquil Areas*. CPRE, London

Declining Genetic Diversity in Agriculture

Farmers of traditional and sustainable agricultural systems have long favoured diversity on the farm. Around the world, many farmers still cultivate a huge variety of mixtures of cereals, legumes, root crops, vegetables and tree crops. In Africa more than 80 per cent of all cereals are intercropped, often with as many as 20 species grown in close proximity. In the Andes in Peru, the anthropologist Robert Rhoades (1987) once recorded 36 potato varieties in one field alone. These were all shapes and sizes, and a variety of colours, including black, red, blue, purple, yellow and white. Altogether some 3000 potato varieties are still grown by Andean farmers.

Although modern varieties and breeds have almost always displaced traditional varieties, it is only in the latter part of this century that fields monocropped to single species have become common. During the 20th century, about three-quarters of the genetic diversity of agricultural crops worldwide was lost. Only about 150 plant species are now cultivated, of which just three (rice, wheat and maize) supply almost 60 per cent of calories derived from plants (FAO, 1993a; Fowler and Mooney, 1990). The trend has been spiralling downwards in most countries. In India, more than 30,000 rice varieties were once grown, but now just ten cover 75 per cent of the whole rice area. In the US, 80–90 per cent of vegetable and fruit varieties have been lost this century, including 6121 apple varieties (85 per cent), 2354 pear varieties (88 per cent), 546 garden pea varieties (95 per cent), 394 field maize varieties (91 per cent), and 383 pea varieties (94 per cent) (Fowler and Mooney, 1990; Pretty, 1995a). Now 71 per cent of the maize area is planted to six varieties; 96 per cent of the pea area to two varieties; and 65 per cent of the rice area planted to four varieties. Similar concentrations have occurred in Europe. In The Netherlands, a single potato variety covers 80 per cent of the potato land, while 90 per cent of wheat is planted to three varieties. In the UK, 68 per cent of early potatoes are planted to three varieties, and four winter wheat varieties account for 71 per cent of winter wheat area. As Cary Fowler and Pat Mooney (1990) put it: 'the losses of fruit and vegetable varieties are staggering'.

Animal variety has also fallen. In Europe, some 750 breeds of domestic animals (horses, cattle, sheep, goats, pigs and poultry) have become extinct since the beginning of the 20th century; and a third of the remaining 770 breeds are in danger of disappearing by 2010. In Britain, 26 breeds of farm animal have been lost this century – five of cattle, two of goat, four of horse, seven of pig, and eight of sheep. Pig farming, for example, is now dominated by just two breeds – the large white and the British landrace.

Livestock raising has undergone great intensification across Europe, with many animals reared according to factory farming methods. This has made animal production more efficient, but it has also initiated many new animal welfare concerns and ultimately food safety problems. Several livestock products have been the subject of massive health concerns in recent years. These include strains of dangerous bacteria, such as *Listeria* in soft cheeses, *Salmonella* in chicken meat and eggs and *E coli* in various meats, as well as BSE in cattle and the potential link to CJD in humans. Welfare continues to be a concern, particularly over the conditions of animals reared inside buildings and their stress levels; and the transport of live animals, especially cattle and sheep. Nevertheless, some of these systems are being changed – pigs, now, are increasingly reared outdoors in extensive and humane systems.

In Spain, traditional breeds of cattle have been steadily replaced by foreign breeds. In 1955, traditional breeds made up three-quarters of the national herd; now they comprise less than a fifth (see Table 2.5). These changes were encouraged by the Ministry of Agriculture during the 1970s and 1980s. But, recently, many farmers have started to move back to traditional breeds, as Friesians have been unsuitable for much of the Spanish grazing lands. What we have lost is not only genetic diversity, but also some of the basic building blocks of sustainability. The narrowing of the genetic base makes agriculture more vulnerable to pest

Table 2.5 *Changes in numbers of different cattle breeds in Spain, 1955–1986*

Breeds	*1955 ('000 of cattle)*	*1986 ('000 of cattle)*
Friesian	338	1374
Swiss Brown	104	194
Charollais	0	22
Rubia Gallega	311	189
Asturiana	62	37
Retinta	39	137
Other native breeds	730	10
Total	1584	1963
Number of traditional breed animals as a proportion of all cattle	72%	19%

Notes: Rubia Gallega is a typical Galician breed.
Retinta is a typical breed of the *dehesa* grassland/woodland system.
Source: IEEP and WWF. 1994. *The Nature of Farming: Low Intensity Systems in Nine European Countries.* Institute for European Environmental Policy, London, and World Wide Fund For Nature, Geneva

and disease attack. Equally important, it also restricts consumer choice and welfare.

Soil Erosion

It is well established that agricultural intensification increases soil erosion. Yet it is only recently that the problem has been recognised as serious in the UK (RCEP, 1996; Evans, 1990a,b; 1996). It also clearly illustrates the cost of pursuing productivity increases without accounting for environmental damage. Erosion is now known to be greatest when there is little vegetative cover, such as during winter when winter cereals are being grown; when slopes are long, such as in big fields; and when farmers cultivate up and down slopes, rather than across the contour. Some 30 to 95 tonnes per hectare of soil can be lost from fields if field boundaries and hedges have been removed.

A major cause of soil erosion has been the shift in recent years towards the cultivation of winter cereals, driven by production-oriented policies. Since the late 1960s, the area of land sown to winter cereals has tripled, largely at the expense of grassland and spring cereals. The high price of wheat has encouraged winter cultivation in fragile environments, and this has led to a massive increase in soil erosion. Bob Evans, a former government soil scientist, estimates that some 620,000 hectares (4.4

per cent) of land in Britain is now at high risk, with a third of this at very high risk (Evans, 1990a,b). In the very high risk category, erosion affects 5 to 10 per cent of fields each year, and during two years in five as many as 20 to 25 per cent are affected. Altogether, it is thought that some 40 per cent of arable land in England and Wales has been eroded.

Wind erosion is mainly a problem in the drier eastern areas. Flat open expanses of intensively farmed arable land on sandy or peaty soils in East Anglia and the Vale of York are particularly at risk. Here, too, the costs can be high, as a 1994 story about a large arable field in Suffolk illustrates: 'for 200 years, the field had been cultivated, but our generation... removed the hedges and created a dust bowl... This year the road was effectively impassable for five days...and it took a crew of workers from the county's highways department most of a month to clear... to the tune of £10,000, most of which came from public funds' (*Daily Telegraph*, 1994).

Both water and wind erosion lead to losses of organic matter from the soil. First identified as a matter of concern in the 1970 Strutt Report (MAFF, 1970), organic matter levels have continued to fall. Peter Bullock of Silsoe College has warned that 'soil organic matter has fallen to dangerous levels in about 30 per cent of UK soils' (McLaren, 1994). The risk of erosion increases as soil organic matter falls because it improves the permeability of soil to water, increasing infiltration and reducing runoff and surface capping. Without organic matter, there are serious implications for soil stability, water retention and nutrient-holding capacity – problems that affect farmers directly as well as producing off-site costs. The advisory level is 2 per cent or higher for organic matter in soils, but many have now less than 1.5 per cent. As a result, yields on some 15 per cent of soils are thought to be affected by reduced organic matter levels. The Royal Commission on Environmental Pollution (RCEP) reports that further decreases in organic matter are expected well into the next century (DoE, 1997).

On the chalk downland of southern England erosion was uncommon until winter cereals were cultivated widely. In the late 1970s, only 5 per cent of these chalk downs were under winter cereals, but this leapt to 65 per cent by the early 1990s. As a result, soil loss accompanied by flooding has caused great damage to housing and farms. About 60 incidents of flooding have occurred on the South Downs alone in the past 15 years, including at Mile Oak, where £105,000 of damage was caused to housing and a further £150,000 spent on flood alleviation works; at Rottingdean, where £420,000 of damage occurred to 40 houses, gardens and roads; and at Breaky Bottom, where £81,000 of damage was caused to a valley-bottom vineyard.

These types of incidents are increasingly occurring elsewhere and appear most common in the West Midlands, Somerset, Dorset, Suffolk

Table 2.6 *Annual cost of soil erosion in the UK (£ million)*

On-farm costs (£ million)		*Off-farm costs (£ million)*	
Lowlands			
water erosion	2.14	Roads and property	3.3
wind erosion	1.60	Accidents (5 per year)	0.1
Uplands	0.08	Water pollution	
Loss riparian land	3.46–4.22	uplands	3.6–30
		lowlands	2.0
		Foot paths	0.5
		Stream channels	7.0
Total	7.28–8.04	Total	16.53–42.9
Grand Total		23.81–50.94	

Source: Evans, R. 1996. *Soil Erosion and its Impact in England and Wales.* Report to Friends of the Earth, London

and the Isle of Wight. The costs of clearing soil from roads and housing can be substantial. In Kent, it takes 7200 person-hours a year to clear washed soils from the roads, putting the total cost per 100 hectares of chalk and greensand soils at £121 (Evans, 1996). This is slightly above the average cost of £96 per 100 hectares incurred by local authorities. The greatest costs are in the sandy soils of the Isle of Axholme (£186 per 100 hectares) and the fens of East Anglia (£474 per 100 hectares). These costs are borne by local authorities, individual households and insurance companies, as well as by neighbouring farmers (Evans, 1996; Boardman and Evans, 1991; Boardman, 1990; Robinson and Blackman, 1990). Soil erosion is therefore costly both to farmers and to the wider public. On the farm, farmers experience costs due to the loss of the soil's productive capacity and the direct loss of nutrients, seeds and other inputs. Off the farm, soil erosion can cause flooding, damage to property and traffic accidents, and substantially contribute to water contamination and pollution. A recent comprehensive survey has put the cost of soil erosion in England and Wales at £23.8 to £50.9 million per year (see Table 2.6). These off-farm costs are similar to those in North America. In England and Wales, off-site costs are US$164 to $426 per 100 hectares; in the US they are $350 to $1070 per 100 hectares (Evans, 1986; Faeth, 1993).

The Modernist Policy Context

It is clear that the type of agriculture we have developed has been able to produce remarkable amounts of food. It is also clear that this has come at some considerable cost – both to the environment and to people. For many years, the nature of these external costs was not well understood. Even now, new problems are emerging. Some are threatening the very nature of whole sectors of agriculture – such as BSE in cattle and the effects of organophosphates on farmers' health.

If there were no alternatives, then we would be truly in a fix. We need the food, so we have to pay the full cost. Fortunately, viable alternatives to industrialised agriculture do exist. As we shall see, sustainable agriculture offers the opportunity for agriculture to be productive while at the same time protecting food quality, the environment and rural communities. It has, however, been short-term policies that have helped to make modern agriculture so favourable.

Fifty Years of Production Policies in Europe

The principal goal of agricultural policy throughout Europe this century has been increased productivity. Financial support from the state, and later the European Community and then the Union, has been tied to output, and markets for produce have been guaranteed. In Britain, this began in the 1940s when provisions were made under various acts. The historic 1947 Agriculture Act was a landmark, since its principal objectives were raising food production and combating the chronic balance of payments deficit. Provisions were made in this and later acts for ploughing grants, for price subsidies of crop and livestock products, for grants for field drainage and other investments in fixed assets, for subsidies of fertilisers and lime, for per capita payments for beef calves, and for hedgerow removal. An annual price review guaranteed prices so that farmers would have 'at least a modest prosperity and insulation from economic factors' (Bowers and Cheshire, 1983).

Intensification was actively sought. The 1952 Agriculture (Ploughing Grants) Act set two rates for ploughing up grassland; the higher rate of £30 per hectare (equivalent to £430 at current prices) was for the removal of at least 12-year-old grassland. Rates of uptake of the grant were greatest in eastern and south-western counties where grassland was rapidly converted to cereal production (Bowler, 1979). The 1957 Farm Improvement Grant further favoured the development of capital intensive cereal and dairy farming, since land drainage and hedgerow removal qualified for subsidy as well as buildings and machinery. These grants and subsidies continued through the 1960s, with new provisions to encourage the amalgamation of small farms into larger units and the early retirement of farmers. It was not until after Britain entered the European Community that many of these direct grants and subsidies were discontinued, such as those for fertilisers and lime in 1974. Nonetheless, the Common Agricultural Policy (CAP) continued to support agricultural prices, protect markets and provide for export subsidies. The guaranteed prices have generally been well above world market prices.

But the policy climate began to change in the early 1980s. Food commodities were beginning to accumulate at an alarming rate in the European Community, producing the first 'food mountains'. It was increasingly apparent that something must be wrong with a system that produces too much food and which therefore necessitates great expenditure on both storage and subsidising exports to other parts of the world. By the early 1990s, these surpluses were absorbing 20 per cent of the Common Agricultural Policy budget for storage alone. A further 28 per cent was expended on export subsidies. By 1997, the total CAP budget had grown to 41 billion ECU, of which about half was for direct payments to farmers (EC, 1995). In the UK, total expenditure under the CAP and on national grants was £3.03 billion during 1995 and 1996 – equivalent to £52 for each citizen.

These direct payments or subsidies are, however, seen to be unjust, with the largest farmers capturing the greatest proportion. In England, seven farms received arable area payments in excess of £500,000 in the mid 1990s. The largest 651 arable farmers (1.3 per cent of the total) received 10 to 15 per cent of all support, whereas the smallest 30,000 farmers (58 per cent of the total) received a third of all support. No farmers in Wales received more than £100,000. In both England and Wales, there were more than 30,000 farmers who each received less than £10,000 (Waldegrave, 1995).

Support under the Common Agricultural Policy

Despite growing recognition today of the real or potential social and environmental benefits of farming, the principal policy framework of the European Community still enshrines the notion of increased food production at all costs. The Common Agricultural Policy objectives, as outlined in Article 39 of the Treaty of Rome, are to increase agricultural productivity, secure EU food supplies, stabilise prices, and ensure a 'fair standard of living' for the community's farmers. The 1992 MacSharry reforms of the CAP did not change these objectives. Nonetheless, the reforms did begin to weaken the link between production and farm income. The reforms introduced a system of direct payments to farmers and a move away from market support as a means of securing farm incomes. To qualify for these payments, farmers had to comply with a range of specific controls which are intended to restrain production.

Arable production is restrained by set-aside, and livestock production by quotas and headage payments. Incentives were put in place for farmers to comply with new practices, and so reduce food production. Sustainable technologies and practices represented only a very small element of these compliances.

One of the most controversial elements of the reforms was set-aside. This encourages farmers to remove from production a fixed amount of their land (beginning at 15 per cent in 1992 but falling to 5 per cent by 1997) if they wished to continue to qualify for area payments, which in turn compensated them for the reductions in support prices. Compensation payments are based on the average historic yield for regions, rather than on individual farms; they therefore result in producers across Europe receiving different levels of compensation depending upon whether they are in a high or low yielding region. As with earlier attempts to reform the CAP, such as the introduction of milk quotas in 1984 and in 1988, legally binding ceilings on market support and voluntary set-aside, these reforms failed to address the underlying problems within the agricultural sector. With regard to the impacts on rural employment, these reforms reduced employment as farmers sought to reduce costs in response to falling prices. The Rural Development Commission (RDC) estimated that some 5400 full-time agricultural workers lost their jobs in Britain, with a further 2680 seasonal and part-time workers affected (RDC, 1996).

Set-aside has been the logical solution to overproduction under the current policy paradigm. But set-aside is unhelpful in sustainable agriculture because it both encourages intensification on the remaining land and undermines many of the basic principles of farming. It is true that set-aside has positively helped some wildlife, particularly birds. But to many farmers, it still feels fundamentally wrong to leave land unmanaged, or to spray herbicides on vegetation rather than plough it into the soil to help restore fertility.

The Gradual Greening of Policy

The CAP reforms lightly 'greened' agricultural policy by including, for the first time, policy measures designed to fulfil environmental objectives. Regulation 2078/92 requires member states to implement an agri-environmental programme. This obliges governments to offer farmers voluntary incentive schemes for adopting environmentally friendly forms of land management. The amount of farmland designated under these agri-environmental measures varies across member states. The average for the EU as a whole is about 13 per cent, but ranges from

Table 2.7 *Farmland under agri-environmental regulations, 1995–1996*

Country	*Amount of land designated under agri-environment regulations (per cent)*
Austria	100%
Finland	74.2%
Germany	28.6%
France	16.7%
Portugal	12.1%
UK	7.8%
Ireland	4.5%
Italy	4.2%
Denmark	2.4%
Netherlands	1.0%
Spain	0.4%
Belgium	0.2%
EU average	13.1%

Note: Spain has introduced ESAs over 10 per cent of its farmland, but take-up by farmers has been very low.
Source: Cuff J and Rayment M. 1997. *Working with Nature: Economies, Employment and Conservation in Europe*. RSPB and Birdlife International, Sandy, Bedfordshire

a low of 0.2 per cent in Belgium to a high of 100 per cent in Austria (see Table 2.7). These designations, however, do not guarantee that all farmers have adopted conservation friendly farming.

The Environmentally Sensitive Areas (ESA) scheme preceded these measures. It was first set up in 1986, following various protests about the draining and ploughing up of the Halvergate marshes in Norfolk. In the early 1980s, farmers were paid to drain marshland and convert it to cereal production. The damage caused to Halvergate led directly to the establishment of the Broads Grazing Marsh Conversion Scheme in 1985 – the forerunner of ESAs. This offered farmers a flat rate hectarage payment in return for cutting stocking rates and reducing pesticide and fertiliser use. Some 5000 hectares were designated, and 90 per cent of farmers came into the voluntary scheme. When ESAs were then mooted for sites across the country, a wide range of bodies were involved in the design process, choosing 160 sites. But these were then cut to only ten covering 738,000 hectares, the management prescriptions were simplified, and the final design centrally imposed (Blunden and Curry, 1988).

Under the current system, farmers enter ten-year voluntary management agreements in return for annual payments. These ESAs cover specified areas of designated high landscape or ecological value. There are currently 22 ESAs in England, covering some 1.7 million hectares, plus a further 21 in Scotland, Wales and Northern Ireland. This is about

14 per cent of total agricultural land. However, only a small number of farmers within ESAs have elected to join the schemes. Uptake within ESAs was 483,000 hectares in 1997 – about 28 per cent of the designated area (MAFF, 1997).

The Countryside Stewardship Scheme was established in 1991 and is available only in England outside ESAs. It aims to protect and enhance valued landscapes and habitats, and to improve the public enjoyment of the countryside. The scheme targets chalk grassland, waterside landscape, lowland heaths, coastal land, uplands and historic landscapes, such as orchards and meadows. Again, farmers receive payments for entering ten-year management agreements. Some 6500 management agreements with farmers have been made, with 107,600 hectares covered by the scheme in 1996.

The Welsh equivalent is the Tir Cymen scheme administered by the Countryside Council of Wales, which had some 900 agreements to the end of 1996. This scheme has had a substantial positive effect on farmers' incomes, on the environment and on local job creation (see Chapters 7 and 8). In Scotland, the Countryside Premium Scheme, launched in 1997, seeks to integrate all environmental schemes and is open to all 15,000 farmers and crofters currently not in designated ESAs (some 80 per cent of land is not covered by ESAs). A prior conservation audit is required with a five-year plan, and a range of payments available for both management options and capital works.

The Organic Aid Scheme is open to all farmers and offers incentives over a five-year period to convert to organic production. The payment levels are, however, so low that few farmers appear able to risk conversion. Payments for the first two years of conversion of cereals in Britain are 82 ECU per hectare per year (about £65) – the lowest in the EU. Over the whole of the EU, the average is 190 ECU, with four countries exceeding 275 ECU (Lampkin, 1996). The result is, of course, that there are more farms under organic agriculture elsewhere. Austria has the most, with 18,000 farms on 252,000 hectares.

Nitrate sensitive areas (NSAs) have been designed to limit nitrate leaching to aquifers used to supply drinking water. Farmers are offered voluntary incentives for following strict practices that limit leaching. Prior to regulation 2078/92, ten pilot NSAs were set up in 1990 on 10,700 hectares. Now a further 22 have been designated, covering 25,000 hectares of agricultural land. In addition to the NSAs, 68 nitrate vulnerable zones (NVZs) covering 600,000 hectares have been designated under the EC Nitrates Directive (EEC/91/676), where mandatory uncompensated measures apply. NVZs apply to any catchment where drinking water abstractions exceed 50 milligrammes of nitrate per litre. Denmark, Germany and The Netherlands have indicated that all their land will be designated as NVZs and will be subject to compulsory measures, and France intends to designate ten million hectares, some 50 per cent of agricultural land.

There is a range of other schemes of a more targeted nature, including:

- the Moorland Scheme, which pays farmers outside ESAs for each ewe removed, so as to encourage restoration and conservation of heather and other shrubby moorland;
- the Countryside Access Scheme, which is restricted to guaranteed set-aside land which is suitable for new or increased access;
- support for Sites of Special Scientific Interest (SSSIs) in which English Nature, Scottish Natural Heritage, the Countryside Council of Wales and the DoE in Northern Ireland provide payments for the adoption of management practices to protect habitats and species on 45,000 hectares;
- the Arable Incentives Scheme, in which farmers are able to apply for funding to test methods for protecting wildlife, especially birds, during arable farming;
- the Farm and Conservation Grant Scheme and the Farm Woodland Premium Scheme, both run by MAFF; the Woodland Grant Scheme run by the Forestry Commission; the Landscape Conservation Grants run by the Countryside Commission; and various National Nature Reserve agreements administered by English Nature.

Expenditure on all these schemes amounts to about £100 million, the largest proportion being for ESAs (£64.5 million in 1995 and 1996). But this remains a very small proportion of the total expenditure on the CAP. Taken together, these schemes cover some 1.86 million hectares of land in the UK. There is clearly a long way to go before all farming becomes environmentally sensitive.

Another problem is that these schemes do not necessarily support more sustainable agriculture. The piecemeal action that focuses on individual aspects of a farm, such as a riverside meadow or chalk grassland, does not encourage integrated farming on the whole farm. As a result, farmers may be encouraged to manage sustainably one particular field, but not the rest of the farm. In this way, the internal linkages and processes essential for sustainable agriculture are not necessarily promoted. However, there is a trend towards whole farm agreements.

This great diversity of schemes is seen by many farmers as both bureaucratic and unjust. It may be a matter of luck as to whether a farmer is inside or outside a designated scheme, and some miss out altogether.

Others, though, have many possible schemes to chose from, each with different rates and conditions. The House of Commons Select Committee on Agriculture recognised these problems when it reported in early 1997, indicating that all farmers should have access to all schemes.

Summary

Agricultural and rural development policies have in the latter part of the 20th century successfully emphasised external inputs as the means to increase food production. This has produced remarkable growth in global consumption of pesticides, inorganic fertilisers, animal feedstuffs, and tractors and other machinery. As a result, there have been spectacular increases in production.

But these achievements have brought many environmental and social costs. Several attempts have been made to put a cost on polluting activities associated with agriculture. But it is difficult to place a market value on environmental goods. How much is a landscape or clean water worth? Where data do exist, and they are mainly for the costs incurred in cleaning up or replacing lost resources, the costs are substantial. In the UK, it costs drinking water users £7.57 every time a farmer applies a kilogramme of pesticide active ingredient, and £0.03 for each kilogramme of nitrogen fertiliser. Put another way, every hectare of farmland produces external costs to drinking water of between £23 and £26 per year. This amounts to a free transfer of resources to farmers who pollute water.

Pollution problems arising from intensive agriculture have been widely documented over recent decades. Many, though, are on the increase. Pesticides continue to give cause for concern. Between 0.5 per cent and 5 per cent of those applied are lost to the environment. As a result, humans are still being poisoned. Globally, between three and 25 million agricultural workers are poisoned each year, with some 20,000 dying. Organophosphates are an increasing problem, with studies from Brazil, Egypt, the Philippines and Taiwan showing that 30 to 50 per cent of farmers are affected by a wide range of chronic symptoms. In the UK, sheep farmers are clearly suffering ill-health from the use of OPs in sheep dips.

Habitats and wildlife continue to be under threat across Europe. These are valued components of the countryside, produced mainly by farming. It is 'modern' farming, though, that is taking them away too. The mixed landscapes of trees, flower-rich grasslands, crops and traditional livestock in Spain, Portugal, Italy, Sweden, France and Germany are being lost. In the UK, every habitat is in decline. Some 13 to 16,000 kilometres of hedgerows disappear every year. Wildlife is disappearing too, particularly farm birds. Many of the most familiar and best-loved birds are being lost at an alarming rate; the corncrake, snipe, yellow wagtail and corn bunting are now rarities. The number of skylarks, a symbol of summer for so many, has fallen by nearly 60 per cent. Studies from Spain and Scotland show that diverse landscapes are the key to rich and diverse wildlife.

At the same time as these changes are occurring in wildlife, farming is losing genetic diversity. In The Netherlands, a single potato variety covers 80 per cent of the potato land, while 90 per cent of wheat is planted to three varieties. In the UK, 68 per cent of early potatoes are planted to three varieties, and four winter wheat varieties account for 71 per cent of winter wheat area. Animal variety has also fallen. In Europe, some 750 breeds of domestic animals (horses, cattle, sheep, goats, pigs and poultry) have become extinct since the beginning of the 20th century; and a third of the remaining 770 breeds are in danger of disappearing by 2010. What we have lost is not only genetic diversity, but also some of the basic building blocks of sustainability. The narrowing of the genetic base makes agriculture more vulnerable to pest and disease attack. Equally importantly, it also restricts consumer choice and welfare.

Soil erosion is a growing problem. Since the late 1960s, the area of land sown to winter cereals has tripled, largely at the expense of grassland and spring cereals. The high price of wheat has encouraged winter cultivation in fragile environments, and this has led to a massive increase in soil erosion. Altogether, it is thought that some 40 per cent of arable land in England and Wales has been affected by erosion. The natural capital of the soil, its organic matter, structure and invertebrates, has fallen year on year since at least the 1970s. Soil erosion is costly both to farmers and to the wider public. On the farm, farmers experience costs due to the loss of soil productivity and the direct loss of nutrients, seeds and other inputs. Off the farm, soil erosion can cause flooding, damage property, cause traffic accidents, and substantially contribute to water contamination and pollution. A recent survey has put the cost of soil erosion in England and Wales at £23.8 to £50.9 million per year.

The principal goal of agricultural policy throughout Europe this century has been increased productivity. Financial support from the state, and later the European Community and then the Union, has been tied to output, and markets for produce have been guaranteed. By 1997, the total CAP budget had grown to 41 billion ECU, of which about half was for direct payments to farmers. In the UK, total expenditure under the CAP and on national grants was £3.03 billion during 1995 and 1996 – equivalent to £52 for each citizen. The 1992 CAP reforms lightly 'greened' agricultural policy by including, for the first time, policy measures designed to fulfil environmental objectives. Regulation 2078/92 requires member states to implement an agri-environmental programme. This obliges governments to offer farmers voluntary incentive schemes for adopting environmentally friendly forms of land management. The amount of farmland designated under these agri-environmental measures varies across member states, varying from a low of 0.2 per cent in Belgium, to 7.8 per cent in the UK, to 100 per cent in Austria.

In the UK, there are many schemes to support the gradual greening of farming, including environmentally sensitive areas, Countryside Stewardship, Tir Cymen, the organic aid scheme, nitrate sensitive areas, arable incentive schemes and many others. But despite these schemes, costing about £100 million per year (about 3 per cent of the UK budgetary support to farming), there is still a long way to go before all farming becomes environmentally sensitive or sustainable. Many farmers do not enter schemes that are seen as too top-down in design; many schemes are piecemeal, focusing on individual habitats rather than on whole farms. Nonetheless, there are many advances in sustainable agriculture, both on research stations and on farms, that show that system-wide change is profitable for farmers and good for the environment and rural economies.

Chapter 3

Sustainable Agriculture in Europe

A close connection exists between soil fertility and health... the health of humans, beast, plants and soil is one indivisible whole; the health of the soil depends on maintaining its biological balance... such health as we have almost forgotten should be our natural state.

Eve Balfour, 1943, *The Living Soil*

What is Sustainable Agriculture?

Basic Principles of Sustainable Agriculture

In the latter part of the 20th century, external inputs of pesticides, inorganic fertilisers, animal feedstuffs, energy, tractors and other machinery have become the primary means to increase food production. These external inputs, though, have substituted for free natural control processes and resources, rendering them more vulnerable. Pesticides have replaced biological, cultural and mechanical methods for controlling pests, weeds and diseases; inorganic fertilisers have substituted for livestock manures, composts, nitrogen-fixing crops and fertile soils; and fossil fuels have substituted for locally generated energy sources. What were once valued local resources have all too often become waste products.

The basic challenge for sustainable agriculture is to make better use of available biophysical and human resources. This can be done by minimising the use of external inputs and by utilising and regenerating local or internal resources more effectively. A more sustainable agriculture systematically pursues five goals (Pretty, 1995a,b):

- a thorough integration of natural processes such as nutrient cycling, nitrogen fixation, soil regeneration and pest–predator relationships into agricultural production processes, ensuring profitable and efficient food production while increasing natural capital;
- a minimisation of the use of those external and non-renewable inputs that damage the environment or harm the health of farmers and consumers, and a targeted use of the remaining inputs used to minimise costs;

- the full participation of farmers and other rural people in all processes of problem analysis, and technology development, adaptation and extension, leading to an increase in local self-reliance and social capital;
- a greater use of farmers' knowledge and practices in combination with new technologies emerging from research, including innovative approaches not yet fully understood by scientists or widely adopted by farmers;
- the enhancement of both the quality and quantity of wildlife, water, landscape and other public goods of the countryside.

Resource-Conserving Technologies

There is an increasing number of proven and promising resource-conserving and regenerative technologies for pest, nutrient, soil, water and energy management (see Table 3.1). These have emerged from many sources: from traditional agricultural systems; from both the long history of organic agriculture and from recent experiences with 'precision farming'; and from farms in both developing and industrialised countries. Many of these have arisen on farms where steps have already been taken to reduce both the costs and the adverse environmental effects. Natural processes are favoured over external inputs, and by-products or 'wastes' from one component of the farm are emphasised, as they can become inputs to another. In this way, farms remain productive while reducing negative impacts on the environment.

These technologies do two important things. They conserve existing on-farm resources, such as nutrients, predators, water or soil. And they introduce new elements into the farming system that add to the stocks of these resources, such as nitrogen-fixing crops, water harvesting structures or new predators, substituting for some or all external resources. Most represent low- or lower-external input options.

Many of the individual technologies are multifunctional. This implies that their adoption will result in favourable changes in several aspects of the farming system and natural capital at the same time. For example, hedgerows encourage wildlife and predators and act as windbreaks, so reducing soil erosion. Legumes are introduced into rotations to fix nitrogen and also act as a break crop to prevent carry-over of pests and diseases. Clovers in pastures can reduce fertiliser bills and lift sward digestibility for cattle. Grass contour strips slow surface runoff of water, encourage percolation to groundwater, and are a source of fodder for livestock. Catch crops prevent soil erosion and leaching during critical periods, and can also be ploughed in as a green manure. The incorpora-

Table 3.1 *Promising and proven resource-conserving and regenerative technologies for European agriculture*

Primarily for Pest and Weed Management

- reduced and low-dose applications of pesticides, fungicides and herbicides;
- patch spraying of pesticides, herbicides and fungicides;
- releasing natural enemies – mainly for glasshouse crops – such as of *Encarsia* wasps for whitefly control or fungal parasites (eg *Verticillium*) for aphid and whitefly control;
- beetle banks and flowering strips in fields to encourage predators;
- rotational farming to keep weed levels low and restore natural fertility;
- conservation headlands unsprayed to encourage weeds and insects that are good for birds;
- bio-pesticides and bio-fungicides;
- mechanical weeding;
- resistant varieties and breeds;
- pest monitoring to identify and assess levels before treatment (such as sticky, suction, bait and pheromone traps).

Primarily for Soil and Nutrient Management

- legumes and green manures in mixed rotations, such as clovers in pastures and undersowing of clover in cereals;
- undersowing of grass in maize to reduce soil erosion, soak up excess nitrogen and provide early spring grazing next year;
- targeted inputs using computer-aided technologies and global positioning systems;
- recycling of livestock manures and sewage sludge;
- deep placement and/or slow release fertilisers;
- sub-soiling of pastures.

Primarily for Soil and Water Conservation

- contour cropping for soil conservation;
- low, zero or non-inversion tillage for soil conservation;
- irrigation scheduling and on-farm reservoirs.

Primarily for Energy and Feed Management

- energy crops of poplar and willow in long-term rotations, burned on-farm for heat and electricity;
- regular maintenance and servicing of machinery;
- time switches and energy-efficient bulbs in buildings;
- rotational grazing to improve grass productivity and reduce bought feed requirements.

tion of green manures not only provides a readily available source of nutrients for the growing crop but also increases soil organic matter and hence water retentive capacity, further reducing susceptibility to erosion. Low-lying grasslands managed as water meadows that are good for wildlife also provide an early-season yield of grass for lambs.

The principle of integration, a central theme of sustainable agriculture, implies a focus on increasing the number of technologies and practices, and the positive, reinforcing linkages between them. If a technology, such as a nitrogen-fixing legume, is taken by farmers and adapted to fit their own cropping systems, and this leads to increases in output or reductions in costs, then this is the strongest evidence of success. What is clear is that resource-conserving and regenerative technologies are spreading. In Denmark, some 150 farms have in-field weather stations to help predict disease outbreaks in potatoes, leading to cuts in fungicide use, with some growers able to postpone first applications for five or more weeks. In the UK, some 150,000 hectares of cereal farms were computer-mapped by 1997, enabling inputs to be targeted more precisely and the total use of pesticides and fertilisers to be cut. In the UK, three-quarters of crops grown in glasshouses also use natural predators to control pests rather than pesticides. In France, there are 700 farms in the national network researching and implementing 'agriculture durable'. In the state of Baden-Württemberg in southern Germany, 102,000 farms use sustainable practices and technologies, though not all integrated at the whole farm level. The organic revolution also continues, with demand from consumers growing. Now the number of farmers converted entirely to organic practices in the EU is about 50,000, including 18,000 farms in Austria, 10,300 in Italy and 6000 in Germany.

The Sustainability Dividend – More from Less

This chapter will show that resource-conserving technologies and practices can be hugely beneficial for both farmers and rural environments. They make a positive contribution to natural capital, either returning it to predegraded levels or improving it yet more. At the same time, direct costs are reduced. There is more from less: more natural capital from fewer external inputs; more social capital from involving farmers and rural communities in change; more food output from fewer fossil-fuel derived inputs.

The best evidence of redesigning farming systems to incorporate sustainability comes from countries in Africa, Asia and Latin America (Pretty, 1995a; Hinchcliffe et al, 1996; Pretty, 1996b). Here a major concern is to increase food production in areas where farming has been largely untouched by the modern packages of externally-supplied technologies. In these lands, farming communities which adopt regenerative technologies have substantially improved agricultural yields, often using few or no external inputs. In the higher-input systems where the Green Revolution of the 1960s and 1970s had such an impact on food output, particularly of rice, the dividend comes from a reduction in pesticide use. Most of these successes are community-based activities that have involved the complete redesign of farming and other economic activities at local level with the full participation of farmers and local people in the process.

In the 20 countries recently examined by the International Institute for Environment and Development (a total of 63 projects and programmes), we found that there were at least 1.93 million households farming more than four million hectares with resource-conserving technologies and practices (Pretty and Thompson, 1996). What is remarkable is not so much the numbers – at one level they are large, but in comparison with total agricultural area in the world, they are still tiny – but that this happened in the previous two to five years. Moreover, many of the improvements are occurring in difficult, remote and resource-poor areas that were commonly assumed to be incapable of producing food surpluses. Highlights from these initiatives include:

- Some 223,000 farmers in Paraná and Santa Catarina in southern Brazil using green manures and cover crops of legumes and livestock integration have seen yields of maize and wheat increase from two to three, to four to five tonnes per hectare.
- Some 45,000 farmers in Guatemala and Honduras have used regenerative technologies to triple maize yields to some two to 2.5 tonnes per hectare and to diversify their upland farms, which has led to local economic growth that has in turn encouraged migration back from the cities.
- More than 300,000 farmers in southern and western India farming in dryland conditions, and now using a range of water and soil management technologies, have tripled sorghum and millet yields to some two to 2.5 tonnes per hectare.
- Some 200,000 farmers across Kenya, as part of various government and non-government soil and water conservation and sustainable agriculture programmes, have more than doubled their maize yields to about 2.5 to 3.3 tonnes per hectare and substantially improved vegetable production through the dry seasons.
- 100,000 small coffee farmers in Mexico have adopted fully organic production methods and yet increased yields by half.

- Close to 800,000 wetland rice farmers in Bangladesh, China, India, Indonesia, Malaysia, Philippines, Sri Lanka, Thailand, and Vietnam have shifted to sustainable agriculture, where group-based farmer-field schools have enabled farmers to cut pesticide use while still increasing their yields by about 8 to 12 per cent.

There are many smaller-scale successes too. In countries of Sub-Saharan Africa, such as Burkina Faso, Ethiopia, Senegal, Uganda and Zambia, many thousands of community-level initiatives are now showing that if farmers are involved in technology development, they can substantially improve the food outputs from farming without diminishing natural capital. These successes share important common characteristics. They have made use of resource-conserving technologies in conjunction with group or collective approaches to agricultural improvement and natural resource management. They have made participatory approaches and farmer-centred activities the focus of their agenda, and so these activities occur on local people's own terms. They have not used subsidies or food-for-work to 'buy' the participation of local people, or to encourage them to adopt preselected technologies; improvements, therefore, are unlikely simply to disappear at the end of projects or programmes. They have generally supported the active involvement of women as key producers and facilitators of change. And they have emphasised 'adding value' to agricultural products through agroprocessing, marketing, and other off-farm activities, thus creating employment and income-generating opportunities and retaining the surplus within the rural economy.

Increased Yields Are Only Part of the Picture

These findings suggest that the widespread adoption of sustainable agriculture would have a massive impact on the productive capacity of most countries. Indeed, the currently low- and medium-yielding countries, where yields are in the range of 0.5 to two tonnes per hectare and which are also the poorest, would benefit most in terms of food production than the high-yield countries. But yield data provide only a partial picture of the improvements that have occurred, or could occur, through sustainable agriculture and rural development. Sustainable agriculture also offers farmers the opportunity to diversify their strategies. For this reason, cereal production may not increase in the short term following the adoption of sustainable agriculture, since many farmers respond to increased yields by diversifying into new crops, thereby reducing the area under cereals.

One pattern is as follows: soon after transition to sustainable agriculture, farmers get excited by the extra production; in the following year, everyone is producing so much more that the local market price declines or collapses and producers get lower returns. In the next year, they reduce the area under cereals and diversify into new crops. This cycle has been observed in a number of contexts. In the Honduras, for example, farmers who used to grow only maize now cultivate upwards of 25 crops per farm; in Gujarat, India, farmers have diversified away from sorghum- and millet-based systems to grow many types of vegetables; in Bolivia, upland potato farmers have reduced field sizes (by up to 90 per cent) to save on labour while producing the same amount of food; in Kenya, some farmers have diversified without much change to cereal production, introducing traditional staples of sweet potato, arrowroot and bananas, together with vegetables and fruit trees (Bunch and López, 1996; Shah, pers comm; Ruddell, 1997; World Neighbors, 1996; Pretty, 1997).

Nonetheless, output of basic staples can increase. There are many examples of food deficit areas, such as parts of Ethiopia, Kenya, Uganda, Zimbabwe, Honduras and Guatemala, becoming food surplus areas, following the adoption of sustainable agriculture (Hinchcliffe et al, 1996; Bunch and López, 1996). Clearly, changes in infrastructure, markets and policies may have been important, but sustainable agriculture played a key role. As we shall see later in this chapter, there are also many remarkable changes occurring throughout Europe and North America as farmers begin to take steps towards sustainability. As agriculture becomes more knowledge, management and labour intensive, so farmers can return to their traditional role of making contributions to both natural and social capital.

Sustainability as a Contested Concept

Sustainability is a word that has entered common use in recent years. Since the Brundtland Commission put sustainable development on the map in the mid to late 1980s, more than 100 different definitions of sustainability have been developed and published (WCED, 1997). Each emphasises different values, priorities and practices. Clearly, no reasonable person is opposed to the idea. But what does it mean in practice?

To some it implies the capacity of something to persist unchanged for a long time. To others, it means the capacity to adapt in the face of uncertainty or to bounce back from stresses and shocks. To some, it implies economic growth while protecting natural resources. To others, still, it is just accounting for the environment while continuing on a business-as-usual track. Does any of this help in the context of farming?

We all know sustainability represents something good, but what exactly? And, more importantly, has the notion of sustainable agriculture contributed to better farm practices, or is the term too easily hijacked? A vigorous debate has recently emerged in Europe around these issues.

A wide range of more sustainable forms of agriculture is now emerging and spreading in Europe and North America. Many different terms are being used including, in no particular order, organic, sustainable, alternative, integrated, regenerative, low-external input, balanced input, precision farming, targeted inputs, wise use of inputs, resource-conserving, biological, natural, eco-agriculture, agro-ecological, biodynamic, and permaculture. Some of these have precisely defined standards; most do not. Some are making substantial contributions to natural capital; others very little.

The emerging problem is that some stakeholder groups take the stance that their type is the only form of sustainable agriculture. Some argue, for example, that precision farming (involving yield mapping and targeted use of inputs) is sustainable, as it is efficient and productive (MF, 1995; DowElanco, 1994). We even have some arguing, with convoluted logic, that high-input farming is sustainable as its very high productivity protects uncultivated forests and wildernesses from coming under the plough (Avery, 1995).

In the UK, integrated crop management (ICM) has emerged as a popular concept in British farming in recent years. ICM, though, is another contested concept (Pretty, 1996a; Redman, 1995; Farmer, 1995; Wise, 1996; Leake, 1996). Some believe it offers great promise, indicating that steps can be taken towards more sustainable farming. Others say that it is just rhetoric, representing nothing more than an attempt by large farmers, supermarkets and chemical companies cynically to claim the environmental high ground while not significantly changing farming practices. In an editorial for the Minnesota Institute for Sustainable Agriculture's newsletter *CornerPost*, Tim King (1996) warned that

> *...precision agriculture, as it relates to satellite and computer technology, is squarely and firmly boxed in the packaging of the industrial agriculture paradigm. Like all industrial agriculture technologies preceding it, precision agriculture will further alienate farmers from their fields, the food they produce and their communities.*

But there is also good evidence to show that precision farming can substantially reduce input use and can therefore contribute to natural capital. Some, however, say precision farming means using more inputs. What is clear, however, is that these are mostly all forms of *more* sustainable agriculture. Nevertheless, it is difficult to say how much more sustainable one is than another unless we know much more about the specific local conditions and technologies in use (see Three Steps of Sustainability later in this chapter).

Common Misconceptions about Sustainable Agriculture

Some important misconceptions persist about sustainable and regenerative agriculture (Pretty, 1995a). The most common is that sustainable agriculture represents a return to some form of low technology, backward or traditional agricultural practices. This is untrue. Sustainable agriculture incorporates recent innovations that may originate with scientists, with farmers or both. Some are very high technology; some are practices that are proven over thousands of years. Sustainable agriculture can represent economically and environmentally viable options for all types of farmers, regardless of their farm location, and their skills, knowledge and personal motivation.

A common misconception relates to what exactly organic agriculture is. Organic farming uses external inputs where they are needed. In general, it seeks, first, to maximise the use of internal and biological cycles, but utilises external inputs including inorganic fertilisers (such as phosphorus in the form of rock phosphates and potassium in the form of sulphate of potash) and naturally derived pesticides as appropriate. Many of the goals of sustainable agriculture correspond closely to those set out in organic standards. It is also commonly stated that any farming using low or lower amounts of external inputs can only produce low levels of per-hectare output. Again, this is known to be untrue. Many sustainable agriculture farmers show that their crop yields can be better than or equal to those of their more conventional neighbours. Most show that their costs can be reduced. In developing countries, this offers new opportunities for economic growth for communities that do not have access to, or cannot afford, external resources. In the industrialised world, it means that farmers can remain profitable while contributing more to natural and social capital. Either way, this shows that sustainable farming can be compatible with small or large farms, and with many different types of technology.

Nowhere were these misconceptions more clearly stated than in the Barber Report, the investigation sponsored by the Royal Agricultural Society of England and chaired by Sir Derek Barber in 1991. This 'commission' concluded that any 'criticism of modern farming is unsubstantiated and much public concern is misconceived... The image of a simple "green" agriculture with a contrived non-intensive output is incompatible with the aim of maintaining a competitive position in the market place.' Barber and his colleagues could not have been more wrong.

One reason is that they sought the views of just 85 individuals and organisations, and indicated that: 'the evidence is clearly defined in our report and is derived from the foremost authorities in the land'. Yet no meetings were held with environmental, community or consumer groups. What these would say, along with many farming organisations, is that sustainable agriculture is competitive, making farmers better off while helping to rebuild natural capital.

The emerging empirical evidence is perhaps more contested in the US. Some 82 per cent of conventional US farmers still believe that low-input agriculture will always be low output, even though there are tens of thousands of sustainable agriculture farmers who are showing, year after year, that this is untrue (Hewitt and Smith, 1995). Influential politicians continue to reinforce these beliefs. In 1991, the former US Secretary of Agriculture Earl Butz said:

> *We can go back to organic agriculture in this country if we must – we once farmed that way 75 years ago. However, before we move in that direction, someone must decide which 50 million of our people will starve. We simply cannot feed, even at subsistence levels, our 250 million Americans without a large production input of chemicals, antibiotics and growth hormones (quoted in Schaller, 1993).*

The myth-makers are well supported by Dennis Avery and colleagues from the Hudson Institute in Indianapolis. The 1995 book *Saving the Planet with Pesticides and Plastic* and subsequent publications seek to distinguish between high-yield and low-yield agriculture (Avery, 1995; 1996). Their narrative states that high-yield agriculture is good since it takes the pressure off wildernesses and wildlife; it is good because it can feed the world; it is good because evidence for any negative impacts on the environment is greatly exaggerated. High-yield agriculture can only be delivered with high levels of pesticides, fertilisers and plastics. Low-yield agriculture, by contrast, involves too few inputs and so is a threat to world food security (see Box 3.1).

What all this ignores is that sustainable agriculture can be high yield with low levels of external inputs. This is not pie in the sky. It is already happening on many thousands of farms worldwide. During the 1990s in Africa, Asia and Latin America, some 1.12 million farmers of rainfed agriculture, where yields are typically only 800 to 1500 kilogrammes per hectare, more than doubled their yields with sustainable agriculture; and a further 0.79 million farmers of irrigated rice systems have improved their yields by 8 to 12 per cent while substantially cutting pesticide use.

Box 3.1 The Myth-Makers

The now infamous Knutson study investigated the consequences of two potential agricultural policies in the US: a banning of all pesticides, and a banning of all fertilisers and pesticides. More than 140 agricultural scientists were called upon to provide estimates of how such cuts would affect yields and costs. The authors concluded that a 'no chemical' policy approach would lead to sharply lower yields – down 35 to 50 per cent coupled with higher crop prices, higher erosion, and higher inflation in the national economy. Food price inflation was highlighted: 'food price inflation arising from no chemical use would exceed 10 per cent annually and approach the 14 to 15 per cent level that existed during the world food crisis of 1973–1974'.

Many, though, have questioned the methodology and the potential conflicts of interest on the grounds that:

- a total ban of all chemicals is unrealistic as zero-option policies are unlikely ever to occur;
- the study failed to account for potential and real increases in resource-conserving technologies to substitute for cuts in fertilisers and pesticides;
- the study failed to account for induced research and development by farmers, researchers and businesses.

Source: Knutson R D, Taylor L R, Penson J B and Smith E G. 1990. *Economic Impacts of Reduced Chemical Use*. Knutson and Associates, College Station, Texas

Existing Forms of Sustainable Agriculture in Europe

Low-Intensive Farming and Traditional Systems

Despite the expansion of modern agriculture documented in Chapter 2, there are still millions of hectares of land farmed in environmentally sensitive or positive ways throughout Europe. These systems tend not to damage the environment. They also contribute to local communities and economies in other important social ways. Most are relatively unintensive, unchanged or unimproved. In a recent study of low-intensive agriculture in nine countries, Eric Bignall and Davy McCracken calculate that there are some 56 million hectares of farmland in this category, representing some 38 per cent of all utilisable agriculture land (Bignall and McCracken, 1996; IEEP and WWF, 1994; see Table 3.2). Many are under threat from modern farming methods and from rural abandonment (see Chapters 2 and 6).

There is diversity of both farmed products and wildlife in these habitats and systems. The wet grasslands in France, particularly in the

Camargue and Normandy, are important for cattle, goat and sheep grazing, and provide an important habitat for wild horses, cattle and many birds. About 1.25 million hectares are classified as important bird areas (IBAs) worthy of protection under the EU Birds Directive. The drier unimproved grasslands, such as the 1.4 to 1.7 million hectares of *maquis* and *garrigue* in France, the *machair* dune-grazing in coastal west Scotland, and the 500,000 hectares of lowland grasslands of Hungary, are also valuable and highly diverse.

The high mountain pastures of southern Europe are characterised by transhumance and other seasonal movements of livestock, and the grazing is essential to prevent encroachment and maintain floristic diversity – Mediterranean grasslands, for example, can have as many as 120 to 180 different species per 100 square metres. There are still ten million hectares in Spain and five million hectares in Greece under low-intensive livestock systems at high altitude, with several million animals still making annual summer migrations. The wood-pasture systems of Spain, Portugal, central Italy and parts of Hungary involve livestock grazing on permanent pastures with scattered tree cover. Small amounts of cereals are often cultivated, producing a diverse and mosaic-style landscape. The five to ten million hectares of *dehesas* in Spain and the 1.3 million hectares of *montados* in Portugal are characterised by species-rich grassland, mixtures of cork and holm oak trees, and sheep, pigs and cattle, and a remarkable abundance of wildlife, including many rare and now threatened species of birds. In Italy, there are some 250,000 hectares of hill pasture and coppiced woodland grazings for the traditional Maremmana breed of cattle.

Traditional mixed systems of subsistence or part-time farming with integrated crop and livestock production still exist too, including crofting in Scotland; Hungarian *tanya* which comprise small mixed holdings of two to 25 hectares with cereals, legumes, cattle, sheep, vineyards and orchards; *minifundia* in central and northern Portugal; *coltural promiscura* in Italy; and low-intensity mixed farms in Ireland. Low-intensive dryland arable on the steppes of Spain and Portugal are also valuable, and include 700,000 hectares in Alentejo in Portugal and four to nine million hectares in Spain. Cereal yields in these systems rarely exceed 800 kilogrammes per hectare. Traditionally managed permanent crops of fruit trees, olives and vines, primarily some 600,000 hectares in Greece, one million hectares in Italy and two to three million hectares in Spain and Portugal, also provide diverse habitats for wildlife. Olive groves provide important feeding for overwintering and migrating birds, which is no loss to farmers since they arrive after the harvest. But when they are intensified, many of the micro-habitats and food sources disappear, rendering them much less beneficial to birdlife.

Table 3.2 *Extent and characteristics of traditional and low-intensive agricultural systems in seven European countries*

Country	*Farmland under low-intensive systems (million ha)*	*Low-intensity farmland as a per cent of utilised agricultural area*	*Examples of low-intensive systems*
France	7.75	25%	wet and dry grasslands (eg *garrigue* and *maquis*)
Greece	5.6	61%	migratory livestock systems, olive groves
Hungary	1.5	22%	traditional grasslands, mixed *tanya* holdings
Ireland	2.0	31%	traditional mixed farms
Italy	7.1	14%	mountain pastures, traditional olive groves, lowland steppes, hill pastures, coppiced woodland grazing, traditional arable
Poland	2.74	60%	traditional mixed farming
Portugal	2.74	35%	low-intensive cereal on steppes, *montado* tree, livestock, cereals, sheep and goat grazing in mountains
Spain	25.0	82%	*dehesa* grasslands and open woodland, dryland arable, olive groves, extensive livestock
UK	2.0	11%	upland low-intensive grasslands and moors, grasslands by rivers, marshes and dunes
Total	56.4	38%	

Sources: Bignall E M and McCracken D I. 1996. 'Low intensity farming systems in the conservation of the countryside'. *Journal of Applied Ecology* 33, 416–424. IEEP and WWF. 1994. *The Nature of Farming: Low Intensity Systems in Nine European Countries.* Institute for European Environmental Policy, London, and World Wide Fund for Nature, Geneva

Organic Farming

The aim of organic farming is to create integrated, humane, environmentally and economically sustainable agriculture systems. Maximum reliance is put on self-regulating agro-ecosystems, locally or farm-derived renewable resources, and the management of ecological and biological processes. The use of external inputs, whether inorganic or organic, is reduced as far as possible. In some countries, organic agriculture is known as ecological or biological agriculture.

The term organic, as Nic Lampkin of the Welsh Institute for Rural Studies has put it, is 'best thought of as referring not to the type of inputs used, but to the concept of the farm as an organism, in which the component parts – the soil minerals, organic matter, micro-organisms, insects, plants, animals and humans – interact to create a coherent and stable whole' (Lampkin, 1996). What currently distinguishes organic farming from other forms of sustainable agriculture is the extension of legal and voluntary standards and certification procedures, such as the British standards laid down under the UK Register of Organic Food Standards (UKROFS) and EC standards under Regulation 2029/91.

In western Europe, there have been dramatic increases in organic agriculture in recent years. The extent has increased tenfold from about 120,000 hectares in 1985 to 1.2 million hectares in 1996 (Lampkin, 1996). Nearly 50,000 farmers are now engaged in certified organic farming (see Table 3.3). The greatest growth has been in Austria (from 500 to 18,000 farms), in Germany (from 1597 to 6068 farms), in Italy (from 600 to 10,300 farms) and in Sweden (from 150 to 3000 farms). In Austria, organic farming now covers 250,000 hectares, representing about 7 per cent of all farmland. In the UK, there are about 820 organic farms, with an area of 47,900 hectares. In the US, there are 4060 government-certified organic farms, but they do cover 450,000 hectares.

The most rapid expansion of organic production has occurred in Austria. The principal reason is the high level of funding and support from national policies and the public at large. Organic agriculture has become mainstream, with organic produce no longer selling at a premium but as a staple for many. In some locations, whole villages have shifted into organic production using box schemes for direct marketing (see Chapter 5). Labour use is 10 to 50 per cent higher on organic farms, and therefore this agriculture can help to keep communities economically viable. Finland is another country that has seen a very rapid growth in organic farming, from less than 7000 hectares in 1990 to more than 44,000 hectares on nearly 2800 farms in 1996. Signed agreements already indicate that the total area reached 100,000 hectares in 1996. Since the Finish government established fertiliser taxes in 1990, nitro-

Table 3.3 *Number of organic farms in the EU (1996)*

Countries	*Number of certified farms*	*Land area organic or in conversion in 1996 (ha)*
Austria	18,000	252,000
Belgium	200	5000
Denmark	1200	42,180
Finland	2780	44,730
France	3700	97,000
Germany	6068	310,480
Greece	1000	4500
Ireland	300	7000
Italy	10,300	202,210
Luxembourg	21	625
Netherlands	590	12,240
Portugal	325	10,190
Spain	1200	28,130
Sweden	3000	105,000
UK	820	47,900
EU total	49,504	1,169,185

Note: these numbers are continually changing, and not just upwards. In Austria, for example, many farmers withdrew from organic farming in 1995–1996 to take up alternative agri-environment options which had less stringent requirements than organic farming. But Nic Lampkin's 1997 estimates for both certified and in-conversion organic farms in the EU and central Europe put the total at more than 60,000 farms on 1.8 million hectares: Welsh Institute for Rural Studies at www.aber.ac.uk.

Sources: Lampkin N. 1996. *Impact of EC Regulation 2078/92 on the development of organic farming in the European Union*. Welsh Institute of Rural Affairs, University of Wales, Aberystwyth, Dyfed. Rundgren G. 1996. 'The northern lights of the organic world'. *Ecology and Farming*, 12, pp 12–14. Inger Källandar, pers. comm. 1997

gen consumption has fallen by 15 per cent, potassium by 30 per cent and phosphorous by 30 per cent (Brueninghaus, 1996).

The role of positive public policies has been vital in these transitions to organic agriculture. Proponents of organic agriculture have long argued that they deserve considerable public support to compensate for the conversion or transition costs. In addition, organic agriculture internalises many of the environmental and social costs that 'modern' agriculture actively externalises (and so passes to other people or sectors of the economy). The payments in the UK to 1997 were derisory compared with some offered in other EU countries. The irony is that the demand for organic food in Britain exceeds internal supply – some three-quarters of organic food consumed in the UK is imported.

Under EC Regulation 2078/92, Austria offers 335 ECUs per hectare per year for conversion of cereal systems to organic farming; this compares with 365 in Finland; 235 to 275 in The Netherlands and Ireland; and 120 to 180 for all other EU countries, except for the UK, which offers 82 ECUs (as of 1997). For vegetable conversion, the disparities are even greater, with Finland and The Netherlands offering more than 500 ECUs per hectare; Austria 450; Greece, Belgium and Ireland between 300 and 370; Spain, France, Italy and Portugal between 210 and 250; and the rest below 200, with only the UK below 100.

Another myth may need exposing when it comes to organic farming: it does not always mean lower yields. A recent study by the Institute of Grassland and Environmental Research of organic milk production found that grassland production fell by 15 per cent in the first year of conversion, but returned to preconversion levels (at eight tonnes of dry matter per hectare) as the clover became established. Average milk yields were maintained, despite a 15 per cent cut in concentrate use (IGER, 1996). In Norway, organic milk production is lower cost and so gives higher returns (Rosvold et al, 1996). Many individual farmers will support this evidence. They say that although yields do fall on conversion, they increase over time.

Integrated Farming Systems

Integrated farming systems have emerged in recent years as another more environmentally friendly approach to farming. Once again, the emphasis is upon integrating technologies to produce site-specific management systems for whole farms, incorporating a higher input of management and information for planning, setting targets and monitoring progress. There are important historical, financial and policy reasons why still relatively few farmers have taken the leap from 'modern' high-input farming to organic agriculture. But it is possible for anyone to take a small step which can, in theory, be followed by another step. Integrated farming in its various guises represents a step, or several steps, towards sustainability.

What has become increasingly clear is that farmers can cut inputs with the adoption of integrated crop management (ICM) or integrated farming system technologies without losing out on profitability. Some of these cuts are substantial; in others they are relatively small. But all contribute to natural capital. It is difficult to generalise about what happens to inputs and outputs, since much depends on locally specific conditions. It used to be thought that more sustainable agriculture, whether organic or integrated, would mean substantial reductions in both crop and livestock yields and economic returns (Pretty and Howes, 1993). However, this generalisation no longer stands.

It appears that farmers can make some small cuts in input use without negatively affecting gross margins. By adopting better targeting and precision methods, there is less wastage and so the environment benefits. Yields may fall initially but will rise over time. They can then make greater cuts in input use (20 to 50 per cent) once they substitute some regenerative technologies for external inputs, such as legumes for inorganic fertilisers or predators for pesticides. And, finally, they can replace some or all external inputs entirely over time once they have learned their way into a new type of farming characterised by new goals and technologies. But if too many changes are made too quickly – such as before natural capital in the soil is rebuilt or beetle banks established for predator management, then integrated farming can result in lower yields and lower gross margins.

Since the mid 1980s, a range of experimental studies and on-farm analyses across Europe and North America have shown what kind of dividend can be released from the adoption of more sustainable agriculture. The most notable of these studies have been in Germany, The Netherlands, France, Switzerland, the UK and the US. Figure 3.1 summarises the findings from most of the studies. It maps the change of yield following the adoption of sustainable agriculture technologies against the change in gross margins. The graph contains four possible routes from the 100:100 per cent starting point:

- top right sector: where both yields and gross margins increase with sustainable agriculture;

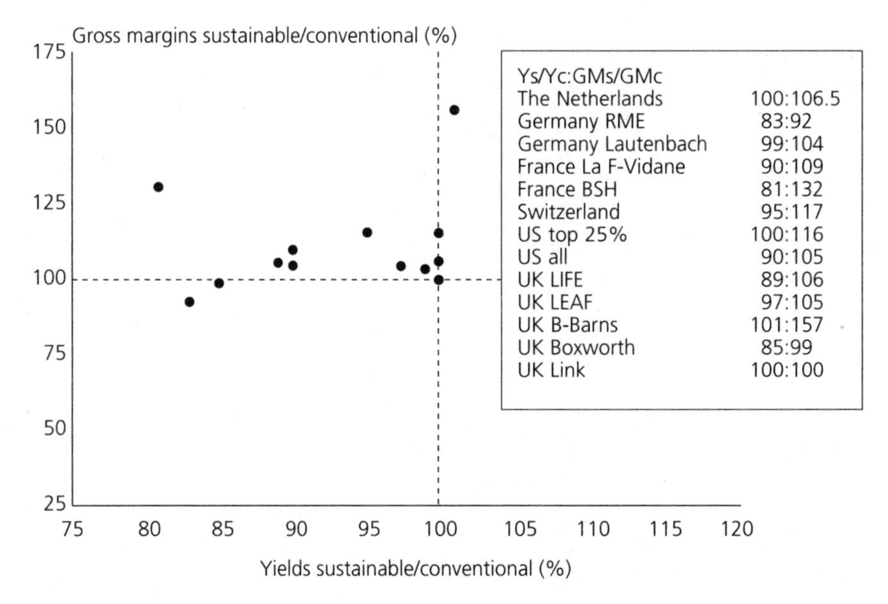

Figure 3.1 *Transition routes for sustainable agriculture systems in Europe and North America – changing yields against changing gross margins*

- top left sector: where yields decline, but gross margins are still higher;
- bottom left sector : where both yields and gross margins fall;
- bottom right sector: where gross margins decline, but yields increase.

The desirable outcome for individual farmers is clearly anywhere above the 100 per cent gross margin line. The antagonists of sustainable agriculture (the low input is low output lobby) assume that all sustainable agriculture falls into the bottom left sector. The bottom right is unlikely to occur, as it is irrational for farmers. The graph shows that most of the empirical data falls into the top right sector – where yields slightly decline, but economic returns increase, owing in the main to the sharp cuts in input costs. There are cases, however, where the cuts are too great for the system to bring greater returns, such as for some of the research in Germany and the UK (Boxworth).

Figure 3.2 maps the same yield changes against changes in pesticide and fertiliser use. This time the sector preferences differ:

- top right sector: where yields increase with increased use of pesticides and fertilisers – this has been the route followed by modernist agriculture during the course of this century;
- top left sector: where yields decline, but input use increases – this is a wasteful and uneconomical route for farmers;
- bottom left sector: where both yields and input use fall – which is good for farmers if cost declines exceed the forgone productivity;
- bottom right sector: where yields increase, but input use falls – which would be desirable for all.

As the graph shows, most of the transitions have occurred in the bottom left sector – when yields fall but inputs fall too. These studies show that at almost every site the capacity to cut pesticides is greater than for fertilisers. The research also shows that insecticide use can be cut more easily than herbicides and fungicides in the early stages of transition (the graphs do not show this since all pesticide types are combined). In the Lautenbach experiment, for example, insecticide use was cut over 12 years by 83 per cent, fungicides by 50 per cent and herbicides by just 28 per cent, putting the total reduction at 33 per cent (El Titi, 1992).

It is also important to note that the differences between countries are probably not significant. Time spent on developing and innovating sustainable agriculture technologies is more important that spatial factors. In the early stages, farmers are able to make some savings by cutting down on wastage. But the big dividends come from adopting

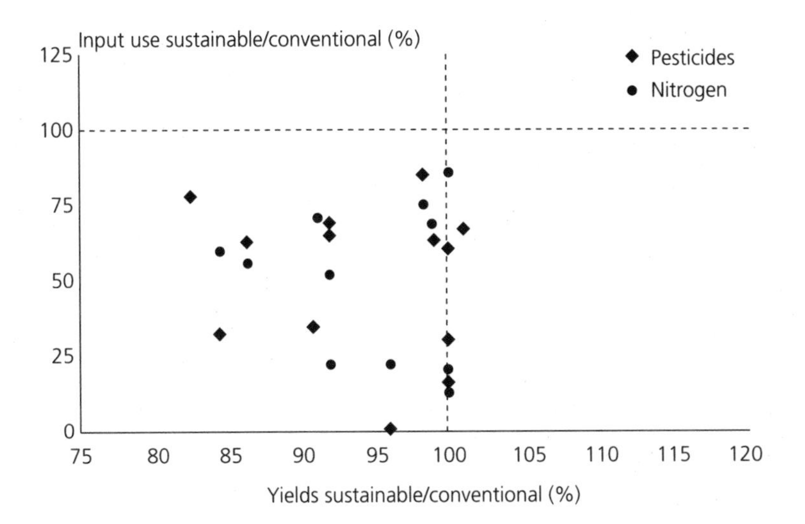

Figure 3.2 *Transition routes for sustainable agriculture systems in Europe and North America – changing pesticide and nitrogen fertiliser use*

regenerative technologies and eventually the comprehensive redesign of farming (see the Three Steps of Sustainability). In the UK, for example, there has generally been less research and fewer farmers are engaged in experimentation on sustainable agriculture. However, empirical evidence is proving critical in encouraging the transition towards sustainable agriculture. Research by the LIFE project at Long Ashton, by the Scottish agricultural colleges, by a range of English agricultural colleges, by the Boxworth project, at Boarded and Bundish Hall farms, by Rothamsted, and most recently by the LEAF network of farmers has shown that more sustainable agriculture can result in substantial cuts in inputs, with the maintenance, or even improvement, of gross margins. The greatest changes have been in reductions to pesticide use, with cuts of 20 to 50 per cent possible. There appears to be room for manoeuvre on fertiliser use – the range of cuts so far possible, with no impact on margins, appears at the moment to be of the order of 15 to 30 per cent (Jordan and Hutcheon, 1994; Jordan et al, 1993; Ogilvy et al, 1995; Rew and Cussans, 1995).

It is important to note that these findings do not represent endpoints. Evidence from elsewhere shows that continuous improvements are possible over long periods. To begin with, farmers clearly need to pay for information, services and technologies for integrated and precision farming. But money tends to be spent locally on knowledge and management skills. Over time, farmers increasingly take on the role of their own service provider as they become better. As learning increases, farmers get to keep more of the value.

These changes are again being made by farmers supported by research, education, extension and input-supply systems largely geared almost exclusively towards 'modernist' and high-input farming. Their choices for alternative, workable technologies are few, and so their transition and learning costs are high. The on-farm biological processes that make sustainable agriculture productive also take time to become established, including the rebuilding of depleted natural buffers of predator stocks and wild host plants; increasing the levels of nutrients; developing and exploiting micro-environments and positive interactions between them; and establishing hedges and trees. These higher variable and capital investment costs must be incurred before returns increase. Examples include labour for planting of trees and hedgerows; for pest and predator monitoring and management; for fencing of paddocks; and for purchase of new technologies, such as manure storage equipment or global positioning systems for combines.

Empirical evidence does tell us that performance improves over time. One remarkable set of data from 44 organic farms in Baden–Württemburg, Germany, has shown that wheat, oats and rye yields steadily increased over a 17-year period following transition (Dabbert, 1990). Elsewhere, at the Lautenbach estate, gross margins and yields steadily improved over the years, rising from an average of 98 per cent of conventional in 1979 and 1981 to 115 per cent by 1985 and 1986. This coincided with an increase in soil organic matter content from 1.2 per cent to 1.7 per cent (El Titi and Landes, 1990). In the US, the top quarter of sustainable agriculture farmers now have better yields and much higher gross margins than the top quarter of conventional farmers. But these changes have taken five to ten years.

Summary of Recent Field Evidence

Germany

Germany has probably the largest number of farmers in any EU country engaged in attempts to shift towards sustainable agriculture. This is because of positive policies enacted largely by Länder governments. In the state of Baden–Württemburg, there are more than 100,000 farmers engaged in various forms of organic production, low-input integrated practices and extensification.

It is also because the first fully integrated farming system experiment was established at Lautenbach in 1978 by Adel El Titi of the State Institute for Plant Protection in Stuttgart. This led to other experiments being set up in The Netherlands, Switzerland and the UK. These became part of a network coordinated by Pieter Vereijken of the DLO Research

Institute for Agrobiology and Soil Fertility. In Baden–Württemburg, activities started small with a group of 15 farmers, but over the years this grew to the active involvement of 50 members on some 5000 hectares. All are testing and experimenting with integrated farming technologies and practices that have shown dramatic improvements to natural capital with the switch to sustainable agriculture. Yields remained about the same over nine years, but gross margins were 3.5 per cent higher, fertiliser use down 25 per cent and pesticide use down 36 per cent. Soil faunas were considerably more diverse and present in greater numbers under the integrated system – with more earthworms and beneficial insect species. There was also about 0.5 per cent more organic matter in the soil than in the conventional fields. Excess nitrogen in the soil profile fell from 85 kilogrammes per hectare to just ten kilogrammes per hectare (El Titi and Landes, 1990; El Titi et al, 1993).

Five years of experimental evidence from the University of Göttingen, however, shows that it is sometimes difficult to maintain both yields and gross margins using mechanical weed control, reduced tillage, varied timing of sowing, reduced nitrogen and pesticides, and no use of PGRs (Georg-August-Universität Göttingen, 1996). Although inputs use and costs substantially fell, gross margins were reduced by 8 per cent too. But these savings on inputs and transfers to labour are significant. Farmers are saving 298 Deutschmarks (DM) per hectare from their fertiliser and pesticide bill, and spending almost 12 DM per hectare more for labour. Clearly if all 653,000 German farmers shifted to sustainable agriculture, then this would represent massive savings for them and benefits for rural economies at large.

The Netherlands

In The Netherlands, the first experimental site was established at Naegele in 1979. This was followed by on-farm testing of integrated and ecological farming systems, first with a group of 38 farms and now with an expanded countrywide network of 550 farmers who are part of the Arable Farming 2000 Project (van Weperen and Röling, 1995; Wijnands et al, 1995; Proost and Matteson, 1997). Farmers using a variety of resource-conserving technologies, including rotations, mechanical weed control, precision spraying, soil testing and resistant cultivars, have reduced their levels of inputs by 30 to 50 per cent, almost cutting out insecticides and stopping plant growth regulators. Yields have also been sustained at former levels and profitability increased. Importantly for water pollution, there have been dramatic reductions in the levels of nitrogen and phosphorus, both in the soil and drainage water (down 40 to 80 per cent).

Farmers have become increasingly aware of new problems and how to solve them. Their motivation has increased over time. On the core pilot farms, the use of active ingredients has been cut year on year from 18.9 kilogrammes per hectare in 1989 to 2.9 kilogrammes per hectare in the mid 1990s: sustainable agriculture farmers get better with time (Wijnands, 1992). The integrated arable farming systems approach has had no negative effect on farm economic results. Indeed, gross margins are 5 to 8 per cent higher, and savings are of the order of 250 to 500 guilders per hectare per year (roughly £100 to 200 per hectare). For the farmers' part of the Arable Farming 2000 Project, this translates into an annual saving of £0.75 to £1.5 million.

France

Like Germany and The Netherlands, France has many more farmers engaged in sustainable agriculture than in the UK. The leading research has been conducted at the Institut des Céreales et des Fourrages (ITCF) by Philippe Viaux and colleagues. This has shown that although yields of 'agriculture durable' are 15 to 25 per cent lower for wheat, oil seed rape and barley, margins are up to 32 per cent higher. These results have encouraged researchers to expand the experimentation to larger numbers of farmers. In the La Ferté–Vidame watershed that provides drinking water for Paris, farmers working with ITCF have been able to reduce variable costs from 1900 to 1500 French francs (FF) per hectare (from £210 to £165) by reducing input use. On these farms, gross margins have risen from 5800 to 6300 FF per hectare per year (Viaux and Rieu, 1995; Ansay and Viaux, 1996). The public benefits twice – from more environmentally friendly farming and also from the reduced contamination of the groundwater.

Such research has led the Ministère de l'Agriculture to develop Plans de Développement Durable, which is a wider attempt to foster sustainable rural development (Ministère de l'Agriculture de la Pêche et de l'Alimentation, 1996). Some 700 farmers in 59 regions are organised into groups, with a technical steering committee and very broad representation from many local organisations. The hope is that this will lead to a much greater understanding of the differing technical requirements that farmers have, and how sustainable technologies can better be fitted to particular locations (see Chapter 8).

Switzerland

Following positive research results in the 1980s, a network of 200 farms was set up in 1990. Results from these farms have clearly shown that low-input sustainable agriculture brings better returns for farmers. Research led by David Dubois has shown that wheat, barley, potato, grain maize and oat yields fall by 5 to 15 per cent because of the substantially reduced pesticide and fertiliser inputs (pesticides cut to zero and fertilisers to 20 to 25 per cent of intensive levels), but economic returns are consistently better (up 8 to 17 per cent).

When these sustainable agriculture technologies are combined into five-year rotations, then the number of grain crops makes an important contribution to whether the system is economically viable. A rotation with maize, wheat, clover-pasture and potatoes, using 45 per cent of the nitrogen and only about half the pesticides, has slightly lower yields and higher gross margins, with substantial benefits to soil structure and organic matter. But if the rotation is changed to increase to three cereal years out of five, then gross margins are 3 per cent lower. Nonetheless, this rotation sees the pesticide treatments down by 78 per cent (Dubois et al, 1995; Media Villa et al, 1995).

These results were so convincing to policy-makers and the public that new laws and regulations were introduced in 1996, shifting the whole of Swiss farming substantially towards sustainable agriculture. A three-tiered system offers farmers financial support to change their practices (see Chapter 8).

United Kingdom

A range of different research groups and farmers have been investigating the potential dividend from sustainable agriculture in the UK. Perhaps the best known is the LIFE project (less intensive farming for the environment) at Long Ashton where, since 1990, on-farm research is coordinated by Vic Jordan and colleagues. This shows that yields in low-input systems are in a band from 11 to 18 per cent down, to 3 to 7 per cent up. Since 13 to 19 per cent less energy, 25 to 36 per cent less nitrogen (about 40 kilogrammes per hectare), and 20 to 80 per cent less insecticide, herbicide and fungicide active ingredient are used, the variable costs are down about 40 per cent. The result is that gross margins are up 2.3 to 5.8 per cent and net margins up by 20 to 26 per cent (Jordan and Hutcheon, 1995; 1996). Such changes in input use, some 40 per cent off the fertiliser bill and 20 to 80 per cent off the pesticide bill, would mean annual savings to UK farmers of £530 to £820 million.

The Boxworth experiments were conducted over five years in the 1980s on 50-hectare blocks to assess the effect of reducing pesticides in arable systems (Greig-Smith et al, 1993). At the time, the comparisons were between conventional, supervised and integrated systems. The more sustainable systems were characterised by considerably less pesticide applied to all crops (six applications per crop compared with 14). For wheat, yields were down by 15 per cent, though gross margins were the same; for oil seed rape, both yields and gross margins were up. There were considerably more predatory insects in the integrated fields, leading to greater predation of pests, mainly aphids, by beetles and spiders within the crop. Over the five-year period, beetles and spiders in the conventional crop declined in number.

At Boarded Barns and Bundish Hall Farms, Rhône-Poulenc Agriculture has conducted seven years of experiments comparing conventional with organic systems, with an ICM comparison added in 1995. Organic margins for the whole rotation are considerably lower because of the need for two years of legumes and leys to build soil fertility. But the ICM margins outperform their conventional counterparts by £200 per hectare. ICM wheat far outperforms conventional wheat: the costs are lower (65 per cent) and the yields are higher (104 per cent) (Rhône-Poulenc, 1991–1997).

The LEAF (linking environment and farming) network of farms was set up in 1991 and has been promoting a whole farm approach to sustainability, combining the best of traditional methods with appropriate modern technologies. At mid 1997, there were 24 demonstration farms, with a further 400 individuals as members of the network. LEAF recently compared some ICM farms with non-ICM equivalents (LEAF, 1996). The first comparison was of dairy-arable holdings. A wide range of resource-conserving technologies and practices were used. For animal nutrition, the focus was on welfare, putting maize into integrated rations and increasing concentrate efficiency. For crop rotations, the focus was on disease-resistant varieties; a five-year rotation pattern; careful use of livestock manure; targeting and timeliness of fertiliser applications (down 11 per cent); regular agronomy visits and soil testing; and efficient machinery use through employing skilled contractors. For wildlife and habitats, the focus was on hedgerow, woodland and pond enhancement. For energy, the focus was on dairy-plant savings and regular machinery maintenance. The outputs per cow were the same for the ICM and non-ICM farms. But on the ICM farm, the variable costs for concentrates and forage were £38 lower per cow.

The second comparison was of ICM and non-ICM arable farms. The resource-conserving technologies used on the 700-hectare integrated farm in the Cotswolds included a maximum of two cereals per rotation, break crops, disease-resistant varieties, efficient machinery use with trained operators; targeting of fertiliser applications, incorporation of crop residues, soil compaction management, and reduced treatment with pesticides. The gross margins for the whole farm were 5 per cent greater, and profits 13 per cent up. Labour costs were up by 5 per cent, mainly because farm staff were better skilled and so more motivated. The farmer, because of reduced costs and increased gross margins, was also spending more on improving wildlife and landscape features. This amounted to £9.65 per hectare (£6700 for the whole farm) of his own money on building up the natural capital.

These kinds of impacts are repeated throughout the LEAF network. Even a new LEAF farm, at Whitsome Hill in Scotland, after using global positioning systems (GPS) and precision farming technologies, and applying fertilisers just on a per hectare basis rather than on the whole field, was able to cut its fertiliser use by 10 to 15 per cent. Again, this is not making the kind of contribution to natural capital that organic farming does, but it represents achievable early steps that large numbers of farmers could take. The winners are farmers, the countryside, local jobs and businesses. The losers are the agrochemical companies.

In recent years, the growth in the use of precision farming technologies has been dramatic. The basic principle is that there is much greater diversity in nutrient reserves and weed and pest density within fields than has been commonly supposed. Inputs can, therefore, be targeted more precisely, cutting down on waste to the environment. Most precision farming technologies make use of global positioning systems to map resources at great accuracy. Some of these maps are yield maps – with flow meters on combine harvesters able to convert rates of harvest of a crop to infield yields. Others are soil maps, with soils sampled at regular intervals, analysed for various nutrients, and then mapped. By mid 1997, at least 150,000 hectares of farmland were mapped in this way. On average these are producing a 10 to 15 per cent cut in fertiliser and pesticide use, with up to 45 per cent in some cases (pers comm with Jim Potter, Simon Parnington, Steve Garner, Ann Willington, Matt Ward).*

This includes yield mapping with Massey-Ferguson systems. Some 130 commercial systems are in use (40 systems were sold in early 1997 alone), each capable of 400 hectares, which suggests that 56,000 ha have been mapped. Another system is soil mapping with the SOYL developed system. Some 80,000 ha have been mapped for P, K, Mg and lime, half of which have had variable rate application. The SOYL system allows for spatially adjusted fertiliser spreading, and these variable applications can bring savings of £24/ha in fertiliser costs alone. Andrew Blanchard of New Barn Farm in Hants has been using mapping and variable rate fertiliser applications for four years on his 625-ha farm. The amount of P and K used has been reduced over this period from 250 kg/ha to 150 kg/ha, saving him some £11,000 per year for the whole farm, and £24 per hectare. Other systems include those run by Law Fertilisers, who have sampled 12,000 ha; by Optimix, who have sampled 12,000 ha on 100 farms; and by Willington Services, who have sampled 6000 ha on 50 farms, resulting in savings of £40–50 per ha from the fertiliser bill, though at the moment this is mostly balanced by the costs of the precision farming service itself.

(continued overleaf)

The United States

Research, extension and non-governmental organisations (NGOs) in the US have been working on sustainable agriculture issues for longer than most in Europe. Sustainable agriculture was first noted as an emergent and increasingly relevant goal for farming in the 1985 Farm Bill. A range of policies have been put in place to discourage polluting agriculture and to encourage resource conservation. And many institutions have established an excellent track record of research and development partnerships with farmers and local communities. These include the Center for Rural Affairs in Nebraska, the Minnesota Institute for Sustainable Agriculture, the Kansas Rural Center, the Center for Sustainable Systems in Kentucky, the Center for Agroecology of the University of California at Santa Cruz, and the Leopold Center for Sustainable Agriculture at Iowa State University.

It is very difficult to say how many farmers are now engaged in forms of sustainable agriculture, owing to the size of the country and the number of institutions involved. A conservative estimate puts the numbers at somewhere between 50 and 60,000. Again, research shows that most of these farmers are doing better than their conventional counterparts. A recent multistate study funded by the North West Area Foundation of farming in Minnesota, Iowa, North Dakota, Montana, Oregon and Washington compared several thousand sustainable agriculture farmers with conventional equivalents (NAF, 1994). The results are extraordinary (see Table 3.4). Sustainable agriculture farmers are more diverse in terms of crops and livestock and the amount of land under woodland and wetland. They spend less on pesticides and fertilisers and more on labour. Their soils have better structure. They spend more in

---

(footnote carried over from page 105)

These are by no means the full extent of researching activities on sustainable and integrated farming. A selection of results include trials run by CWS Agriculture (which farms some 20,000 ha in all) on Stroughton Farm on a seven-year rotation of two silage leys, three winter wheat, one bean and one set-aside, which show that the gross margins at 1997 prices were £436 compared with £429 for conventional. Research on a number of sites in Scotland by the Scottish Agricultural College shows that the profitability of many crops can be maintained even with 75 per cent reductions in the use of herbicides and fungicides; ADAS research as part of the TALISMAN project at Boxworth and High Mowthorpe found no negative impact from 50 per cent cuts in recommended doses of fungicides and herbicides across a wide range of crops, provided better management is substituted. Research at Rothamsted shows that herbicide use can be cut by 30 per cent when weeds are accurately mapped and patch spraying technology used. Research by the Institute of Grassland and Environmental Research in Devon on including clover in pastures has shown that fertiliser costs can be cut by £50 per hectare, as well as cutting concentrate use as sward digestibility increased. And LINK-integrated farming systems research conducted on six sites across the country shows that crop yields and gross margins can be maintained with 15 per cent less nitrogen and 35 to 40 per cent less use of

Table 3.4 *Differences between sustainable and conventional farms in six states of mid-west and north-west US*

Indicators of difference (and location if specific)	*Conventional farms*	*Sustainable farms*
Farm diversity and conservation		
• Proportion of land to maize-soybean (Iowa) (%)	94%	61%
• Number of different crops grown per farm (Montana)	2.3	5
• Proportion of farmland under woodlands, wetlands and conservation uses (Minnesota) (%)	7%	25%
• Per cent of farms with livestock (Minnesota)	37%	91%
• Impact on soil structure	depleted	enhanced
Input use		
• % farmers applying less than 110 kg N/ha	1%	90%
• % farmers applying more than 155 kg N/ha	67%	0
• Amount spent on pesticides ($/ha)	$15.7	$0.02
• Amount spent on fertilisers ($/ha) (North Dakota)	$23.9	$2.72
• % farmers relying exclusively on inorganic fertilisers	62%	2%
• % farmers relying only on organic sources (Minnesota)	6%	50%
Total expenditure on pesticides and fertilisers ($/ha)		
Montana	$19.2	$6.2
Iowa	$257	$170
Yields and income		
• Maize yields of top quarter farmers	same	same
• Maize yields of all farmers in survey	318 bu/ha	281 bu/ha
• Gross income ($/ha)	$237	$188
• Net income ($/ha)	$30	$35
Expenditure on labour and other local goods and services		
• Expenditure on goods and services purchased from farmers and local businesses (Iowa)	$78/ha	$128/ha
• Extra labour required for sustainable agriculture	na	+22–52%
• Additional hours of labour for sustainable agriculture		
Iowa	na	+19 hours/week
Montana	na	+70 hours/week

Source: Northwest Area Foundation. 1994. *A Better Hoe to Row*. Minnesota, US

local communities on goods and services from local businesses. The yields of the top quarter farms are better, though for the whole sample they were 10 per cent down. Their net income was higher.

What these changes could mean if whole communities made the transition to sustainable agriculture is very significant. One community in the mid-west situated in the middle of a ten-by-ten kilometre square of sustainable agriculture would get £82,000 more spent in their local community each year than if all the farms were conventional. Labour demand would be greater, input use lower, and yields perhaps 5 to 10 per cent lower, though net income for the block would be up by some US$5 million. But these are not the only advances that have been made (Hewitt and Smith, 1995). There are 500 sustainable agriculture farmers in Iowa who are part of the Practical Farmers of Iowa network (see Chapter 5). There are 2300 cotton farmers in Alabama now using no pesticides. There are 30,000 hectares of potato farms in Wisconsin, with US$5.9 million saved by farmers through the use of IPM. Apple growers in Georgia have cut pesticide costs from $600 to $244 per hectare with no reduction in yields. And the internationally renowned wine producers, Gallo, in California, now use no external inputs on 2500 hectares, but do not label their 45,000 cases sold each year as organic: 'Gallo produces as many grapes for less money per hectare than they did in the chemical intensive days' (Hewitt and Smith, 1995).

The Three Steps of Sustainability

If all these good things are already happening across Europe and North America, what can be done to encourage further transition to sustainable agriculture? One way of reconciling some of the differences is to regard sustainable agriculture as a series of steps (Pretty, 1996a). The transition towards sustainable agriculture is often seen as requiring a sudden shift in both practices and values. But not all farmers or land managers are able or willing to take such a leap. However, everyone can make one small step. And small steps added together can bring about big change in the end. Small steps also reduce the threat of the unusual and the risks of the unknown. According to some psychologists, changes in attitudes and values follow changes in behaviour and practice, and so once individuals begin to engage in more sustainable agriculture, they may be provoked into thinking differently about the environment and so, in turn, make further changes in behaviour and practices.

Drawing on the work of Rod MacRae, Stuart Hill and colleagues in Canada, and the Federation of Swedish Farmers, four steps have been devised in this process, three of which are on the path to sustainability

(MacRae et al, 1993; Federation of Swedish Farmers, 1997). Step 0 is, of course, conventional modern farming. The other three incorporate changes in economic and environmental efficiency; changes in the integration of regenerative technologies; and wholesale redesign with communities.

Step 0: Conventional Modern Farming

This is farming with modern crops or livestock breeds, supplemented with pesticides, inorganic fertilisers, machinery and irrigation, with information coming from off-farm and outside the community (from extensionists, private suppliers of inputs, researchers). The basic principle is to emphasise external solutions and technologies in order to overcome internal productivity constraints; farming systems are simplified to maximise production; farming systems are expected to adapt to the technologies; economic efficiency is improved by cutting labour costs; and natural resources are said to be conserved as high-output farming decreases the need to cultivate more land.

However, the leakage of external inputs away from the farm damages both natural capital (water, air, soil, plants and wildlife) and human health (farmers and food consumers); the simplification of farming systems results in loss of on- and off-farm biodiversity and impoverishes landscapes; jobs and farms are lost; and there is increased impoverishment of social capital.

Step 1: Improved Economic and Environmental Efficiency

This step comprises adoption of information-intensive technologies and practices for precision farming: it includes targeted inputs, patch spraying, deep placement and slow-release fertilisers, global positioning systems and satellite mapping, low-volume and minimal dose pesticides, soil testing, weed maps, no-till or non-inversion farming, mechanical and weed harrowing, and pest and disease resistant crops. Efficiency is increased as inputs are no longer wasted and therefore lost to the environment. Costs fall and yields and gross margins remain the same or are improved.

However, the goals of farming remain essentially the same, and therefore existing values and principles are not fundamentally challenged.

Step 2: Integrating Regenerative Technologies

This step incorporates regenerative technologies and drops some conventional ones: it incorporates crops and trees that fix nitrogen (legumes and clovers, green manures, cover crops); uses alternative pesticides (biological, bacterial and viral); manipulates habitats to encourage predators; employs natural enemy release; integrates animals (sheep, goats, cattle, pigs, fish, bees) into cropped systems; and emphasises technologies that conserve soil and water. These regenerative technologies make the best use of all locally available biological and human resources, and there is more reliance on local management skills and knowledge. Demand for labour increases as ecological and landscape diversity improves. External resources are targeted where they will add value – provided none degrade or damage the environment or human health, or lead to the loss of local resources.

However, rural communities may remain relatively uninvolved in farming and food matters, and farmers may not have been motivated to act collectively by forming new institutions or groups.

Step 3: Redesign with Communities

This step has agriculture as a central part of community economic and social activities, with sustainability as an emerging property of whole communities, catchments and landscapes. Natural and social capital are regenerated.

Attitudes and values are now completely different as new philosophies and participatory practices emerge; resource-conserving technologies are locally specific and varied by farmers; local people have greater self-reliance and cohesion; local groups and institutions are strengthened for both natural resource and financial resource management; external institutions are reformed since professionals work as facilitators and enablers of local change; and agriculture as a whole is structured to emphasise local economic regeneration. The basic principle is that there is much that local people can do by themselves with the resources at hand – but perhaps never realised they could.

However, this will need widespread reform of institutions and the development of supportive policies if redesign is to spread through whole nations.

Investing in Learning

How can more farmers be encouraged to move through the three steps of sustainability? The first thing to note is that sustainable agriculture should not prescribe a concretely defined set of technologies, practices or policies. This would only serve to restrict the future options of farmers. As economic, ecological, climatic and social conditions change, so farmers and communities must be encouraged and allowed to adapt too. Sustainable agriculture is, therefore, not a simple model or package to be widely applied or fixed with time. It is more a process of social learning, with targets to measure progress (Pretty, 1995b). As Don Vetterman, a successful sustainable agriculture farmer from Nebraska, put it in 1995: 'this is farming you learn every year. You don't get it all at once' (in Klinkenburg, 1995).

This is backed up by recent research on the changes made by 550 farmers in The Netherlands and some 2800 in the US (NAF, 1994; Somers, 1998). It appears that once farmers begin to make the change, taking one small step, they tend to keep going. As farmers use new, more environmentally sensitive technologies, this appears to provoke more thinking about the environment. This in turn leads to more curiosity about what can be done. Further changes in behaviour provoke further changes in values, and further steps are taken. In The Netherlands, farmers in the integrated arable farming systems network steadily reduced their pesticide losses to the environment over a six-year period by four to fivefold. All the benefits cannot be expected to occur in the first year (Vereijken et al, 1995). But it is also true that not all farmers make the transition. Some adopt the economically and environmentally efficient stance using precision farming, and then go no further. Part of the problem is that adopting resource-conserving technologies and practices is not a costless process. Farmers cannot simply cut their use of fertilisers or pesticides and hope to maintain outputs. They have to substitute management skills, new knowledge and new technologies in return.

Farmers must therefore invest in learning. Lack of information and management skills is a major barrier to adopting sustainable agriculture. We know much less about these resource-conserving technologies than we do about the use of external inputs in modernised systems. As external resources and practices have substituted for internal and traditional ones, knowledge about the latter has been lost. In addition, much less research on resource-conserving technologies is conducted by research institutions. It is clear that the process by which farmers learn about technology alternatives is crucial. If they are enforced or coerced, then they may adopt for a while, but will only continue if their incomes are dependent on the technology. Cross-compliance and green conditionality of this type may only buy grudging support, or indeed none at all (see Chapter 8).

However, if the process is participatory and enhances farmers' capacity to learn about their farm and its resources, then it appears that the foundation for redesign is laid. There are many good examples of the effectiveness of this participatory learning in developing countries (Pretty, 1995b). However, a notable case from the industrialised country context is from south Queensland, Australia, where Gus Hamilton and other extensionists from the Department of Primary Industry developed simple learning tools that enabled farmers to investigate the impact of rainfall on their soil. These encouraged more than four out of five farmers to switch to conservation technologies. Many of these have gone on to develop new and different technologies for their own farms, and they now fully support the values and principles that once they would have opposed (Hamilton, 1995).

The importance of maintaining social capital has been shown very clearly in a recent comparison by Gérald Assouline and colleagues of the conversion to organic farming in Denmark (Lemvig, West Jutland) and in France (Drôme Department in south-east France) (Assouline, 1997). They found several key factors in the transition process. The first was the presence of good local pioneers who could demonstrate that their alternative farming works and pays; as one farmer put it, 'if they dare, so do we'. There were also effective consultants and extensionists who could give back-up support and provide economic data and technical advice when needed.

Furthermore, those engaged in the transition deliberately stayed in touch with conventional farmers in order to prevent the emergence of ideological divisions. The authors indicated that 'one of the Danish converts thinks that it is important for organic farmers not to criticise conventional farmers too much... They go out of their way to maintain good relations with their colleagues... respecting each other's professionalism as organic and conventional farmers'. These attitudes are similar to some of the partnerships that have emerged between ecological and conventional farmers in Sweden (see Chapter 8). In the mountainous area of Diois, farmers chose to work together in groups; this is greatly advancing the shift towards sustainable farming. By contrast, in nearby Val de Drôme, where farming is more intensive and entrepreneurial, farmers tend to work separately and the spread of sustainable technologies is slower. As Gérald Assouline said: 'it takes great courage to begin to make the transition from conventional towards organic farming'. This course is given strength when others are engaged in the process too – when the social capital is present.

The North-West Area Foundation's study of sustainable and commercial farming in the US found that:

> *sustainable agriculture farmers learn most from other farmers and from their own research. Large majorities report never having used local extension agents, state specialists, university scientists or soil conservationists... only one in four consider these expert institutions very useful to them. These institutions are perceived as too oriented to conventional agriculture and its needs (NAF, 1994).*

Sustainable agriculture farmers have to rely more on themselves. And it works – in Iowa, sustainable agriculture farmers who belong to groups have the highest yields over both conventional farmers and non-group sustainable agriculture farmers.

The central policy challenge, as we shall see in later chapters, is to find methodologies and mechanisms to make positive changes to farming while not alienating farmers themselves. Rural communities and their institutions can play a positive role in supporting the transition towards sustainability.

Measuring the Transition to Sustainability

How will we know when progress along these steps is being made? It is clear that most agriculture is currently in an unsustainable state. There are many potential areas for improvement on both the input and output side of farming, including increasing positive inputs, such as through greater use of regenerative and resource-conserving technologies; decreasing negative inputs, such as using fewer pesticides, fertilisers, energy, land, labour and water; increasing the amount of positive outputs, such as of wholesome food, energy crops, clean water, better diets, more wildlife and their habitats; and reducing the amount of negative outputs, such as pesticide pollution, fertiliser contamination, farm wastes, health damage of farmers and the general public, polluted water, carbon dioxide and other gas emissions, and wildlife losses.

It is clear that there is a need for indicators to assess progress. These must be easily measured, and therefore not costly; non-contestable and therefore convincing to internal and external stakeholders; prone to immediate management action; and able to create value for farmers and rural communities, both in terms of natural and social capital. In Table 3.5, ten classes of indicators are listed with examples of resource-conserving technologies and management practices that would help to deliver the expected improvements. There are four key issues that must be addressed when considering indicators to improve performance: targets; timescale; prioritisation; and measurement.

The immediate target for each of these indicators should be 50 per cent improvements compared with the past 20-year running average. These would deliver significant benefits both in terms of reduced costs and improved returns, and would act to protect natural and social capital. A 50 per cent improvement is also chosen to represent an achievable but tough target. For the first class of indicators, for example, this would imply that the amounts of pesticide active ingredient used in agricultural production should be reduced by 50 per cent, that leakage to water would be reduced by half, and that the proportion of narrow-spectrum, wildlife-safe products has increased by half. For the third class, it would imply an increase in organic matter content of soils by 50 per cent and a reduction by 50 per cent in soil eroded from fields.

Another key question is the timescale. Clearly, the sooner the targets are met, the sooner the benefits are delivered. Some changes can be made almost immediately: inserting clovers into pastures allows an immediate cut in inorganic fertiliser use; building up organic matter, however, takes much longer. A reasonable target timescale for achieving the 50 per cent improvements would be a maximum of five years.

The third issue is prioritisation. How do we decide on the best place to begin? There is neither an objective starting point nor endpoint for sustainability. Choosing any of these indicators over another is, therefore, a matter of judgement. What we can say is that if fewer pesticides are used, then this is a step towards sustainability; if there are fewer people suffering ill-health from pesticide pollution, then this too is a step towards sustainability; if more legumes are used, then this is another step to sustainability. But what we cannot say is whether focusing attention on energy consumption is more or less important than focusing it on pesticide consumption and IPM alternatives. We know that resource-conserving technologies will deliver agricultural systems that are less energy intensive and less reliant on pesticides.

The fourth issue is measurement. How will we know if progress is being made? The tendency is to go for the easily measured. This is what policy-makers in The Netherlands and Scandinavia have done in seeking to reduce pesticide use. But even measuring pesticide use is contested by many different groups. Cutting pesticide use by half in The Netherlands (where 20 kilogrammes of active ingredient per hectare were used in 1987) means something quite different in Sweden (where 1.1 kilogrammes of active ingredient per hectare were being used at the same time), or in the US (1.8 kilogrammes per hectare) and Denmark (2.6 kilogrammes per hectare).

In the UK, pesticide use on cereals has fallen in the past decade by 40 per cent (from 18.4 to 11.5 million kilogrammes of ingredient between 1982 and 1994), putting applications at about 2.5 and 2.9 kilogrammes per hectare. Yet the area treated has increased by 20 per cent. Over the 1990 and 1994 period, the total average annual use of pesticides in Britain was 29.2 million kilogrammes of active ingredient – some 20 per cent less than the 1982 level of 35.5 million kilogrammes. (In the US, however, the use of pesticides reached an all time high of 550 million kilogrammes in 1995 – over twice as much as when *Silent Spring* was published in 1962, and an increase of 40 million kilogrammes over 1994. Expenditure on pesticides was $10.4 billion in 1995.) But the falls are almost entirely due to the use of more biologically active products and the lower doses required, rather than the adoption of alternatives to pesticides. As the UK Royal Commission on Environmental Pollution put it: 'pesticides should be used as sparingly as possible and in conjunction with other control measures' (RCEP, 1996).

Policy Discrimination Slows the Transition

Although there are several important national initiatives and EU programmes that promote more sustainable agriculture, most policies still currently act as powerful disincentives against sustainability. In the short term, this means that farmers switching from modernist to resource-conserving technologies can rarely do so without incurring some transition costs. In the long term, this means that sustainable agriculture will not spread widely beyond localised success. The main problem is that the external costs of modern farming, such as soil erosion, health damage or polluted ecosystems, are not incorporated into individual decision-making by farmers. In this way resource-degrading farmers bear neither the costs of damage to society, nor those incurred in controlling the polluting or damaging activity.

In principle, it is possible to imagine pricing the free input to farming of the clean, unpolluted environment. If charges were levied in some way, then degraders or polluters would have higher costs, would be forced to pass them on to consumers, and would therefore be encouraged to switch to more resource-conserving technologies. This notion is captured in the polluter pays principle, a long stated principle of national policies in the OECD. However, beyond the notion of encouraging some internalisation of costs, this has not yet been of practical use for policy formulation in agriculture.

The effect of policies is strikingly clear in the US, where commodity programmes inhibited the adoption of sustainable practices by artifi-

Table 3.5 *Ten classes of indicators for sustainable agriculture*

Key indicators	*Examples of resource-conserving technologies and management practices that deliver improvements*
Pesticides	
• Total active ingredient used per hectare and per tonne of net output • Leakage to groundwater and surface water • Proportion of products used that are narrow spectrum and wildlife-safe	• Adoption of integrated pest management technologies – rotations, pheromone traps, resistant varieties and breeds, biological control, in-field forecasting, beetle banks and flowering strips, field mapping to target inputs
Fertilisers	
• Total N, P and K used per hectare and per tonne of net output • Leakage of nutrients to ground and surface water • Proportion of N used on farms that has been fixed from the atmosphere	• Field mapping to target inputs • Use of legumes and nitrogen-fixing crops in arable and pasture systems • Soil testing for nutrients • Slow-release and deep-placement fertilisers
Soils	
• Organic matter content of soil (per cent) • Amount of soil eroded per hectare and per tonne of net output	• Legumes and nitrogen-fixing crops grown • Organic matter returned to soil (eg green and animal manures, sewage sludges) • Hedgerows and grass-barriers to control surface water flows, and contour cropping
Energy	
• Direct and indirect energy used (machinery, electricity, embodied energy in fertilisers and pesticides) per hectare and per tonne of net output • Product miles or km per tonne of net output	• Local energy crop production, eg from arable coppicing, and on-farm electricity and heat generation • Reduced use of fertilisers • Local sourcing of inputs and services
Water	
• Irrigation water used per hectare and per tonne of net output	• Irrigation scheduling • Winter storage reservoirs
Wastes	
• Amount of livestock waste incorporated on-farm • Emissions of methane, nitrous oxide and carbon dioxide per	• Proportion of mixed farms with livestock and crops • Adoption of fertiliser practices that reduce emissions

hectare and per tonne of net output • Proportion of wastes recycled (plastic bags, polythene)	• Detailed recording of inputs and outputs of non-renewables
Wildlife	
• Number of species and populations of birds per hectare of farmland • Insect diversity in non-crop habitats • Population size of key predators	• Increased rotations and diversity of crops • Beetle banks and flowering strips in crops • Hedgerow and woodland/ wetland protection and management
Local innovations	
• Rate of innovation and adaptation of technologies (no. experiments per farm) • Investment in training and learning	• Support to farmers' research and discussion groups for technology testing, experimentation and exchange
Local community	
• Number of jobs per farmed hectare and per farm • Proportion of farm inputs and environmental services bought from local businesses (within 15 km radius) • Number of local businesses supported through adoption of more sustainable agriculture	• Substitution of management and knowledge-intensive technologies for non-renewables • Audit of expenditure patterns and explicit policies for buying from local businesses and suppliers
Partnerships and learning	
• Number of stakeholder consultations and discussion forums • Investment in and support for group-based activities	• Participatory processes and forums for decision-making in rural communities

cially making them less profitable to farmers. In Pennsylvania, the financial returns to monocropped and continuous maize were comparable to those in sustainable agriculture involving mixed rotations in the early 1990s (see Table 3.6). But the continuous maize attracted about twice as much direct support in the form of deficiency payments. And these farms required more nitrogen fertiliser, eroded more soil and caused three to six times as much damage to off-site resources. Clearly, a transition to the resource-conserving rotations would have benefited both farmers and the national economy.

Many countries have also encouraged modernist farming by subsidising pesticides, fertilisers, credit and irrigation. These subsidies have quite

Table 3.6 *Comparison of the impact of conventional and alternative rotations on off-site damage and yields in Pennsylvania, and the subsidies received by farmers for each rotation*

	Continuous maize	*Sustainable agriculture rotations (maize, barley, beans, oats, clover, grass)*
• Crop sales	267 $/ha	203–220 $/ha[1]
• Production costs	247 $/ha	188–198 $/ha
• Net returns	20 $/ha	15–22 $/ha[2]
• Additional deficiency (subsidy) payments	145 $/ha	62–89 $/ha
• Total returns	165 $/ha	77–111 $/ha
• Nitrogen fertiliser use	168 kg/ha/yr	0 kg/ha/yr
• Soil erosion (tonnes/ha)	23 t/ha/yr	8–15 t/ha/yr
• On-farm depreciation (–) or appreciation (+)	–62 $/ha/yr	+7–20 $/ha/yr
• External costs of soil erosion	230 $/ha	42–71 $/ha

1 This rises to $309 after transition period (a period of conversion from high to low input).
2 This rises to $121 after transition period.
Source: Faeth P (ed). 1993. *Agricultural Policy and Sustainability: Case Studies from India, Chile, the Philippines and the United States*. WRI, Washington, DC

clearly increased the use of these inputs. But they also tend to encourage excessive use and wastage. Subsidised water and electricity, for example, have aided the expansion of irrigated agriculture in parts of Asia, which has contributed to a decline in groundwater levels and accelerated accumulation of salts in the soil. The financial cost is great: Peter Hazell and Ernst Lutz (1998), in a new World Bank book, put the annual cost of subsidising inputs in developing countries at US$20 to $25 billion for water, $6 billion for fertilisers, and $4 billion for pesticides. As they put it: 'many governments have also kept agricultural output prices too low through export taxes and overvalued exchange rates, reducing farmers' profits and their returns to investing in the conservation and improvement of natural resources'.

These policies have reduced the economic viability of sustainable agriculture technologies for pest management. In Indonesia, for example, it was only the removal of pesticide subsidies in 1986, coupled with the banning of 57 rice pesticides, that has allowed farmer-field schools to flourish and to allow farmers to make the transition to pesticide-free or pesticide-low rice farming. In Costa Rica, however, public support for IPM is offset by a whole raft of policies and processes that stimulate pesticide use, including the promotion of pesticide intensive systems, education in crop protection, lack of implementation of legislation, public funding of pesticide research, tax and duty exemptions for pesticides, and information transmitted by chemical companies. Over five years in the early 1990s, the value of pesticide imports increased from US$56 million to $84 million – despite the apparent promotion of IPM and sustainable agriculture (Agne and Waibel, 1997).

There remain many contradictions under the CAP and national policies in the EU. Livestock headage payments, for example, encourage overstocking on upland pastures in Britain, Ireland and the steppes of Spain. Forestry payments encourage farmers to replace species-rich grasslands with tree monocultures. High cereal prices have encouraged Portuguese farmers to grub up 160 to 200,000 hectares of holm oaks in the past 30 years (Viera and Eden, 1995).

Many traditional agricultural systems throughout Europe that are good both for people and for the rural environment are also under threat. In some cases, policies have discriminated against these systems. Gaston Remmers's study (1996) of the diverse tree, crop and animal farming systems in Alpujarra, Spain, has shown how both agricultural policies seeking to increase output, and forestry policies encouraging the planting of conifer plantations, are threatening existing systems. Subsidies for almond cultivation are on a per hectare basis, but to qualify fields must contain no more than 10 per cent of another crop. This prohibits farmers from using the bean *moruna* (*Vigna articulata*) as a green manure. In the past, whenever farmers considered land as 'weak', they planted *moruna*. It is considered a better fertiliser than both animal manures and inorganic fertilisers, is resistant to drought, boosts almond yields for three years, and is also valued as animal fodder. Yet its use is in decline.

Summary

The basic challenge for sustainable agriculture is to make better use of available natural and social resources. This can be done by minimising the use of external inputs, by utilising and regenerating internal resources more effectively, or by combinations of both. Sustainable agriculture offers many opportunities to integrate a wide range of economic, social and environmental concerns in the countryside. Farming does not have to produce its food by damaging or destroying the environment: resource-conserving technologies allow farmers to be productive and earn a living while protecting the landscape and its natural resources for future generations. Farming does not have to be dislocated from local rural communities: sustainable agriculture, with its need for increased knowledge, management skills and labour, offers new upstream and downstream work opportunities for businesses and people in rural areas.

The best evidence of redesigning farming systems to incorporate sustainability goals comes from a wide range of countries in Africa, Asia and Latin America. Here a major concern is to increase food production in the areas where farming has been largely untouched by the modern packages of externally supplied technologies. In these lands, farming communities which adopt regenerative technologies have substantially improved agricultural yields, often using few or no external inputs. Most of these successes are community-based activities that have involved the complete redesign of farming and local economic activities with the full participation of whole communities in the process.

A wide range of more sustainable forms of agriculture are now emerging and spreading in Europe and North America. Some are making substantial contributions to natural capital. But there are important misconceptions about sustainable and regenerative agriculture. The most common characterisation is that sustainable agriculture represents a return to some form of low-technology, backward or traditional agricultural practice. This is manifestly untrue. Sustainable agriculture incorporates recent innovations that may originate with scientists, with farmers, or both. It is also commonly stated that any farming using low or lower amounts of external inputs can only produce low levels of output. This is untrue for two reasons. Firstly, many sustainable agriculture farmers show that their crop yields can be better than or equal to those of their more conventional neighbours. Most show that their costs can be reduced. In developing countries, this offers new opportunities for economic growth for communities that do not have access to, or cannot afford, external resources. Secondly, sustainable agriculture produces more than just food – it significantly contributes to natural and social capital, both of which are sometimes difficult to measure and cost.

Despite recent changes to modern agriculture, there are still millions of hectares of land farmed in environmentally sensitive and low-intensive ways throughout Europe. Their extent dwarfs recent transitions to sustainably intensified agriculture. Organic farming is one form of sustainable agriculture in which maximum reliance is put on self-regulating agro-ecosystems, locally or farm-derived renewable resources, and the management of ecological and biological processes. In Europe, there have been dramatic increases in organic agriculture in recent years. The extent increased tenfold to 1.2 million hectares in 1996. Nearly 50,000 farmers are now engaged in certified organic farming. In the UK, there were about 820 organic farms in 1997, with an area of 47,900 hectares.

Integrated farming systems have emerged as another type of more environmentally friendly approach to farming. Once again, the emphasis is on incorporating a higher input of management and information. Integrated farming in its various guises represents a step or several steps towards sustainability. It is difficult to generalise about what happens to outputs. It used to be thought that more sustainable agriculture, whether organic or integrated, would mean reductions in crop and livestock yields. However, this generalisation no longer stands. It appears that farmers can make some cuts in input use (at least 10 to 20 per cent) without negatively affecting gross margins. By adopting better targeting and precision methods, there is less wastage and therefore the environment benefits. Yields may fall initially but will rise over time. They can then make greater cuts in input use (20 to 50 per cent) once they substitute some regenerative technologies for external inputs, such as legumes for inorganic fertilisers or predators for pesticides. Sustainable agriculture farmers get better over time.

One way of reconciling some of the differences is to regard sustainable agriculture as a series of steps. The transition towards sustainable agriculture is often conceived of as requiring a sudden shift in both practices and values. But not all farmers are able or willing to take such a leap. However, everyone can take one small step. And small steps added together can bring about big change in the end. Four steps have been devised in this process, three of which are on the path to sustainability. Step 0 is, of course, conventional modern farming. The other three incorporate changes in economic and environmental efficiency (step 1); changes in integrating regenerative technologies (step 2); and wholesale redesign with communities (step 3).

How can more farmers be encouraged to move through the three steps of sustainability? The first thing to note is that sustainable agriculture should not prescribe a concretely defined set of technologies, practices or policies. This would only serve to restrict the future options of farmers. As economic, ecological, climatic and social conditions change, so must farmers and communities be encouraged and allowed to adapt. Sustainable agriculture is, therefore, not a simple model or package to be widely applied or fixed with time. It is more a process of social learning with targets to measure progress.

Farmers must therefore invest in learning. Lack of information and management skills is a major barrier to adopting sustainable agriculture. We know much less about these resource-conserving technologies than we do about the use of external inputs in modernised systems. As external resources and practices have substituted for internal and traditional ones, knowledge about the latter has been lost. In addition, much less research on resource-conserving technologies is conducted by research institutions. It is clear that the process by which farmers learn about technology alternatives is crucial. If they are enforced or coerced, then they may adopt for a while, but will only continue if their incomes are dependent on the technology. Cross-compliance and green conditionality of this type may only buy grudging support, or indeed none at all. However, if the process is participatory and enhances farmers' capacity to learn about their farm and its resources, then it appears that the foundation for redesign is laid. The importance of maintaining social capital has been shown clearly by recent work in France, Denmark and the US. How will we know when progress along these steps is being made? Today, most agriculture is currently in an unsustainable state.

It is clear that there is a need for indicators to measure progress. These must be easily measured, and therefore not costly; non-contestable and therefore convincing to internal and external stakeholders; prone to immediate management action; and able to create value for farmers and rural communities, both in terms of natural and social capital. This chapter contains ten classes of indicators, and discusses the four key areas that must be addressed when considering indicators for improved performance, namely targets; timescale; prioritisation; and measurement. Policies, though, still slow the transition, and there is need for fundamental reform to increase the incentives for more sustainable agriculture and to reduce the distortions that give an unfair advantage to modernist farming. As the next chapter discusses, reforms in the food system will help farming become more sustainable.

Part 2

Towards Sustainable Food Systems

Chapter 4

The Food System: Hunger and Plenty

There have been oppression and luxury,
There have been poverty and licence,
There has been minor injustice.
Yet we have been gone on living,
Living and partly living.
Sometimes the rain has failed us,
Sometimes the harvest is good,
One year is a year of rain,
Another a year of dryness,
One year the apples are abundant
Another year the plums are lacking
Yet we have gone on living
Living and partly living.

T S Eliot (1888–1965),
Murder in the Cathedral

The Global Challenge

A Generation of Progress?

In the early 1960s, the spectre of global famine spurred agricultural researchers and policy-makers to give a very high priority to agricultural improvement. This gave rise to the so-called Green Revolution. As a result of the modernisation of rice, wheat and maize varieties, and increased consumption of fertilisers and pesticides, food production per capita has, since the mid 1960s, risen by 7 per cent for the world as a whole, with the greatest increases in Asia – up by about 40 per cent. Nitrogen consumption has increased from two to 75 million tonnes in the last 50 years, and pesticide consumption in many individual countries has increased by 10 to 30 per cent since the 1980s alone.

According to the World Bank, some 70 to 90 per cent of these recent increases in production have been due to increased yields rather than agriculture expanding into previously uncultivated lands. Some scientists have described this as 'the greatest agricultural transformation in the history of humankind' (Plunkett, 1993). But a major problem is that these benefits have been poorly distributed. Many people have missed out, and hunger still persists in parts of the world. In Africa, for example, food production per capita has fallen by a fifth since the 1960s. Worldwide there are close to one billion people hungry today, with several hundred million children malnourished. The causes are complex and not entirely the fault of modern agricultural technologies. These have had an undoubted positive impact on the overall availability of food. Nonetheless, agricultural modernisation has contributed, in that the technologies were more readily available to the better off.

The process of modernisation has produced three distinct types of agriculture: these are industrialised, Green Revolution, and all that remains – the low-external input, traditional and unimproved (Chambers et al, 1989; Pretty, 1995a). The first two types have been able to respond to the technological packages, producing high-input, high-output systems of agriculture. Their conditions were similar to those where the technologies were generated, or else their environments could easily be changed and homogenised to suit the technologies. These tend now to be endowed with access to roads and urban markets; with this comes ready access to modern crop varieties and livestock breeds, inputs, machinery, marketing infrastructure, transport, agroprocessing facilities and credit, and good soils and stable rainfall or irrigation water supply.

Most agricultural systems in industrialised countries count as high-external input systems, save for the relatively small number of organic farmers. In developing countries, high-external input systems are found in the large irrigated plains and deltas of South, South-East and East Asia, and parts of Latin America and North Africa, but also in patches in other regions. They tend to be monocrop or monoanimal enterprises, geared for sale, and therefore include lowland irrigated rice, wheat and cotton; plantations of bananas, pineapples, oil palm, and sugar cane; market gardening near urban centres; and intensive livestock rearing and ranching. These are the lands of the Green Revolution, the success of which lies in simplicity. Agricultural scientists bred new varieties of cereals that matured quickly, permitting two or three crops to be grown each year; these cereals were day-length and could be extended to farmers at many latitudes. They also produced more grain at the expense of straw. These modern varieties were distributed to farmers, together with inputs of inorganic fertilisers, pesticides, machinery and credit. Technical innovations were implemented in the best-favoured agroclimatic regions and for those classes of farmers with the best expectations of, and means for, realising the potential yield increases.

The third type of agriculture comprises all the remaining agricultural and livelihood systems. This is a largely forgotten agriculture, located in drylands, wetlands, uplands, savannas, swamps, near-deserts, mountains, hills and forests. Farming systems in these areas are complex and diverse, agricultural yields are typically only 0.5 to one tonnes per hectare. They are remote from markets and infrastructure; located on fragile or problem soils; and are unlikely to be visited by agricultural scientists and extension workers or studied in research institutions. The poorest countries have higher proportions of these agricultural systems.

The extraordinary thing is that by the mid 1990s, some 30 to 35 per cent of the world's population, about 1.9 to 2.1 billion people, were still directly supported by this third and 'forgotten' agriculture (based on estimates from FAO and World Bank production and distribution data; see Pretty, 1995a). A US Office of Technology Assessment study said, in the late 1980s, that 'most agricultural development assistance...has emphasised external resources' (OTA, 1988). But these people can rarely afford to use external resources. Their only immediate alternatives lie in making the best use of systems with resource-conserving technologies.

Hunger and Plenty: the Current Situation

Despite the progress that has been made in the last generation, notwithstanding those who have missed out, there is still too much food and too little food. The rich over-consume, and the poor are hungry and continue to be ignored. By the end of the first quarter of the 21st century, the world will have some eight to 8.5 billion people. Today, even though enough food is produced in aggregate to feed everyone, and world prices have been falling in recent years, some 800 million people are suffering from chronic hunger, of whom 180 million are underweight children suffering from malnutrition. As many as two billion people subsist on diets deficient in vitamins and minerals essential for normal growth and development (UNICEF, 1993).

Even as countries get more wealthy, the poor continue to lose out. In the past three decades, global income has doubled, yet the number of people living in poverty has continued to rise from 944 million to 1300 million. More than a billion have access neither to basic health services, nor to clean water. Just under a billion adults cannot read, and 120 million children cannot attend primary school. Twelve million children die before the age of five every year. Worse than more people living in absolute poverty, the gap between the wealthy and the poor has also widened. The poorest fifth of the world's population have seen their share of global income fall from 2.3 to 1.4 per cent since the 1960s, whereas the richest fifth have increased their share from 70 to 85 per cent. In some industrialised countries, the disparities have been even greater. In Britain, from 1979 to 1992, average incomes rose by 36 per cent. But for the top tenth, they rose by 60 per cent; for the lowest tenth, they actually fell by 17 per cent (Jacobs, 1996). In addition, there are increasing numbers of hungry people in industrialised countries. This does not fit the popular myth in wealthier countries – the poor are elsewhere, and can be helped by well-meaning charities and overseas development aid. But as Chapter 6 describes, poverty and deprivation have become substantially more common in rural areas in Europe and North America.

The global problem is now common to us all. We produce enough food to feed everyone. But hunger remains both in developing and industrialised countries. We seem to be getting richer. But poverty persists. We have productive agricultural technologies, but not everyone can afford them. They also damage the environment. So where do we go from here?

Schools of Thought

It is clear that agricultural production will have to increase substantially in the next quarter to half century if some lasting impact is to be made on global and local food insecurity (IFPRI, 1995; Crosson and Anderson, 1995; Leach, 1995; CGIAR, 1994; FAO, 1993b; 1995). Regrettably, there is simply no prospect of resources being sufficiently evenly shared in the near future to make a difference for the hundreds of millions of poor and excluded families.

Nevertheless, the views on how to proceed vary hugely. Some are optimistic, even complacent; others are darkly pessimistic. Some indicate that not much needs to change; others argue for fundamental reforms to agricultural and food systems. Some indicate that a significant growth in food production will only occur if new lands are taken under the plough; others suggest that there are feasible social and technical solutions for increasing yields on existing farmland. There are five distinct schools of thought (Pretty and Thompson, 1996; McCalla, 1994; Pretty, 1995a).

Business-as-Usual Optimists

The business-as-usual optimists, with a strong belief in the power of the market, say supply will always meet increasing demand and therefore growth in world food production will continue alongside expected reductions in population growth (Rosegrant and Agcaolli, 1994; Mitchell and Ingco, 1993; FAO, 1993).

Since food prices are falling (down 50 per cent in the past decade for most commodities), this indicates that there is no current crunch over demand. Food production will continue to rise as the fruits of biotechnology research ripen, boosting plant and animal productivity, and as the area under cultivation expands, probably by some 20 to 40 per cent by 2020. Some say this means an extra 79 million hectares of uncultivated land will have to be converted to agriculture in Sub-Saharan Africa alone. It is also assumed that population growth will slow, and that developing countries will substantially increase food imports from industrialised countries. It is unfettered markets that are said to be the best way to enhance efficiency. Referring to arable farming in East Anglia, Sir Derek Barber famously put it this way in the 1991 Royal Agricultural Society of England study on agriculture and the environment: 'why clutter up such landscapes with thin green threads of new hedges? Why not let this type of highly efficient grain country get on with its job of producing a tonne of wheat at the very lowest cost?' (Barber, 1991).

However, the optimists do not recognise the environmental costs of production. If they do, they say these are overstated or are simply small in comparison with the pressing need for more food. They do not address who will be engaged in increasing production, or draw attention to who has missed out. In particular, they do not explain how the poor will acquire the purchasing power to buy the food they need, when some 800 million people are already hungry today.

Environmental Pessimists

The environmental pessimists contend that ecological limits to growth are being approached or have already been reached (Brown, 1994; Brown and Kane, 1994; Ehrlich, 1968). These pessimists follow a neo-Malthusian line of argument by claiming that populations continue to grow too rapidly, while yields of the major cereals will stop growing or even fall. This is because of growing production constraints in the form of soil erosion, land degradation, forest loss, pesticide overuse, and fisheries overexploitation. Dietary shifts, especially increasing consumption of livestock products, are seen as an emerging threat, since this means an even greater consumption of cereal products. They do not believe that new technological breakthroughs are likely. Solving agriculture and food problems means putting population control as the first priority. Says Lester Brown of the Worldwatch Institute in Washington: 'I don't see any prospect of the world's fishermen and farmers being able to keep up with the growth in world population' (Brown, 1994).

However, like the optimists, the pessimists do not account for economic arguments that say current supply simply reflects current demand and price conditions. They also fail to incorporate the latest empirical evidence of production increases in many poor countries, and cost reductions in many industrialised countries.

Industrialised World to the Rescue

The industrialised world to the rescue lobby believes that developing countries will never be able to feed themselves, for a wide range of ecological, institutional and infrastructural reasons, and so the looming food gap will have to be filled by modernised agriculture in industrialised countries (Avery, 1995; Carruthers, 1993; Knutson et al, 1990). This group sees highly intensive agriculture coexisting with wildlife, though in different places. By pushing up production in large, mechanised operations, small and more 'marginal' farmers will be able to go out of business, taking the pressure off natural resources. These can then be conserved as protected areas and wildernesses. The larger so-called high-yield producers will be able to trade their food with those who need it, or have it distributed by famine relief. It is also vigorously argued that any adverse health and environmental consequences of chemically-based agricultural systems are minor in comparison with those wrought by the expansion of low-output agriculture into new lands. The best representation of this school of thought is put by Dennis Avery of the US think-tank the Hudson Institute, whose recent publications include the book *Saving the Planet with Pesticides and Plastic.*

However, opponents of this group argue that no account is taken of the impact that such intensification of agriculture in the Europe and the US, combined with food dumping, would have on the rural economies of developing countries. The contention that high-yield agriculture can only be sustained by large inputs of fertilisers and pesticides conveniently ignores a very large body of alternative empirical evidence, as does the view that agrochemicals are safe.

New Modernists

By contrast, the new modernists argue that biological yield increases are possible on existing farmland in countries of the South, but that this food growth can again only come from 'modern' high-external input farming (Borlaug, 1992; 1994a; 1994b; Sasakawa Global 2000, 1993–1995; World Bank, 1993). This group argues that farmers, particularly in Africa, simply use too few fertilisers and pesticides, which are again said to be the only way to improve yields and keep the pressure off natural habitats. This model approach replicates the Green Revolution with so-called science-based agriculture; the objective is to increase farmers' use of fertilisers and pesticides. Curiously, it is also argued that high-input agriculture is more environmentally sustainable than low-input agriculture, since the latter represents the intensive use of local resources which may be degraded in the process.

Some of the new modernists' views are extreme. Norman Borlaug, Nobel laureate for his contribution to international agricultural development, said in 1992 that agriculturalists:

> *...must not be duped into believing that future food requirements can be met through continuing reliance on ... the new complicated and sophisticated 'low-input, low-output' technologies that are impractical for the farmers to adopt.*

In 1995, he then said: 'Over the last decade, extremists in the environmentalist movement in the affluent nations have created consumer anxiety about the safety of food produced using agricultural chemicals.'

Nevertheless, the opponents of this group point out that many farmers cannot afford or even get access to external inputs and technologies when they need them. They also indicate that a sole focus on external inputs undermines the potential productivity of locally available resources.

Sustainable Intensification

Others are making the case for the benefits of sustainable intensification on the grounds that substantial growth is possible in currently unimproved or degraded areas while at the same time protecting and regenerating natural resources (Pretty, 1995a;1996b; McCalla, 1994; NAF, 1994; Hewitt and Smith, 1995). Empirical evidence now indicates that regenerative and low-input (but not necessarily zero-input) agriculture can be highly productive, provided farmers participate fully in all stages of technological development and extension to others. This evidence also suggests that the productivity of agricultural and pastoral lands is as much a function of human capacity and ingenuity as it is of biological and physical processes. Such sustainable agriculture seeks the integrated use of a wide range of pest, nutrient, soil and water management technologies. It aims for an increased diversity of enterprises within farms, combined with increased linkages and flows between them. By-products or wastes from one component or enterprise are inputs to another. As natural processes increasingly substitute for external inputs, so the impact on the environment is reduced and regenerative capacity is increased.

However, opponents of this group still argue that external inputs will always be necessary, and so sustainable agriculture represents only a low-output agriculture that cannot increase crop and livestock production sufficiently to feed the world.

The Globalisation of the Food System

Whichever school of thought eventually prevails, it will have to do so in an increasingly globalised food system. Some see a globally linked system as providing extra efficiencies and benefits for all. Others say this will simply mean that the most powerful will prevail, with the poorest missing out again. The term globalisation refers to the growing integration of the world economy. Since the 1940s, world trade has grown twelvefold – much faster than the fivefold growth of output. Now imports and exports make up a much larger proportion of economic activity than ever before, and international trade flows annually amount to some US$4000 billion. The current orthodoxy is that the liberalisation of trade, through the reduction of tariff and non-tariff barriers, increases competition and raises efficiency, releasing productive resources for growth.

The trend to greater trade liberalisation will have the most profound impacts. On 15 April 1994, trade ministers from 117 governments met in Marrakesh, Morocco, after seven years of negotiations under the General Agreement on Tariffs and Trade (GATT). They signed the Uruguay Round, which entered into force on 1 January 1995. Comprehensive agreements were reached in investment, intellectual property rights, trade in services, and, for the first time, agriculture and textiles. The Uruguay Round replaced the GATT with the more powerful and now permanent World Trade Organisation (WTO). To join the WTO, members have to accept every part of the Uruguay Round Agreement, treating it as a single undertaking (LeQuesne, 1996).

Huge financial gains are expected as a result, with early estimates varying from US$200 billion to $500 billion by 2005. According to then director-general of the GATT, Peter Sutherland, all countries will emerge as winners. But is this true? Caroline LeQuesne has documented for Oxfam the winners and losers. The winners are likely to be the countries of the North; the losers to be those in the South. A World Bank–OECD study put the gains for the South at one third of the total, with Sub-Saharan Africa actually losing out by an estimated $2.9 billion. But as she points out: 'the corresponding social and environmental costs of the agreement have not figured in any of the official calculations' (LeQuesne, 1996). Others have argued that globalisation poses considerable threats to food security, environmental protection and consumer choice.

There are two vital consequences for the food system. Farming will become more concentrated on exports, with developing countries encouraged to produce and sell high-value crops and to purchase staples in the world market. At the same time, organisations concerned with food and input trading, manufacture and sales will have to get larger in order to compete in the global market. They will have to find ways to force down costs and capture greater shares of markets. This implies that the gradual globalisation of trade will concentrate rather than open up markets. This is already happening. The 1996 merger of two of the largest agro-chemical companies, Ciba and Sandoz, produced the massive Novartis. On the production side, between 60 and 90 per cent of all wheat, maize and rice is now marketed by just six transnational companies: Cargill, Continental, Ferruzzi, Louis Dreyfus, Bunge, and Born and Cook. One North American company, Cargill, supplies over 60 per cent of the world trade market. Its revenues from coffee sales alone are greater than the gross domestic product (GDP) of any one of the African countries from which it buys coffee. It operates from about 800 locations in 60 countries, employing some 72,700 people. Even though its annual turnover is some $51 billion, Cargill's stated aim is to double in size every five to seven years (Tansey and Worsley, 1995).

Processors have also become larger and fewer in number. The global production of processed food and drinks amounts to some $1500 billion per year. Production in the OECD region is about $800 billion, in the former Soviet Union about $200 billion, and in developing countries about $150 billion (OECD, 1992). This makes this business sector one of the world's largest, with several very powerful companies (see Box 4.1). Concentration has occurred rapidly in the past 20 years. In Britain, there were 5600 food and drinks firms 20 years ago, of which the top 30 accounted for 60 per cent of the employment and value in the industry. By the late 1980s, the sales of just eight firms in Europe made up 70 per cent of the $250 billion food and drink market. In the 1990s, most product markets were each dominated by the top three suppliers, taking 50 to 80 per cent of these markets throughout the EU (Tansey and Worsley, 1995).

This concentration may not always be bad. Companies with strategies to ensure they take environmental and social responsibility for their activities, such as Unilever, have a significant mitigating effect. However, most still argue that environmental and social costs will continue to be ignored as the global food system continues its search for the lowest possible costs.

Box 4.1 Two food giants

Unilever

Founded in 1930, Unilever operates through about 500 companies in 80 countries. The turnover is US$28 billion. It is engaged in oils and dairy based foods; meals and meal components; ice-cream, beverages and snacks; and supplies for professional caterers. It is also engaged in animal feed production, fish farming, plantations and plant breeding. Some $500 million, or 2 per cent of turnover, is spent on research and development, and it employs 4000 scientists and technicians.

Nestlé

Founded in 1866, Nestlé has sales exceeding $36 billion, some 90 per cent of which is derived from food and beverages. It operates 480 factories in 69 countries, and employs 218,000 people. It is engaged in producing drinks, particularly coffee and chocolate based; milk products; chocolate and sweets; and prepared dishes and sauces. Some 669 million Swiss francs are spent on research and development.

Farm to Plate: the Environmental and Social Costs

Hidden Transport Costs

Just as during the production of food on farms, there are many hidden environmental and social costs associated with getting food from the farm gate to our plates. One that has grown significantly in recent years is the transport cost. Each item of food now travels on average 50 per cent further than it did in 1979. According to the former Department of Transport, the same amount of food by weight, some 300 million tonnes, is being transported now as in the late 1970s. But today it travels 36 billion tonne-kilometres per year compared with 24 billion tonne-kilometres (DoE statistics, in Raven and Lang, 1995). However, transport costs still only account for a small portion of the food costs. This is because non-renewable fossil fuels and road transport are cheap. Rail freight uses a quarter the amount of energy per tonne-kilometre as road transport; yet only 6 per cent of goods in the UK are transported by rail (Raven and Lang, 1995). In the US, it is said that each item of food now travels an average 2000 kilometres between grower and consumer. All this extra travel is not costless. One estimate puts the total external costs of all road transport in the UK at close to £14 billion annually (Pearce, 1995). These are costs whole societies pay.

One reason for this increase is that large retailers want the regularity of supply that can only be given by centralised distribution, whereas independent, small producers tend to source their food locally. Three English examples illustrate what has been happening (Paxton, 1994):

- Tomatoes sold in one chain of shops are brought by road from Pilling, Lancashire, to Lancaster, then by road to Blackpool, then to Yorkshire to a distribution depot, then by road to shops all over Britain, including back to Lancaster.
- Bananas sold by an international banana company are imported through Southampton, then taken by road to Lancashire for ripening, sent back to a Somerset warehouse, and then distributed all over Britain, including of course back to Southampton and Lancashire.
- In Evesham, two supermarkets sell organic produce, some of which is grown on farms 1.5 kilometres away. But to get the vegetables from the farm to the shelf, they must first travel to a cooperative in Herefordshire, then to a pack house in Dyfed, Wales, then to a distribution depot south of Manchester, and finally back to Evesham.

A now classic study conducted by the Wupperthal Institute in Germany calculated the transport costs of getting yoghurts to consumers (Böge, 1993). It takes a lot to make a yoghurt, and each component comes from different parts of Europe. To the distribution outlets in southern Germany come strawberries from Poland, yoghurt from northern Germany, maize and wheat flour from The Netherlands, jam from west Germany, sugar beet from the east of Germany, and aluminium cover from 300 kilometres away. Only the milk and glass jars are sourced locally. To produce one truckload of 150-gram yoghurts for the distribution outlet, the contents of the yoghurt travel 1005 kilometres, resulting in the burning of huge amounts of diesel fuel.

Food imports by air have also grown – more than doubling during the 1980s (Paxton, 1994; Prest and Bowen, 1996; Clunies-Ross and Hildyard, 1992). In the winter of 1995–96, for example, five freighters were chartered each week to transport vegetables from East Africa alone. Between the farm gate and our plate, food now goes on some remarkable journeys: apples from South Africa, prawns from Bangladesh, green beans from Kenya, and mangetout and baby corn from Zimbabwe all travel between 6500 and 8000 kilometres to reach Britain. Asparagus from Zimbabwe remarkably takes only 60 hours from the field to our supermarket shelves. In 1994, the UK imported 36,000 tonnes of air-freighted fruit and vegetables – including 7200 tonnes from Kenya and 6300 tonnes from the US (14,400 kilometres). In 1996, 12,360 tonnes of strawberries were imported from New Zealand, Australia, Kenya, Zimbabwe and the US. This is not just to supply demand in our off-season; in the middle of the UK's summer, some 1,150 tonnes travelled 4800 kilometres from the US. UK-grown strawberries are only 50 per cent of all sales, down from 70 per cent in 1985 (SAFE Alliance, 1997).

It is important to note that this global trade provides important foreign exchange for countries such as Kenya and Zimbabwe. High-value vegetables have already replaced tea and coffee as the main foreign exchange earners for Kenya, a remarkable shift over a very short period. But this success may be difficult to sustain. Mobile capital that has helped to promote in-country production could just as easily leave to promote production in another country where costs are lower. Some would say this is exactly the model desired – large companies are encouraged to promote production exactly where the direct costs of labour and environmental protection are lowest. If either of these rise, then they can simply move elsewhere.

Another problem is that many developing countries have been encouraged, particularly via structural adjustment policies attached to loan agreements from the World Bank and IMF, to move into high-value export crops, and then to use the cash generated to import cheap wheat from North America. This again undermines local capacity for food production and increases the environmental costs associated with transport.

Who Pays for the Full Costs?

The full costs of these foreign-exchange earning operations can be substantial. The resources used are generally not fully costed against earnings. In Kenya, for example, one British supermarket chain introduced high-yielding bean varieties from the UK. But it found that these beans did not flower without the long hours of daylight of a temperate summer. Rather than shift to low-resource systems, it extended daylight by introducing energy-intensive floodlights.

In South and South-East Asia, prawn farming expanded massively in the 1980s (Briggs, 1994). In the mid 1990s, the UK market for frozen prawns from India, Bangladesh, China and Thailand alone was annually worth £47 million. In Bangladesh, prawn production grew from 1000 tonnes in 1980 to 25,000 tonnes in 1992. In these countries, there are now some 385,000 hectares of coastal habitats under prawn cultivation. Much of this has meant large-scale clearing of mangrove forests and high levels of inputs to sustain the system. As a result, clean surface water has been polluted to the detriment of rice farmers who now have to use salty rather than freshwater. Small farmers have been made landless, as the value of coastal areas has increased. One estimate suggests that 1.2 million residents of Khula in Bangladesh are now landless owing entirely to the expansion of prawn cultivation. To consumers in the industrialised countries, these costs are hidden; to many local people, they are all too real.

Apples from Chile cause different problems. The cost of an apple grown under intensive agriculture and treated with pesticides is still cheaper after being shipped 14,400 kilometres than are organic apples grown in Europe. Again this is because the environmental and social costs of soil, water and air pollution in Chile are not accounted for in the price of an apple. These costs are paid by local people, while consumers in the North pay for the apple alone.

About half of the animal feed requirements in Europe come from grassland and bulky local fodder crops. But intensive livestock rearing currently also depends on the import of protein-rich feeds and cereal substitutes for their feed. Brazil and Thailand provide more than half of the total compound feedstuffs, derived from soya and cassava. Soya meal contains five times more protein than cereals, and so makes a good feed supplement. In Brazil, the area under soya beans has increased from 1.42 million hectares in 1970 to 11.5 million hectares in the mid 1990s. Much is produced in large-scale, mechanised operations, especially in the fragile Cerrado. Here forests have been cut down; soil erosion has increased; and fertilisers and pesticides contaminate water. Again local people pay for these costs. They consume fewer beans themselves – just 14 kilogrammes per person, which is half the 1981 average consumption of 28 kilogrammes per person (Paxton, 1994).

Some of this globalisation of the food system is undermining traditional systems in the UK, leading to a decline in genetic diversity (see Chapter 2). Apples are an example. Some 412 million tonnes of dessert apples are consumed each year in the UK, but only a quarter come from domestic orchards. The UK imports 190 million tonnes from France; 85 million tonnes from South Africa; 55 million tonnes from New Zealand; 22 million tonnes from the US; and 35 million tonnes from The Netherlands, Benelux and Italy. Although domestic apple consumption is increasing, most of the market growth is being filled by imports (Paxton, 1994).

By contrast, 90 per cent of apples sold in France are produced domestically. At present, the EU also provides a grubbing grant to encourage fruit growers to remove their orchards so as to reduce surplus supply. Farmers are being paid £4700 per hectare to remove their orchards. As they do, a crucial part of the rural heritage disappears. At the National Collection at Brogdale in Kent, there are some 2000 varieties of apple. At one time, it was possible to find 200 varieties growing in a single orchard.

The Decline of Local Shops

Supermarkets have revolutionised food shopping in recent decades. They provide massive choice: in 15 years the number of products has trebled and can now be ten times as great as the choice in a corner store. Every year, 10,000 new food products are launched on the European market. One in ten survives a year; one in 20 survives two years. Consumers like supermarkets because they are convenient, well lit, warm and safe. And more of us shop in them. In Japan, for example, only 10 per cent of consumers shopped in supermarkets in the early 1980s, but by the early 1990s this had risen to 60 per cent (Raven, 1996).

An inevitable consequence of this growth is the sharp decline in corner shops in towns and village shops in rural areas. As a result, the number of specialist grocers, butchers, bakers and fishmongers is falling, with a total loss of some 6000 between 1990 and 1995 (see Table 4.1). Family-run grocer shops now account for only 12 per cent of the vegetable market, whilst the supermarket share has grown from 8 per cent in 1969 to 72 per cent in 1995 (Hughes, 1996).

Table 4.1 *Changes in specialist retailing outlets in the UK, 1990–1995*

Retailers	*1990*	*1995*	*Change (%)*
Butchers	17,044	15,150	–11%
Greengrocers	14,339	12,400	–14%
Bakers	6656	5500	–17%
Fishmongers	2974	2050	–31%
Total	41,013	35,100	–14.5%

Source: Retailing Enquiry, Mintel, March 1996

These changes have been promoted by the trend to out-of-town locations for retail and leisure services. By the mid 1980s, there were some 25,000 out-of-town retail developments in the UK, an increase that has coincided with the closure of some 238,000 (from 577,000 to 319,000) independent shops in villages and high streets between 1961–1992 (DoE/MAFF, 1995; McLaren, 1995; TEST, 1988). By the mid 1990s, just five multiple retailers handled 65 per cent of the retail food trade.

When shops move out of towns and villages, something important is lost. Leominster is a small town of 10,000 people in west England. After the opening of a large out-of-town supermarket in 1992, trading in the town centre fell dramatically. Turnover in some small shops fell a third overnight. Now 35 shops in the town centre are empty, two town centre supermarkets have shut, while three garages, a pub and a petrol station are empty. As Monica Todd, president of the Chamber of Commerce, put it in 1996: 'the whole character of the town has changed – apart from Market Day, there isn't a buzz about the place any more because people are not coming here.' Local traders have responded by launching a Loyal to Leominster campaign to lure shoppers back to the town centre. Discounts, gifts and offers are guaranteed for shoppers with the loyalty card. The hope is that this will help to revitalise the centre (quoted in *The Independent*, 2 March 1996).

Caroline Cranbrook of Great Glemham in Suffolk has recently shown just how important small shops are for social capital (Cranbrook, 1997). She visited 81 food shops in East Suffolk in mid 1997. These employed 548 people, of whom 317 were part time. They were also sourcing locally. She was 'astonished to discover the amount of local produce sold'. The 81 shops were buying from 295 local producers, ranging from large and small farmers, vegetable growers, wine producers, cheese and jam makers, village smallholders, beekeepers, and housewives making pies, soups and cakes. One shop was sourcing over 40 local products.

Such interdependence of small retailers, producers and consumers creates a dense social network that provides employment, good quality food and wider social benefits. The shops are the social centres of communities – they keep an eye on the elderly and infirm; they provide notice boards for advertisements; they keep in touch with local people. They provide diverse foods and connect producers with consumers. Caroline Cranbrook described the social network like this:

> *...a wholesale family butcher buys meat from about 30 local farmers. This business cuts and sells fresh meat to other outlets, cures and smokes bacon, makes sausages and cooked meats, and provides freezer packs. These products are supplied to 21 small shops. In addition, the family runs two butchers shops, which in turn are sourcing other produce, such as eggs, vegetables, cakes and preserves from 24 local producers.*

Crazy Cosmetics

The choices we make as consumers in supermarkets affect both farming and food production. Supermarkets permit shoppers to select fruit and vegetables on the basis of appearance, and so any produce marked by the scars of insects or disease is likely to be rejected. Retailers insist on uniform, visually perfect fruit and vegetables for their shelves, which encourages growers to use large amounts of pesticides if there is any perceived risk of downgrading or rejection. Food that does not meet the specifications is rejected, leading to huge wastage and loss for farmers.

This may be 20 per cent for conventional and up to 50 per cent for organic producers. The producers bear these costs themselves.

In the US, cosmetic standards are particularly strict on citrus and apples. Blemishes on the skins of fruit reduce the returns to farmers, even if there is no evidence that they reduce yields or affect nutrient content or flavour. The citrus rust mite, for example, causes russetting or bronzing on oranges but does not affect taste. In the early 1990s, most Florida oranges were being sprayed for rust mites, costing US$40 to $50 million per year. Oranges from treated orchards sold at a premium, even though yields were the same as in untreated farms. One study of retailers found that one quarter would accept no apples with any blemishes, while an additional four in ten would only accept produce with less than 1 per cent blemishes (Rosenblum, 1994). This means that about one third of pesticides used on apples are entirely for cosmetic control. It is even higher for oranges. In California, some 60 to 80 per cent of insecticides are for cosmetic controls.

Sometimes the requirement of a cosmetically perfect product is taken to an absurd extreme. In the US, more than half the insecticide applied to tomatoes is for control of tomato fruitworm which, like the rust mite, affects appearance only. Processors will not accept batches from farmers with more than 1 per cent damaged, yet 90 per cent of tomatoes are for paste, soups and juices, in which consumers cannot possibly detect cosmetic damage.

The Health Costs of the Food We Eat

What We Eat and Where It Comes From

A turning point for food production in Britain came with the repeal of the Corn Laws in the 1840s (Lang, 1995a). These had protected local markets but, with their repeal, Britain came to rely more on the production of much of its staple foods in distant countries, particularly cheap wheat from the vast prairies of North America. Little changed until the advent of the Second World War, when the lack of self-sufficiency (food self-sufficiency before the war was only about 30 per cent) induced another great change in national policies. These now focused on encouraging increased local production above all other farming goals.

Most decades since then have seen major initiatives to promote domestic production. With the 1947 Agriculture Act, the Attlee government sought to boost home-grown staples, especially cereals and dairy produce. Later agriculture acts in the 1950s and 1960s continued to support home production (see Chapter 2). The 1974 White Paper *Food from Our Resources*, reemphasised the need for domestic production and, in the early 1980s, the government set up Food from Britain, a quango whose brief was to promote British food exports. Yet another decade later (in 1992), the House of Commons Agriculture Committee, in its report *The Trade Gap in Food and Drink*, had still to ask: 'why was the British food industry allowing unnecessary imports?'

After all these policy initiatives, food and drink imports still greatly exceed exports. The total trade gap for food and drink is £5.87 billion, a figure that has remained largely constant during the 1980s and 1990s. Except for Japan, Britain has a negative balance with all trading partners (see Table 4.2). Over the past few decades, the gap has narrowed slightly. In 1971, exports were only 25 per cent of imports in value, but by 1994, they had risen to about 60 per cent of import value. However, most recently, the gap reopened – total food self-sufficiency fell to 53 per cent from 1995 to 1996, while the drop for indigenous food was from 75 per cent to 69 per cent (MAFF, 1996).

Again, the situation is quite different in France and the US. Food exports from France are worth US$36 billion each year, second only behind the US in the world at $47 billion (La Fondation pour le Progrès de l'Homme and Solagral, 1995). France has been a net agricultural exporter for the last two decades. But it has not always been so. In 1960, it imported twice as much food as it exported.

Table 4.2 *Imports and exports of food and drink to and from the UK in 1994 (in £ billion)*

Country or group	*Imports (£ bn)*	*Exports (£ bn)*	*Balance*
Netherlands, Belgium, Luxembourg	2.29	0.97	–1.32
France	2.21	1.42	–0.79
Ireland	1.53	0.99	–0.54
Germany	0.99	0.67	–0.32
Spain	0.66	0.64	–0.02
Italy	0.72	0.44	–0.28
Denmark	0.76	0.14	–0.62
Japan	0.02	0.22	+0.20
US and Canada	0.84	0.73	–0.11
Rest of World	4.82	2.77	–2.05
Total	14.84	8.97	–5.87

Source: Lang T. 1995. 'Local sustainability in a sea of globalisation? The case for food policy.' Paper for the Planning Sustainability conference organised by the Political Economy Research Centre, University of Sheffield, 8–10 September

A significant amount of the food imported to the UK could be produced or processed locally. Should we be asking:

- Why import lettuces, spinach and apples in autumn and summer when they can be grown equally well locally?
- Why import so many apples when they could be grown here, and when we are grubbing up orchards at an increasing rate?
- Why eat so much imported bacon, when domestic pig producers could easily (and would like to) expand production?
- Why bring asparagus all the way from Spain for sale in English supermarkets in June, at the height of the English asparagus season?

Of course, no one conceives of growing bananas on Ben Nevis, or mangoes in Macclesfield, or tea in Tiverton. According to Food from Britain calculations, slightly more than half of the £14.9 billion of imports are unavoidable. This still leaves some £6.5 billion of value that could be produced in the UK. Yet between 1980 and 1990, imports of food that could have been produced in Britain greatly increased: imports of apples rose by 92,700 tonnes (up 25 per cent), fresh vegetables by 493,000 tonnes (up 40 per cent), lettuce by 69,000 tonnes (up 575 per cent), and mushrooms by 26,000 tonnes (up by 467 per cent).

Changing Diets and Consumer Preferences

Most scientists agree on the ideal diet for humans. Sometimes referred to as the Mediterranean diet, in honour of the work of Ancel Keys and colleagues, this diet is high in fruit, vegetables and carbohydrates, low in meat, with some fish (Keys, 1980; Cannon, 1992). Yet now we find that the food we eat is making us unhealthy. As Tim Lang, professor of food policy at Thames Valley University, has put it: 'having conquered hunger in the West, food is now the primary cause of premature death' (Lang, 1995a). With rising affluence, diets tend to contain more fats and sugars, and fewer vegetables, fruits and cereal grains. These diets put people at greater risk of contracting heart disease, some forms of cancer and diabetes. The highest rate of coronary heart disease in Europe is found in Scotland, Finland and Ireland (exceeding 300 per 100,000 population for men); the lowest rates are in Portugal, Spain and France (below 100 per 100,000) (Tansey and Worsley, 1995).

The UK has a poor record of food-related ill-health (HMSO, 1992). Like most industrialised countries, we spend less of our incomes on food than we used to: 18 per cent in the 1990s compared with 33 per cent in the 1950s (CSO, 1994). In the US, the fall was from about 18 per cent in 1960 to just 12 per cent in 1990 (Tansey, 1995). But much of what we do spend is on snack, drink and sweet products. Of the top 50 food brands in the UK, only eight are meat, fish, vegetable or yoghurt foods, accounting for just 13 per cent of the value of these top 50. Most of the rest are non-essentials – drinks, spreads, crisps, sweets, sugar and breakfast cereals. Six of the top ten brands are drink-related, bringing total sales of £900 million per year (a quarter of the total value).

There is an increasing trend towards the purchase of convenience foods. The time available that working people have for food shopping and cooking has been substantially reduced. David Buisson of the University of Otago, New Zealand, suggests that early in the next century, 90 per cent of food will be pre-prepared and that 'it is quite conceivable that a proportion of the population will not know how to cook at all' (quoted in Tansey and Worsley, 1995). Convenience foods are more expensive, but for busy people, time is more important than cost. We now eat more processed, frozen and packaged food, and fewer fresh items. We cook less than we used to and microwave more. We can choose from some 20,000 items on the supermarket shelves, drawn from all over the world. Some of this over-processed, overpreserved food has lower nutritional value – both spinach and asparagus, for example, lose half of their vitamin C value after 24 hours' storage at room temperature.

Alarmingly, we also are eating fewer vegetables. Since 1960, the consumption of fresh green vegetables in the UK has fallen dramatically, from 410 grammes per person per week to just 280 grammes. Fruit consumption has increased, largely because of sharp rises in fruit juice consumption. Fresh potato consumption is falling, though the market for chips and crisps is rapidly growing – one kilogramme of crisps will cost the consumer £3.50 to £9, yet one kg of fresh potatoes should only cost £0.50 to £1, depending on variety and season. Britons now eat fewer vegetables than any other Europeans, about half the amount of the average person in France and Italy, and less than a quarter the intake of people in Spain and Greece. Britain also has the lowest salad consumption rate in Europe – some 15 kilogrammes per year per person, compared with 95 kilogrammes in Spain (Strathclyde University Study, 1997). The British salad industry produces 400,000 tonnes per year worth £224 million, about half of current domestic demand.

Some vegetables are doing better than others – mushrooms, carrots and exotics are up over the past 30 years; but fresh potatoes, cabbage, brussels sprouts and turnips are down (see Table 4.3). The trend is towards greater homogeneity, even where vegetable consumption remains high. In the south-east of France, the Provençal diet used to contain 250 plant species at the beginning of the 20th century; now it comprises just 30 to 60.

Table 4.3 *Changes in UK fresh vegetable and fruit consumption, 1965–1994*

Up		*Down*	
Mushrooms	+183%	Turnips	–11%
Exotic vegetables	+115%	Tomatoes	–19%
Carrots	+36%	Potatoes	–40%
Onions	+9%	Cabbage	–51%
		Brussels sprouts	–69%

Source: Hughes D. 1996. 'Dancing with an elephant: building partnerships with multiples'. Paper presented at The Vegetable Challenge Conference, London, 21 May, The Guild of Food Writers

Perhaps the matter of greatest concern is that the young eat the fewest fruit and vegetables. Only 41 per cent of 16- to 24-year-olds eat fruit regularly (five to six times per week), while this rises to 64 per cent for over 45-year-olds. More 16- to 24-year-olds eat vegetables and salads regularly (some 65 per cent), but still the older generation are higher consumers, with 83 per cent of over 45-year-olds eating vegetables regularly (Hughes, 1996).

The daily battle at the dinner table, where parents try to get their children to 'eat their greens', is being lost. Recent research conducted by Strathclyde University for the Cancer Research Campaign has shown just how serious is the problem. Instead of insisting, parents and schools are increasingly allowing children to choose junk food – jam sandwiches and crisps ahead of vegetables and fruit. As Professor Gerard Hastings said in 1997: 'many parents have simply given up forcing the issue of vegetable consumption because they dislike the stress' (quoted in *The Independent*, 21 January 1997). All of this may be storing up more long-term health problems. Epidemiological studies show that diets rich in vegetables and fresh fruit substantially protect against cancers of the lung, larynx, stomach, colon and pancreas. Of the 300,000 people who get cancer each year in the UK, about one third are diet-related and so potentially preventable. Bowel and stomach cancers alone claim 30,000 lives each year.

There are another 150,000 deaths per year from heart disease, and it is estimated that five portions of fruit and vegetable per day would reduce this by 20 to 30 per cent – a saving of 30,000 lives per year. As Imogen Sharp, director of the National Heart Forum, put it: 'whatever shape or form we eat them – frozen, dried or canned – vegetables and fruit can only do us good' (quoted in *The Independent*, 17 March 1997). But vegetables and fruit have a poor image and are weakly promoted. A total of £71 million is spent annually on advertising sweets; for vegetable and fruit it is £2.9 million.

Diets and Poverty

Increasing poverty has long had important links to low consumption of foods that improve health, particularly vegetables and fruit (Lang, 1997a). One in five parents and one in ten children occasionally go without food because of poverty. Between 1979 and 1991, the number of people in Britain living on less than half of average earnings rose dramatically from five million to 13.5 million. The poorest fifth now earn less than £75 per week. Yet, at the same time, the poorer section of the market is losing connection with fresh produce. Poorer households spend proportionally more of their income on food. In Australia, for example, the poorest tenth of the population spend 56 per cent of their income on food; the richest tenth just 12 per cent. The former know they should eat more fresh fruit and vegetables and would like to be able to do so. But they cannot. They have no access to the right shops; prices tend to be higher in local shops; and the £500 million spent on food advertising in the UK continues to promote foods that are high in fats and sugars (Lang, 1995b).

The National Food Alliance (NFA) has identified some very important reasons why this should be so (NFA, 1995; 1997). Much television advertising on food is targeted at children, and it is dominated by foods high in sugars, fats and salt. It is no surprise, therefore, to find children eating too much of these and too little fibre, vitamins and minerals, which miss out in the advertising. What can be done about these trends? In 1996, the Guild of Food Writers held the Vegetable Challenge, a conference drawing attention to these vital challenges. Linda Brown, one of the organisers, put it like this: 'why don't the British value vegetables more, why don't they care where they come from, why have they been persuaded that vegetables are boring and inconvenient, and why, despite overwhelming medical evidence that vegetables are good for you, is it such an uphill struggle to persuade people to eat more?'

Food Scares and Biotechnology

Food safety has become one of the most potent causes for concern amongst consumers in recent years. The number of reported cases of food-borne diseases or illnesses is increasing. Although food preparers and processors try to eliminate the risks associated with micro-organisms, there remain many strains of dangerous bacteria. These include *E coli* in meats; *Clostridium* bacteria that cause botulism poisoning and are found in canned foods, meats and fish; *Campylobacter* in meats and milk; *Listeria* in soft cheeses and unpasteurised milk and cooked crabs and shrimps; *Salmonella* in raw meats, poultry, milk and dairy products; *Shigella* in milk, dairy and poultry products; and *Vibrio* in seafoods.

In the US, there are said to be some 6.5 million cases of food poisoning per year (Tansey and Worsley, 1995). Some say that no more than 10 per cent of actual cases are reported; others that most bacteria are introduced to foods by consumers themselves during preparation and cooking. Some of these have led to significant food scares in the UK and the rest of Europe – with toxic strains of *Salmonella* in eggs, *Listeria* in soft cheeses, and *E coli* in meats causing ill-health and death amongst consumers. In the US, about 11,000 tonnes of ground beef had to be destroyed in mid 1997 after the discovery of a lethal strain of *E coli*. But perhaps the biggest food quality scare in history has been the advent of BSE in cattle. For a variety of reasons, cattle in Britain in the mid 1980s began to exhibit symptoms hitherto unseen. Loss of body function was traced to a destruction of nervous brain tissue, causing the brain to become 'spongiform' in nature. Infected cattle were removed from the food chain and burned or buried. It was first thought that the cause of BSE was adding to feed the remains of dead sheep and cattle. As a result, this practice was banned in the late 1980s, and it was believed that the disease would be contained and therefore eventually eliminated.

It then became clear that female cattle could pass on the disease to their calves, and that the disease was spreading through the British herd. Stricter regulations were introduced for abattoirs, preventing the use of nervous tissue in human food stuffs. It was not until 1996, however, that concerns about human health were clearly identified. The health minister's speech to the House of Commons in March 1996 marked a turning point, putting BSE at the centre of all consumers' concerns about the food they eat. Evidence pointed to a role being played by BSE in the increasing number of reported cases of CJD, a human equivalent of mad-cow disease. Although relatively few cases were identified, and the numbers are still small, it is not clear how many people in Britain have been exposed to the factors in infected beef that could eventually trigger the disease.

The collapse of the market for beef and the restrictions placed on exports have caused enormous damage to the livestock industry. More than one million cattle and 400,000 calves were removed from the food chain. The compensation scheme for lost revenue cost some £4 billion in the first year, though even this did not fully compensate farmers' losses. The cause of BSE is still unknown at the time of writing. It is clear that the prion protein plays a key role, but how this causes the denaturing of tissue is unclear. It is also unclear why BSE developed in the British herd, and why it has not happened to the same extent elsewhere in Europe and in the US. What is clear is that several factors may be important. BSE cases are now being recognised amongst herds throughout Europe – and there is some suspicion that real numbers are suppressed to avoid the imposition of export bans.

Two important events have happened as a result of the BSE and *E coli* crises. Consumers are now much more concerned about the quality of the food they buy. They realise it could do them harm. They are also aware that some forms of farming appear safer than others. There have been no cases of BSE in organic beef herds that have spent their entire lives on organic farms. Organic farming did not permit the use of concentrate made up with animal matter, nor the use of the organophosphate pesticide phosmet to control warble fly.

Many believe that biotechnology and genetically modified organisms in the food chain represent hidden dangers for consumers. The biotechnology industry is now one of the fastest growing sectors associated with food and farming. There are 1500 to 2000 biotechnology companies in the US, and another 700 in Europe. Most of these are engaged in the production of materials for human health improvements. A significant number, however, are producing new genetically modified organisms (GMOs) for farming. Biotechnology involves making molecular changes to living and almost-living things such as proteins. In agriculture, the first generation changes have been to incorporate genes into crops that confer resistance to a particular herbicide, or that encourage plant cells to produce materials that are harmful to insects. In 1994, there were no genetically modified crops grown anywhere in the world. In 1997, more than 12 million hectares of GMO crops were cultivated in the US, Australia, Argentina, Canada and Mexico, involving a quarter of all cotton, 14 per cent of soya and 10 per cent of maize in the US alone (Pretty, 1998).

The biotechnology goods, particularly in the health sector with 'molecule pharming', could be significant: herbicide-resistant crops could mean a decline in pesticide use; salt- and drought-tolerant crops could be grown on hostile soils; cotton with the *Bt* gene needs fewer pesticide applications; delayed-ripening genes in tomatoes means the shelf life is longer; and genetically modified sheep and pigs are producing human proteins, such as insulin and human blood-clotting factor in their milk. But the biotechnology bads are less well understood and will only emerge long after widespread adoption of GMOs. The greatest concerns are over genetic pollution, with genetically modified crops crossing with wild relatives, transferring the genes into the environment. No one knows what will happen in the long term. Already resistance genes in oil seed rape are found in France to have become incorporated into wild radish. In Scotland, potatoes containing snowdrop genes that confer resistance to peach potato aphid have been found to reduce the lifespan of female ladybirds by half. Other concerns are over the potential spread of antibiotic resistance to humans from the antibiotic markers used in these crops.

There are also many fundamental ethical issues, such as whether insurance companies should have access to data about our own genes that may tell them whether we have a greater likelihood of contracting a particular disease. Other issues include the production of 'self-pigs', which could be genetically tailored to an individual human (a human in pig's clothing), the organs of which could be transplanted in the event of accident or disease; and the whole issue of patenting genes, allowing companies to 'own' genetic material that may be common to most of us.

Some argue that there are no problems. The head of regulation affairs at AgrEvo, Windsor Griffith, said in 1997: 'GMOs are the most tested food we will have ever consumed. It really is very frustrating because we are dealing with a very safe technology.' Many agrochemical companies now see GMOs as their main source of revenue in the future and are pushing very hard to prevent labelling, arguing that it discriminates against their 'new' and 'modern' crops. But Vyvyan Howard of the University of Liverpool captured the concerns of many by pointing out that 'once genetically modified organisms escape into the environment, they are self-perpetuating. No one can predict the long-term results of manipulation.' Norman Ellstrand of the University of California has said: 'Within ten years, we will have a moderate to large-scale ecological or economic catastrophe' (quoted in Lean, 1997).

Towards Integrated Food Policies

Most countries give some guidance to their citizens about the types and diversity of foods they should eat. Healthy diets come from eating mainly cereals, fruits, vegetables, meats and dairy products. Guidelines usually suggest a mix of these, combined with suggested caution over consumption of fats, sugars and salt. One example is the Department of Health's Healthy Plate recommendations, which convey the relative proportions of different types of food that should be eaten, the variety of foods, and the need for moderation in fats, oils and sweets. It recommends that a third of consumption is of the bread, cereal, rice and pasta group; a third is of vegetables and fruit; a ninth is of meat, poultry and fish; a ninth is of milk, yoghurt and cheese; and a ninth is of foods containing fat and sugars (Tara Garnett, pers comm, 1997).

With all these known causes of dietary ill-health, has any country been able to put into place policies that have shifted food consumption patterns? Perhaps the best cases are in Scandinavia. In Finland, for example, two key policies have led to a trebling of vegetable and fruit consumption and a dramatic improvement in health in recent years. Firstly, children are provided with set meals at school, rather than allowed a free choice; and public catering establishments make vegetables an intrinsic part of main courses, always providing fruit as alternatives to dessert (James, 1996). In Britain, it was the removal of statutory standards for school meals in 1980 that many believe has led to a decline in dietary diversity and quality. In 1997, new voluntary guidelines were introduced, and it is hoped that school meals will, in future, provide more health benefits to young people.

In Norway, an integrated food policy was adopted in the mid 1970s. It sought to do four things. It promoted domestic food production and the reduction of food imports, increasing national self-sufficiency from 39 per cent to 52 per cent of calories by the 1990s. It promoted agricultural development in the country's less advantaged and marginal regions while preserving the natural resource base. It encouraged healthy diets; reduction of fat consumption, especially saturated fats; and replacement of fats and sugars with grains, vegetables and unsaturated fats. And, globally, it sought to make a contribution to world food security through supporting agriculture in developing countries. The progress has been significant (Milo, 1990). Average fat consumption has fallen from 40 per cent to 34 per cent of total calories. There has been a fall in deaths due to heart disease; an increase in self-sufficiency; and progress in regional development. The initial opponents of the policy, especially the food and agriculture industry, are now more willing to listen to and work with nutritionists and health professionals.

Tim Lang, Professor of Food Policy at the Thames Valley University, has called for much more trust to be built into food policies (Lang, 1997b). As he says, food policy is too important to be left to a narrow clique. More stakeholders must be involved – all the way from suppliers of all inputs to farming, through farmers themselves, to manufacturers and retailers, and the consumers themselves. He suggests ten core components of a reformed and more integrated national food policy.

1 Develop an explicit national food policy aimed at achieving clear goals for improving the health of the nation.
2 Reform public food institutions to take account of the wider goals of all stakeholders.
3 Encourage the food industry to take more account of environmental and social goods, such as through taxes and subsidies.
4 Broaden research policy goals.
5 Give equal weight to safety and trust in foods by sharply reducing pesticide use and pathogen contamination.

Learning Resources Centre

6 Emphasise public education and reskilling of citizens about food matters.
7 Reorient agriculture towards the production of environmental and public health goods.
8 Encourage more localised food systems, both to reduce transport costs and to increase trust in food systems.
9 Monitor new indicators of national health related to food.
10 Integrate information and labelling on food to permit full traceability of all products.

In addition to the changes that can occur at national level, much can also be done with reform at the EU level. The Common Agricultural Policy frames much agricultural practice in the member states of the EU. CAP reform to much wider than production goals is a prerequisite to achieving some of the sustainability dividends expected from more integrated food systems.

Summary

A generation of progress in agricultural development throughout the world has produced three distinct types of agriculture: industrialised, Green Revolution, and all that remains – the low-external input, traditional and unimproved. The first two types have been able to respond to technological packages, producing high-input, high-output systems of agriculture. Their conditions either resembled those where the technologies were generated, or else their environments could easily be changed and homogenised to suit the technologies.

The third type of agriculture comprises all the remaining agricultural and livelihood systems. This is a largely forgotten form of agriculture, located in drylands, wetlands, uplands, savannas, swamps, near-deserts, mountains, hills and forests. Farming systems in these areas are complex and diverse and agricultural yields are typically only 0.5 to one tonnes per hectare. The poorest countries have higher proportions of these agricultural systems. By the mid 1990s, some 30 to 35 per cent of the world's population, about 1.9 to 2.1 billion people, were still directly supported by this third and 'forgotten' agriculture. Most agricultural development assistance has emphasised external resources, yet these people can rarely afford them.

Despite some progress, there is still too much food and too little food. The rich overconsume and the poor are hungry and continue to be ignored. By the end of the first quarter of the 21st century, the world will have some eight to 8.5 billion people. Today, even though enough food is produced in aggregate to feed everyone, and world prices have been falling in recent years, some 800 million people are suffering from chronic hunger, of whom 180 million are underweight children suffering from malnutrition.

It is clear that agricultural production will have to increase substantially if some lasting impact is to be made on global and local food insecurity. But the views on how to proceed vary hugely. The are five distinct schools of thought. The business-as-usual optimists say supply will always meet increasing demand, and so growth in world food production will continue alongside expected reductions in population growth. The environmental pessimists contend that ecological limits to growth are being approached, or have already been reached. The industrialised world to the rescue lobby believes that developing countries will never be able to feed themselves, and so the looming food gap will have to be filled by modernised agriculture in industrialised countries. The new modernists argue that food growth can again only come from modern, high-external input farming. Others, though, make the case for the benefits of sustainable intensification, on the grounds that substantial growth is possible in currently unimproved or degraded areas while at the same time protecting and regenerating natural resources.

Whichever school of thought eventually prevails, it will have to do so in an increasingly globalised food system. Some see a globally linked system as providing extra efficiencies and benefits for all. Others say this will simply mean that the most powerful will prevail, with the poorest missing out again. There are two vital consequences for the food system. Farming will concentrate on exports, with developing countries encouraged to produce and sell high-value crops and to purchase staples in the world market. At the same time, organisations concerned with food and input trading, manufacture and sales will have to get larger so as to compete in the global market, forcing down costs and capturing greater shares of the markets.

Just as during the production of food on farms, there are many hidden environmental and social costs associated with getting food from the farm gate to our plates. One that has grown significantly in recent years is the transport cost, with each item of food now travelling on average 50 per cent further than it did in 1979. One reason for this increase is that large retailers want the regularity of supply that can only be given by centralised distribution, whereas independent small producers tend to source their food locally. Food imports by air have also grown, with an increasing number of crops travelling 6500 to 8000 kilometres to reach Europe. The full costs of all this travel are borne by the global environment.

Supermarkets have revolutionised food shopping in recent decades. They provide massive choice: in 15 years the number of products has trebled, and every year 10,000 new food products are launched on the European market. Consumers like supermarkets, because they are convenient, well lit, warm and safe. But an inevitable consequence of this growth is the sharp decline in corner shops. About 1000 specialist grocers, butchers, bakers and fishmongers are closing every year in the UK. When small shops close, something important is lost. The interdependence of small retailers, producers and consumers creates a dense social network that provides employment, good quality food and wider social benefits. The shops are the social centres of communities – they keep an eye on the elderly and infirm; they provide notice boards for advertisements; they keep in touch with local people.

Most scientists agree that the ideal diet for humans is one high in fruit, vegetables and carbohydrates, low in meat, and with some fish. Yet now we find that the food we eat is making us unhealthy. With rising affluence, diets tend to contain more fats and sugars, and fewer vegetables, fruits and cereal grains. These diets put people at greater risk of contracting heart disease, some forms of cancer and diabetes. We now eat more processed, frozen and packaged food and fewer fresh items. We cook less than we used to and microwave more. We can choose from some 20,000 items on the supermarket shelves, drawn from all over the world. Alarmingly, we also are eating fewer vegetables. Since 1960, the consumption of fresh green vegetables in the UK has fallen dramatically, from 410 grammes per person per week to just 280 grammes. All of this may be storing up more long-term health problems. Epidemiological studies show that diets rich in vegetables and fresh fruit substantially protect against cancers of the lung, larynx, stomach, colon and pancreas.

Food safety has become one of the most potent causes for concern amongst consumers in recent years. The number of reported cases of food-borne diseases or illnesses is increasing. Although food preparers and processors try to eliminate the risks associated with micro-organisms, there remain many strains of dangerous bacteria. But perhaps the biggest food quality scare in history has been the advent of BSE in cattle. The cause of BSE is still unknown.

Many believe that biotechnology and genetically manipulated organisms in the food chain represent hidden dangers for consumers. The biotechnology goods, particularly in the health sector with 'molecule pharming', already look significant. But the biotechnology bads are less well understood. The greatest concerns are over genetic pollution, with genetically modified crops crossing with wild relatives, transferring the genes into the environment. No one knows what will happen in the long term. There are also many fundamental ethical issues, such as whether insurance companies should have access to data about our own genes and the whole issue of patenting genes.

Most countries give some guidance to their citizens about the types and diversity of foods they should eat. But few have properly integrated policies that seek to make the links between sustainable means of food production with proper diets and the consequent benefits for health. However, there are an increasing number of innovations happening in the food system that are contributing to significant reform. The result of these, if they are spread more widely, will be more value for farmers, better national health, and more natural and social capital in both rural and urban areas.

Chapter 5

Adding Value to Food for Farmers and Local Communities

By purchasing food directly, it enhances cooperation and awareness by keeping consumers in touch with production processes.

Japanese consumer cooperative leaflet, 1997

Taking Back the Middle for Farmers and Communities

Who Gets the Value?

The goods and services produced by the natural and social capital in rural areas are extremely valuable. Chapter 4 documented several of the disparities in food systems that leave some people hungry and some with plenty. We all, to a certain extent, derive benefit from our farming and rural systems. But can we, as a whole, get more?

The food and drinks system is one of the largest industrial and commercial sectors in the world, with global production amounting to some US$1500 billion per year. Half a century ago, at least half of the pound, franc, mark or dollar spent on food found its way back to the farmer and rural community. The rest was spread amongst suppliers of various inputs (feeds, pesticides, fertilisers, seeds, machinery, labour and so on) and amongst manufacturers, processors and retailers. Since then, the balance of power has shifted increasingly away from the middle, with value captured on the input side by agrochemical, feed and seed companies, and on the output side by those who move, transform and sell the food. Food consumers appear to have benefited as the real cost of food has fallen. There is now greater choice, and increasingly processed foods reduce preparation and cooking time.

But for farmers and rural communities, the effect has been largely negative. Farmers get a smaller share, no more than 10 to 20 per cent, and they also pass on less, spending less in rural communities and employing fewer local people. Tens of millions of jobs have been lost in farming throughout western Europe and North America in the past 50 years. And farms continue to be abandoned and farm labour laid off. In the US, the farmers' share at the turn of the century was 44 per cent of every dollar spent by consumers. Production costs were about 16 per cent, and the marketing share some 40 per cent. Yet by the 1990s this had changed dramatically, with the farmers' share now only 9 per cent. Marketing takes 66 per cent, with production costs accounting for about 25 per cent. In the decade between the mid 1970s to mid 1980s, the market share of food sold in the US doubled (from $1100 to $2200 billion), but the amount going to the farmer rose only slightly from $40 to $45 billion (Center for Rural Affairs, 1996; Douthwaite, 1996).

Similar changes have occurred in Europe. Over time, more and more of the value is being captured by manufacturing, processing and retailing. In Britain, food prices in supermarkets rose by about half between 1982 and 1992, yet only a third of this increase got back to farmers (Raven and Lang, 1995). The centralised distribution systems and associated transport costs are expensive, and this is partly where the value is going. In Germany, about a fifth of the food mark goes to the farmer, down from some 75 per cent in the 1950s (Greenpeace, 1992). Richard Douthwaite describes in his 1997 book *Short Circuit* what has happened to the pig industry in Ireland over the past 25 years. In the early 1970s, there were some 36,000 family farms rearing pigs. Bacon factories were spread across the country, and about half of the value went back to the farm and local community. By 1996, only 700 pig farmers and six bacon factories remained, and only a fifth of the price of bacon returns to the now largely factory farms.

The organic sector particularly suffers from this capture of value by certain stakeholder groups in the food system. It is commonly supposed that organic produce commands higher prices because growers find it more expensive to farm in an environmentally friendly and welfare-conscious way. As illustrated in Chapter 3, sustainable agriculture does imply a shift of inputs away from pesticides and fertilisers to knowledge, management skills and labour. However, a surprisingly small proportion of the premium paid by consumers gets back to farmers. In Denmark, for example, organic producers receive a tenth more for their milk compared with conventional producers, but the price in the supermarkets is 32 per cent higher. The same is true for wheat – farmers receive 80 per cent more for organic wheat, but organic flour costs 100 per cent more. The extra value is captured almost entirely by distribution and profits (Douthwaite, 1996).

Taking Back the Middle

Can anything be done about this falling share of the food pound that finds its way back to rural communities? Or are the farmers' and rural communities' shares doomed to fall yet more, with further diminishing of rural natural and social capital?

Sustainable agriculture can mean a cut in some production costs as knowledge and information substitute for external inputs. In addition, rural communities need to find new ways of adding value to the goods and services derived from available natural capital. The Kansas Rural Center and the Center for Rural Affairs in Nebraska call this 'taking back the middle'. There are four ways to take back the middle, helping to spread the benefits more evenly amongst stakeholders.

The first option is for farmers to find ways of selling their produce directly to consumers. The number of farmers already selling direct varies from country to country. In the UK, about 5 per cent of farmers sell directly, accounting for about 9 per cent of fresh produce sales. In Germany and the US, this rises to 15 per cent; in France and Japan, it is 25 per cent. But in all these cases, the volume of produce sold is small, owing to the greater proportion of small farmers engaged in direct sales – only 5 per cent in the US and 14 per cent in France (Festing, 1997). There is a variety of proven mechanisms, including farm shops and direct mailing, farmers' and produce markets, community-supported agriculture and box schemes.

The second theme is to enhance links with urban communities through community cooperatives and community gardens and farms. This can enable communities to take back more of the control of the food system from dominant institutions. Poverty is usually associated with ill-health and poor diets. When money is tight, the cheapest way to get sufficient calories is to purchase sugary, fatty and processed foods. But these usually lack many vital ingredients, such as proteins, vitamins and minerals. Locally grown food can enhance diets as well as contribute to social capital.

The third option is to increase collaboration between farmers' groups, cooperatives and alliances, so as to create social capital by learning and working together. Examples include participatory research and experimenting groups, machinery rings, comarketing groups and community food cooperatives. Sustainable agriculture is knowledge and management intensive and therefore needs timely and relevant information to produce value. Yet farmers commonly lack this vital knowledge. By experimenting themselves, they are able to increase their own awareness of what does and what does not work. Few farmers can engage in new methods alone, and so collective efforts are vital.

The fourth option is to enhance labelling and traceability of foodstuffs in order to increase consumers' confidence about both the source and quality of foods. This can be done through ecolabelling and other assurance schemes, organic standards, and fair trade schemes. Consumers are increasingly asking: can we trust the food we see on the shelves? Labelling is important, as it tells consumers something about the way that the food was produced. It helps to create a bond between consumer and producer, provided that the label is meaningful.

Selling Direct to Consumers

The first option for taking back more of the middle is for farmers to find ways of selling their produce directly to consumers. There are a variety of proven mechanisms: farmers' and produce markets; farm shops; community-supported agriculture; and box schemes for local food links.

Farmers' and Produce Markets

Farmers' shops and pick-your-own (PYO) operations have long been an important means for farmers to sell directly to consumers. When successful, they can make a very significant contribution to individual farm income. There are estimated to be some 1500 to 2000 farm shops and several hundred PYO enterprises in the UK (Chris Emerson, pers comm, 1997). A recent study of 150 of these farm shops found that the most successful and long lasting had a great diversity of produce that was clearly labelled and locally sourced (*Farmers' Weekly*, 1 August 1997). Most of the successful businesses also had another attraction on the farm that helped to entice visitors.

However, farm shops can be costly to establish and run and rely on customers driving to farms to make purchases. Clearly, this is not a viable option for every farm in a particular area since competition would be far too great. One option for farms not located on sufficiently busy routes is to develop direct marketing by mail. This can be extremely important as the two cases from Denmark and England in Box 5.1 illustrate. In the US, the mail order business is valued at US$1 billion, growing at 10 per cent per year. The top ten companies, however, account for nearly half of the market, with around 3000 grossing $50,000 or less (Festing, 1994).

In France, there are some 200 collective farm shops located in cities and rural towns. These are run by small groups of farmers, each supplying a different product, such as vegetables, wine, cheese and sausages. The shops operate under a voluntary charter and seek to inform consumers about farm operations too (Festing, 1995).

Box 5.1 Two cases of direct marketing of meat in Denmark and England

Denmark

Ulrich Kern-Hansen farms pigs and cattle on a 22 hectare farm near Silkeborg, east Jutland. One thousand pigs are finished annually on grass. There is an organic meat processing plant on the farm which buys in beef and pig carcasses from other organic producers. The business employs eight people and markets more than 50 pork and beef convenience foods, including minces, sausages and pâtés. Says the farmer: 'I could sell a hundred times more organic pigmeat than I can get.'

England

Helen Browning runs a 540 hectare mixed organic farm on the downs in Wiltshire. The farm grows milling wheat, oats, triticale, peas and roots. There are two dairy herds of 260 cows, and annually some 1700 saddleback pigs, 1700 lambs and 900 free range hens are marketed. All the livestock are slaughtered nearby and sold under the Eastbrook Farm label. The processing side adds value to the beef, lamb and pork. It also buys in meat from six neighbouring farms. Some 25 per cent of the meat is sold through the farm shop; 20 per cent by mail order; 40 per cent goes to small outlets of butchers, caterers, schools and restaurants. Some two tonnes of boned pork is sold in Denmark each month. The whole business employs 23 staff – 11 on the farm and 12 in the meat business. Before conversion, the farm had two family workers and four full-time staff. Like the Denmark case, demand is outstripping supply. Furthermore, half of the beef and lamb receives no premium.

Sources: Helen Browning, pers comm; IFOAM, Germany

Produce markets, in which growers sell directly to customers, are another option for taking back more of the food pound. Produce markets also offer the opportunity for consumers to buy organic or sustainably produced food without paying a premium. There is a long history of markets in rural towns and centres. Weekly or twice weekly markets are a part of the vitality of many small towns in Europe. In France, there are some 6000 weekly street markets involving 70,000 people. In the UK, there are about 800. These markets are almost entirely dominated by traders selling on wholesale produce. Very few farmers are directly involved, and so there are few direct links with the public (Festing, 1995).

In the US, the situation is rather different, as farmers' markets have emerged on a huge scale in recent years (Festing, 1994). Under the Federal 1976 Farmer-to-Consumer Direct Marketing Act, state extension services have a mandate to promote the development and expansion of direct marketing. Held on a weekly or twice weekly basis, farmers and consumer groups have established new market sites to foster direct selling to the local public. There are at least 2400 farmers' markets in the US, involving more than 20,000 farmers as vendors, one third of whom use them as their sole outlet. Each is unique, offering a variety of farm-fresh and organic vegetables, fruits and herbs, as well as flowers, cheese, baked goods and sometimes seafood. Each week, about one million people visit these farmers' markets, 90 per cent of whom live within 11 kilometres of the market. The annual national turnover is about US$1 billion (Andy Fisher, pers comm, 1997).

The benefits these farmers' markets bring are substantial. Harriet Festing of Wye College has shown that they improve access to local food; they improve returns to farmers; they also contribute to community life and social capital, bringing large numbers of people together on a regular basis. Consumers also perceive the food to be of better quality and cheaper than in supermarkets. One piece of research on 15 farmers' markets in California found that produce was 34 per cent cheaper than in supermarkets (Festing, 1994). The contributions to local economies are substantial. One farmers' market in Madison, Wisconsin, contributes $5 million to the local economy each year; another in Santa Fe, New Mexico, brings an added $0.75 million to the nearby farming system.

The evidence also seems to suggest that farmers' markets have a largely positive impact on other local businesses and enterprises, since they increase foot traffic and visibility. There is no evidence that they remove business from other shops. Farmers' markets also recycle resources into other important community functions, contributing particularly to social capital. In Los Angeles, for example, the Encino market is sponsored by an organisation that provides for the elderly, and part of the revenue from the market goes back into health care. Markets run by the Georgia Hunger Coalition bring black farmers from rural south Georgia into black housing estates of Atlanta to sell their produce to 300 households. And in New Orleans, the Vietnamese market features a wide range of Asian vegetables and ducks raised on 16 hectares of former wasteland (Harriet Festing, pers comm, 1997). Andy Fisher, coordinator of the Community Food Security Coalition in the US, put the impacts this way: 'Farmers' markets personalise the food system, lending a face-to-face connection between farmer and consumer, while educating the consumer to seasonal agricultural rhythms.'

These farmers' markets are very much a part of the growing food security movement in the US, recently given support by the passing of the Community Food Security Act as part of the 1996 Farm Bill. This authorises the US Department of Agriculture (USDA) to provide some $2.5 million of grants per year over seven years to community food security projects, thereby recognising the importance of this sector for improvements to food security, poverty reduction and social capital.

Many farmers' markets have been boosted by linkages to state welfare and hunger programmes. Since 1989, the farmers' market coupon programme has been distributing coupons to women and infants for the purchase of fresh produce. In some states, such as New York, schools are encouraged to purchase directly from farmers through legislation that permits them to buy without going through the usual competitive tendering processes.

In Britain, save for a recent experiment in Bath, the nearest equivalent is the Women's Institutes (WI) cooperative markets.* These were started in 1919 as a way to help WI members, unemployed people, pensioners and ex-servicemen to sell surplus garden produce. They comprise a weekly market for the direct sale of home-grown and home-made produce from gardens and kitchens to the public. There are now 538 markets in England, Wales and the Channel Islands, with a further 74 in Ireland. A conservative estimate suggests that these markets reach at least 30,000 customers each week and involve a further 9000 households in production.

Turnover in the British markets has grown from £1 million in 1972 to £10 million in the early 1990s – an average of £18,600 per market. This represents a substantial cash injection directly from consumers into local communities since all the income goes back to the shareholders of the market groups. Richard Douthwaite describes their value to consumers, who like them for the 'quality of the produce, its freshness and presentation'. But these markets have a vital social as well as an economic function as they help local people develop skills, market their produce by working together, and provide a friendly meeting place.

Community-Supported Agriculture

Community-supported agriculture (CSA) is a partnership arrangement between producers and consumers designed, again, to take back more value and also to provide a guaranteed quality of food. The basic model is simple: consumers provide support for growers by agreeing to pay for a share of the total produce, and growers provide a weekly share of food of a guaranteed quality and quantity. CSAs help to reconnect people to farming. Members know where their food has come from, and farmers receive payment at the beginning of the season rather than when the harvest is in. In this way, a community shares the risks and responsibilities of farming (Clunies-Ross and Hildyard, 1992).

* The first farmers' market in Britain was held in Bath on 27 September 1997 and was supported by the Bath and North-East Somerset Council. There were about 30 vendors, and over 3500 customers visited the market.

Box 5.2 What a CSA looks like on the west coast of the US

The Center for Agroecology at the University of Santa Cruz manages a ten-hectare farm and a one-hectare garden on the university campus, using it to teach students about growing and caring for vegetables, fruit, herbs and flowers. In the 1997 season, 75 shareholders were recruited for a 22-week season from June to October. A share is designed for two to four individuals and costs $500 for the season, some $23 per week. Some low-income shares are available. Members receive 30 different crops, including apples, plums, broccoli, cabbage, leeks, onions, garlic, beans, carrots, beans, kiwis, maize, lettuce, pears, eggplant, peppers, potatoes, spinach, strawberries, salads, and winter and summer squash. They also get a regular newsletter with recipes and farm updates, and opportunities to attend farm festivals and workdays.

Source: Center for Agroecology and Sustainable Food Systems, University of California, Santa Cruz, at http://zzyx.ucsc.edu/casfs

In most instances, a detailed annual budget for the farm is drawn up and the costs are shared by the community. Farmers start to receive payment when the crops are planted, and this income is guaranteed. Since they are growing for people, rather than an abstract market, they tend to produce a greater variety of crop. Most offer a mix of eight to 12 vegetables, fruits and herbs per week; some link up with other CSAs to keep up diversity; and others offer value-added products such as cheese, honey and bread. The central principle is that they produce what people want, instead of concentrating on crops that could give the greatest returns. Farmers are, therefore, more likely to use resource-conserving technologies and practices.

In addition to receiving a weekly share of produce, CSA members often take part in life on the farm. Many CSA farms give out newsletters with the weekly food share, so that members stay in touch and know what crops are expected. Box 5.2 describes a CSA in California.

There are some 600 CSA operations in the US, involving 100,000 consumers, and a turnover of $10 to $20 million per year (this assumes an average share of $550, an average number of 50 members and 600 CSAs). But it is social capital that may, again, be the most significant winner. A recent study by Timothy Laird, and reported by Richard Douthwaite, of 83 CSAs in the US has shown what people most value from them. More than 60 per cent of the farmers said that the most successful aspect of their operations was the strengthened bonds between people, resulting in networks that: 'reconnected people with the land and reconnected farmers with the people who eat the food that they grow'

(Laird, 1995). Only about a third of farmers mentioned the value of financial stability.

Retailing outlets are now beginning to appreciate the value of food produced from CSA farms. In April 1997, Erickson's, a major regional company in Wisconsin, signed an agreement with six CSA farms working with the Land Stewardship Project in Minnesota (DeVore, 1997). Erickson's will buy vegetables from the Hay River Produce Co-operative. In a deliberate move away from produce sourced from large farms with poor environmental and community records, the company want to make the most of sustainably produced foods grown locally by family farmers. As a result, the CSAs, which often have the capacity to produce more than their members need, will benefit, as will their members and consumers at the shops. There are fewer than ten subscription farms operating in Britain. Most of the growth of indirect marketing has been through box schemes.

Organic Box Schemes for Local Food Links

Organic farmers and growers in the UK have recently made great strides in the alternative marketing and sale of their produce directly to consumers. Already making use of many kinds of direct links, such as farm shops, combined ordering schemes, and subscription systems, there has been spectacular growth in box schemes. First set up by the Soil Association in 1992, one third of British organic farmers were involved in marketing directly to about 25,000 households by 1997 (Booth, 1996; Festing and Hamir, 1997).

These box schemes have several important attributes. First, they ensure the good quality of the produce. Food is fresh, often picked the same day, and some eight to 12 varieties of vegetables and fruit are put in each box each week. Farmers contract to supply the basics, such as potatoes, carrots, onions, and one green vegetable, and add other produce depending on the season. Over time, box schemes also increase on-farm biodiversity. In response to consumer demand, many farmers have increased the diversity of crops grown from 20 to 50 varieties. Greater diversity satisfies demand and reduces the risk of complete failure in the face of climatic and market uncertainties. Ray Hunter, who farms near Hereford, began by growing three vegetables; now he grows 18; and Jan and Tim Deane, who farm near Exeter, started with 12 varieties and now cultivate 50 (see Box 5.3) (Ray Hunter, pers comm; Devon County Council, 1996).

Prices are comparable to supermarket prices for conventional vegetables, so consumers do not end up paying premiums. Box schemes also help to develop trust and understanding. Newsletters, farm visits and

Box 5.3 Two farms in Devon marketing produce through box schemes

Northwood Farm, Exeter
Jan and Tim Deane farm a 12.5 hectare holding in Devon, where they grow vegetables to supply 200 customers each week with a boxed selection of fresh produce for a fixed price. In most cases, the contents are picked, packed and delivered within 24 hours. The scheme cuts out the chore of shopping, bypasses intermediaries, and guarantees freshness. It provides vegetables at the same price as conventional suppliers. Over time, the number of varieties of vegetables grown has increased from 12 to more than 50, largely in response to demand from consumers for greater diversity. Consumers get vegetables that are in season: carrots, leeks, potatoes, onions and white cabbage in January, but much more perishables in summer. The key to success is trust. Consumers trust the Deanes to give them a mixture of good quality food; and the Deanes trust consumers to go on supporting their whole farm and its efforts through the box payments.

Riverford Organic Vegetables, Buckfastleigh
Riverford Organic Vegetables was begun by Guy Watson in 1985 on 1.6 hectares of land and has grown to an operation on 60 hectares employing 50 people (20 full-time and 30 seasonal), and distributing 1200 boxes weekly. The boxes account for 30 per cent of the produce; 10 per cent is sold through the farm shop, and the rest to wholesale routes.

Source: Devon County Council, 1995; Ian Hutchcroft, pers comm, 1997

personal contacts increase the contact between farmers, consumers and local communities, and by promoting local employment and care for the environment they also foster community regeneration.

One rationale for these schemes is that they emphasise that payment is not just for food, but for support of the farm as a whole. It is the link between farmer and consumer that guarantees the quality of the food. This encourages social responsibility, increases the understanding of farming issues amongst consumers, and increases the diversity of crops grown over time. These schemes have brought back trust, human scale and a local identity to the food we eat. Farmers' Link in Norfolk (1995) report that customers take 'a very tolerant attitude... they do not tend to haggle about the amount or complain about mixes of vegetables'. More positively, Laura Davis, an organic farmer in Oxfordshire, said in 1997 that 'trust and confidence increases... we don't advertise because we'd be swamped by demand. Markets are not the problem.' Her seven-hectare holding produces 50 to 60 different crops per year, and supplies 80 to 90 households for 46 to 52 weeks of the year. She also employs four full-time staff plus seasonal labour: 'we do want to make a contribution to the local economy'.

Importantly, customers report that box schemes encourage them to eat more vegetables. As indicated in Chapter 4, per capita consumption of vegetables has been falling in recent years, and so this seems an important measure to help offset these alarming trends. It is also true, however, that many of those engaged in box schemes are well educated and in upper income brackets, making them least likely to be at nutritional risk. The interest in box schemes has, however, brought forth new problems. Demand is outstripping supply, and some 40 per cent of the organic produce supplied in boxes is imported from continental Europe. There are three schemes with more than 1000 members; these buy in 95 per cent of their produce, more than a third of which is imported. This, of course, undermines the basic principles of direct producer–consumer links, though it still guarantees customers that they will receive the quality and diversity of foods they expect. A recent survey by Harriet Festing and Arzeena Hamir for Wye College and the Soil Association also found that more than half of farmers did not know who their customers were. Consumers say that the greatest problems are lack of choice of vegetables, collection and delivery, and the need for more personal communication with farmers (Festing and Hamir, 1997).

Enhancing Links with Urban Communities

Community Cooperatives

Food cooperatives are an important way to get good food to urban groups with no direct access to farms and the countryside. The Cooperative Society was established a century ago to do exactly this and has grown to be one of the largest wholesale and retailing companies in Britain. Community cooperatives are still having a significant effect. The Glasgow Healthy Castlemill project has a cooperative serving more than 3000 tenants in estates with high unemployment and high levels of heart disease. The coop buys wholesale and sells to local people with just a 1 per cent mark-up. The Birmingham Organic Roundabout was set up in 1992 to supply organic foods to urban customers. It has grown rapidly, now sourcing from a group of 36 Herefordshire organic farmers, supplying 3000 regular customers, and employing 15 full-time employees. Some payments for delivery and sorting are made by exchange trading through LETS (see Chapter 8). Many farmers have benefited from the direct link – their farms have become more diverse and they have become better off.

Direct links between consumers and farmers have had a spectacular success in Japan, with the rapid growth of the consumer cooperatives, *sanchoku* groups (direct from the place of production) and *teikei* schemes (tie-up or mutual compromise between consumers and producers) (Furusawa, 1994). This extraordinary movement has been driven by consumers rather than farmers, and mainly by women. There are now some 800 to 1000 groups in Japan, with a total membership of 11 million people and an annual turnover of more than US$15 billion. These consumer–producer groups are based on relations of trust, and put a high value on face-to-face contact. Some of these have had a remarkable effect on farming, as well as on other environmental matters.

The largest and best-known consumer group is the Seikatsu Club, a consumer cooperation union and recipient of the Right Livelihood Award in 1989. This has a membership of more than 210,000 households organised into 26,000 *hans*, or local branches, all over Japan. It was set up in 1965 by housewives in Tokyo who wanted to find a way of avoiding the high price of milk. Their idea was to band together and to buy milk directly from farmers. Over the next few years, they also began to purchase a range of food free of pesticides, as well as wholesale clothes and cosmetics. Club members then began to take care of distribution themselves. In the late 1970s, a new headquarters was set up in Setagaya and the first Seikatsu Club housewife was elected to local government the following year. Although 37 members have now entered local politics, the club seeks a much deeper change, claiming to 'seek to empower each and every member with a voice and role in participatory politics' (Festing, 1997). Historically isolated in the home, this has given strength and new opportunities to women. The turnover of the Seikatsu Club alone is now 40 billion yen (US$320 to $350 million) and it employs 905 full-time staff.

Some groups are small ventures in which ten to 30 households link with one farmer, who supplies food of a particular quality, usually organic. One larger group is the Young Leaves cooperative begun by Hiroshi Ohira, who farms in Tokyo. It now has 400 household members and 11 organic farmers who supply vegetables, rice, root crops and fruit. Members buy about 75 per cent of their food through Young Leaves. Others have become highly commercial. Radish Boya was started in 1987 by the Japan Ecology Centre and supplies 25,000 members with boxes of produce from 1100 farmers. These *sanchoku* businesses are rapidly expanding, producing a wide range of organic and fair trade products through mail order catalogues and running their own production and processing facilities, farm shops and other farmer-operated businesses. Harriet Festing describes their key components in this way: '*sanchoku* schemes spring from moral commitments as well as commercialism, but their greatest strength lies in promoting the link between farmers and consumers. Farm walks, demonstrations and harvest festivals are organised, and weekly newsletters contain stories from the farm.'

Community Gardens and Farmers

In countries of the South, it is common for very large numbers of urban dwellers to be directly engaged in food production. It has been estimated that 100 to 200 million urban dwellers are now urban farmers, providing food for some 700 million people (Schwarz and Schwarz, 1998; Cook and Rogers, 1996; Smit and Ratta, 1995). In Latin American and African cities, up to a third of vegetable demand is met by urban production; in Hong Kong and Karachi it is about half, and in Shanghai it is over 80 per cent. But in the industrialised countries, far fewer people grow their own food.

Homegardens and allotments have long been important for home food production. During the Second World War, when the Dig for Victory campaign encouraged greater home production, half of all manual workers in Britain kept an allotment or vegetable garden; domestic hen keepers produced a quarter of the national egg production; and pig clubs were promoted. In 1944, some 120,000 hectares of allotments and gardens produced about 1.3 million tonnes of food, about half of the nation's fruit and vegetable needs (Garnett, 1996). The National Society of Allotment and Leisure Gardeners' figures put the current number of allotments in England, Wales and Scotland at about 300,000 covering some 12,150 hectares, just a tenth of the extent during the war. Nonetheless, these yield some 215,000 tonnes of fresh produce every year (G W Stokes, pers comm, 1997).

The American National Gardeners Association (NGA) estimates that some 35 million people are engaged in growing their own food in back gardens and allotments. Their contribution to the informal economy is huge. The so-called Gross National Home Garden Product, a measure of the value of the food grown, is estimated to be about $12 to $14 billion per year. According to the NGA, private gardeners cultivate mostly to produce better tasting and more nutritious food, but also to save money, and for exercise and therapy. It makes them feel better. This is particularly true of community gardens and farms which, by contrast, seek to enhance both food production and social benefits. In New York, 87 per cent of community gardeners invest their time in gardening in order to improve the neighbourhood; 75 per cent for fresh vegetable production; 62 per cent for fun; and 42 per cent to save money (Weissman, 1995; 1996).

There are now several hundred city farms or community gardens in the UK. The 1996 report *Growing Food in Cities* by Tara Garnett for the National Food Alliance and the SAFE Alliance describes 38 examples of community schemes. They provide food, especially vegetables and fruit, for poorer urban groups, and a range of other natural products such as wood, flowers and herbs. They add local value to produce before sale. They sometimes encourage derelict or vacant land to be transformed into desirable areas for local people to visit and enjoy, resulting in the creation of quiet, tranquil places for the community that can support wildlife. The involvement of schoolchildren can mean a reduction in vandalism, as well providing local children with an educational opportunity to learn about farming and animals. City farms also provide the opportunity for mental health patients to engage in work that builds self-esteem and confidence, and for unemployed people to use their time productively in their own community.

Community Permaculture Garden in Newcastle

The Drift Permaculture Project in Scotswood, Newcastle, is a clear indication of the social and natural benefits that can arise from community gardening (Ed Tyler, pers comm, 1997). Established in 1995 on Newcastle College land, it now comprises 0.8 hectares of garden, two ponds, three multilayered forest gardens, beehives and a wildflower meadow. This has provided a significant and new focus for local people. More than 500 people have attended tree planting, pond puddling and green woodworking events. In addition, more than 1000 local schoolchildren have visited the site. Other users include probation service clients and after school clubs. All this helps to bring together many different groups. Fruit and vegetables from the garden are given to volunteers and children.

Much of the success has been because of the strong partnership between the project, local community and many external organisations, including City Challenge, the city council, Newcastle College, local schools, the Drift Garden Centre, and a range of companies and trusts. Children, when asked what they would like to do with a piece of land, now say they want to plant trees and grow vegetables – something they never volunteered before. Ed Tyler states: 'we are convinced that the active involvement of local school children... has meant we have had virtually no vandalism'.

Elder Stubbs Garden Group, Oxford

This 1.6-hectare garden in the east of Oxford is part of the Restore group (Ralph Raistrick, pers comm; Garnett, 1996). It was started in 1989 on derelict allotments and links food production with developing the self-esteem and confidence of 24 mental health patients working at the gardens. It grows vegetables, flowers, fruit and willow coppice. Produce is sold to buyers in the local community through regular deliveries by horse and cart. Value is added by making baskets from the willow, as well as jams and chutneys. There are positive links with the local community, and it shows what people with mental problems can do. Local schoolchildren come to the site regularly to help with work. The gardens are very productive and a box scheme was begun in 1996. But perhaps the greatest benefits have been in the rebuilding of social capital. Keith Birnie says: 'it is not just about work here, it is about socialising, and learning to get on with each other'.

Springfield Community Garden, Bradford, England

The three-hectare Springfield community garden is situated on the edge of the 30-year-old Holmewood Council Estate that is home to 10,000 people. The area has high unemployment, and vegetables are expensive and rarely fresh in the local shops. The garden was designed by Andy Langford on the principles of permaculture, which he defines as 'a way of designing and creating sustainable environments and systems. It can be used for farms, gardens, and architecture; by communities, businesses and schools; to create healthy and efficient places to live and work'.

The Springfield garden was begun in 1994 and now comprises allotments, formal gardens, ponds, play areas, poly-tunnels, mulch areas, and willow and cash crop areas. It has several different functions (Chris MacKenzie-Davey, pers comm, 1997; City of Bradford Council, 1996). Food is grown organically, and sold mainly in social, church and pensioners' clubs, as well as a small amount to shops. The garden provides training opportunities, especially continuing education for those with learning difficulties. It provides plants for local gardeners. There are food processing facilities for the making of chutneys, jams and sauces, and workshop facilities for woodworking. The site itself is a safe recreation area for youngsters. The 'green' buildings onsite also show how waste recycling and energy conservation work in practice.

The project has been community-led from the start and comprises a partnership of agencies and funders, including Bradford City Council, Bradford City Challenge and the local community. Some 300 local residents are regularly involved in activities, and 1000 to 2000 have direct but less regular contact. The garden has improved social capital. There is increased local knowledge of horticulture and improved diets amongst many local families. There has been a decline in vandalism on the estate, and improved esteem, trust and confidence. The estate looks better too, with flower baskets hanging from the ten-storey blocks of flats.

There is a range of other community and food-related activities in Bradford, mostly coordinated by the Local Agenda 21 office. These include the Porishad group of Bangladeshi families growing vegetables on allotments; the city farm; an Allotment Action Group; an urban wildlife group feeding up into district biodiversity concerns; and the Bradford Environmental Trust residents group. These are all part of wider efforts to encourage social and economic regeneration in Bradford Metropolitan District, which has ambitious plans to make all 370 square kilometres sustainable and to bring benefits to all its 480,000 people. In the whole district there are 900 farms, very few of which supply food directly to Bradford itself. The political vision for the council was put in 1996 by Councillor Keith Thompson: 'we are working towards a sustainable district... Sustainability is particularly important because it brings people of different backgrounds together and links Bradford communities and institutions with others throughout the country, Europe and beyond.'

GreenThumb Community Gardens in New York City

The scale of community gardens is much greater in the US. New York alone has some 700 community gardens, involving about 20,000 households in local regeneration. GreenThumb of the New York City Department has been working since 1978 to transform neighbourhood eyesores. It leases city-owned land to community groups at no charge and trains them in garden design, construction and horticultural techniques. According to Jane Weissman, derelict land has now changed into 'safe, thriving and productive oases of green'. The low cost of making gardens benefits taxpayers as well as local people. The average price of constructing one square metre of city park is some US$500; the price of one square metre of community garden is $50.

Most local people see enhanced social capital as a vital end-point. The names of some of the gardens illustrate clearly their meaning to local people, and include the Garden of Hope, the Garden of Happiness, the Magic Garden, United We Stand, and Striving Together (see Box 5.4). In every case, they have helped to bring people together, allowing them to 'take back their cities'. They are important for food production too: GreenThumb (GT) gardens produce some $1 million worth of fruit and vegetables each year. These schemes have created employment, fostered community pride, produced fresh, nutritious food, and are a meeting place for local youngsters. They help to keep children out of trouble. At Hunts Point in the Bronx, a one-hectare plot was once buried under abandoned cars and rubbish. But now teenagers learn about composting, building garden beds, growing and selling herbs, and cutting and selling firewood. This is not because they are forced to but because they want to. Said 23-year-old Ralph Acebedo, 'it kept me out of trouble... we learned a lot about

Box 5.4 The contribution of community gardens to local social capital: testimony of residents

Ruth Fergus, Madison Community Garden
When we first began our community garden, it meant changing an eyesore of a burnt-out building into something beautiful. Now, each morning I wake up to a dream come true. It also changed our mischievous teenagers to a positive junior block association, learning parliamentary procedure and conducting their own meetings instead of destroying the block.

Bertha Jackson, 127th St Block Association, Central Harlem, 1995
This is the beauty. Yearly we got two or three bushels of peaches from the tree. People have come from near and far for Harlem-grown peaches from our garden tree. The peach that grew in Harlem.

Glenn Bader, Mount Eden, Bronx
No one believed it could be done. Everyone told us that the students were failures. Students that had a history of violence and trouble could add nothing to their community. We fooled them.

Tito Arroyo, Bronx
The landlord to the right of the garden said: 'this makes this block, my building, more valuable and more beautiful'.

Mary Scales, in a letter to the New York Times (30 January 1997)
Our community garden was created by students, staff, neighbours, community workers and environment groups. Together we managed to have the lot cleared, a fence erected and a garden created. The students, along with our neighbours, have improved the environment, which has made East New York a more beautiful place to work and live. Flowers bloom, vegetables are harvested, the smell of barbecues fills the air and the students learn. They enjoy learning outside... and our gardens are an oasis of beauty in the deserts of urban decay.

Sources: Weissman J (ed). 1995. *City Farmers: Tales from the Field.* GreenThumb, New York; Weissman J (ed). 1996. *Tales from the Field. Stories by GreenThumb Gardeners.* GT, NY

agriculture and teamwork' (Raver, 1997). Not only do GT gardens mean more food and more community spirit, they also contribute to reduced drug dealing and greater responsibility taken by local people for their own environment. Some 75 local schools are involved in the Education in the Gardens Programme, with gardens developed on derelict parts of school property or vacant lots.

But there is a threat to these gardens. The City of New York is desperate for housing, and GT expects to lose half of its 750 gardens to development over the coming years. As Deputy Mayor Fran Reiter put it in January 1997: 'do we sacrifice gardens to build housing? You're damn right we do' (Raver, 1997). The sadness that many feel is that social capital, so carefully created over years of effort, is lost at a stroke. Houses are needed – but because the housing and development departments record community gardens as vacant plots, they destroy them without knowing what they are doing. Keith Wright, a resident of Harlem, said in late 1996: 'I know we need housing, but until all of Harlem's vacant buildings are restored, and all the garbage-strewn vacant lots cleaned out, I cannot for the life of me understand why you would destroy this programme [of community gardens]' (Trebay, 1996).

Farmers' Groups, Cooperatives and Alliances

Creating Social Capital by Learning and Working Together

Another way that farmers can create extra local value is to work together in groups. For as long as people have engaged in agriculture, farming has been at least a partially collective business. Farmers and farming households have worked together on resource management, labour sharing, marketing, and a host of other activities that would be too costly, or even impossible, if done alone. Local groups and indigenous institutions have, therefore, long been important in rural and agricultural development.

These may be formal or informal groups, such as traditional leadership structures, water management committees, water-user groups, neighbourhood groups, youth or women's groups, housing societies, informal beer-brewing groups, farmer experimentation groups, burial societies, church groups, mothers' groups, pastoral and grazing management groups, tree-growing associations, labour-exchange societies, and so on. These have been effective in many ecosystems and cultures, including collective water management in the irrigation systems of Egypt, Mesopotamia and Indonesia; collective herding in the Andes and pastoral systems of Africa; water harvesting and management societies in Roman north Africa, India, and south-west North America; and forest management in shifting agricultural systems. Many of these societies were sustainable over hundreds to thousands of years.

During the agricultural revolution of the 18th and 19th centuries, farmers' groups played a vital role in spreading knowledge about new technologies (Pretty, 1991). At a time when there was no ministry of agriculture, no research or extension institutions, farmers organised their own experiments and extended the results to others through tours, open days, farmer groups and publications. Farmer groups and societies were central to the diffusion of new technologies. The first were established in the 1720s and increased in number to over 500 by 1840. These groups offered prizes for new or high-quality livestock, crops and machines; encouraged experimentation with new rotational patterns; held regular shows and open days; bought land for experimental farms run by the group; arranged tours to visit well-known innovators; and articulated farmers' needs to national agencies and government.

However, throughout the history of modern agricultural development, local groups and institutions have rarely been recognised. As a result, external institutions and governments have routinely suffocated local institutions during agricultural modernisation. Local management has been substituted for by the state, leading to increased dependence of local people on formal state institutions. Local information networks have been replaced by research and extension activities; banks and cooperatives have substituted for local credit arrangements; and cooperatives and marketing boards have been replaced by input and product markets. When traditional social institutions collapse or disappear, it is common for natural capital to degrade.

As indicated in earlier chapters, the fundamental challenge for sustainable agriculture is to create and support processes that foster learning. When farmers and other rural people are well linked and trust each other, then it is possible for learning mechanisms to be established. Sustainability needs perpetual novelty and adaptive performance. These are best delivered by local people working together in groups. And the more that information and knowledge is locally generated and locally applied, the more that value is taken back into local systems.

Participatory Research by Farmers

The normal mode of modern agricultural research has been to conduct experiments under controlled conditions on research stations, with the results and technologies passed on to farmers. In this process, farmers have no control over experimentation and technology adaptation. Farmers' organisations can, however, help research institutions to become more responsive to local needs, if scientists are willing to relinquish some of their control over the research process. But this implies new roles for both farmers and scientists, and it takes a deliberate effort to create the conditions for such research-oriented local groups. Nonetheless, there have been successes in both industrialised and developing countries.

Self-learning is vital for sustainable agriculture. By experimenting themselves, farmers can increase their own awareness of what does and does not work. But few farmers can engage in developing the new methods and principles alone, and so any kind of group activities that embody co-learning and study are vital. The transition to sustainable agriculture, therefore, needs networks of farmers who can jointly engage in learning and experimentation. In Norway, the first experimental farmers' groups, or *forsöksring*, were set up in 1937 (Gedde-Dahl, 1992). There are now more than 100, each employing an agronomist. Members contribute to costs, but only two-thirds of the funding comes from the state. The greatest impacts have been on mutual trust and respect, on changed farmers' knowledge and social learning, and on more rapid uptake of technologies developed by other farmers through on-farm research.

In The Netherlands, the 38 pilot farms that were part of the Integrated Arable Farming Systems (IAFS) network showed that it was possible to farm more sustainably without losing money (see Chapter 3). This was scaled up to involve 500 farmers as part of the Arable Farming 2000 Project. This in turn is beginning to have an impact on other organisations. The hitherto conservative Northern Agricultural Farmers' Organisation, with 12,000 members, launched its own scheme to test and spread integrated farming amongst its members, even though it has, in recent years, been a vigorous supporter of only high-input farming. According to farmers, the IAFS network creates an opportunity to experiment with new practices; farmers feel there is greater use of their professional skills, and so they get increased pleasure from their work. And they know there is now a more positive contribution to the environment. This is a common feature of farmers working in groups. They feel less alone and more a part of a joint effort for agricultural improvement that they themselves can influence and control (van Weperen and Röling, 1995).

Recent work in Devon as part of the wider Sustainable Agriculture Partnership has also illustrated the critical role of farmer involvement in experimentation. Designed Visions has been helping farmers to develop regenerative technologies in the Blackdown Hills, a remote and predominantly livestock area. Soil reconditioning has been an important innovation for the rejuvenation of permanent pastures. The process involves machines that cut into the surface of grassland to leave parallel furrows 10 centimetres or more deep. These substantially increase the water-holding capacity, growing season, number of earthworms, and soil organic matter. It also means that farmers do not need to engage in the destruction of traditional grasslands by reseeding or adding large amounts of fertiliser. According to Andy Langford: 'farmers' willingness to experiment and innovate has been crucial – without this the technologies would not have been fitted to highly site-specific conditions'.

In the US, the Land Stewardship Project in Minnesota organises farm families into peer-support and information-sharing groups as part of a Sustainable Farming Association (Kroese and Butler Flora, 1992). These groups encourage farmers to experiment with alternative farming practices on their own farms and at their own pace, and facilitate the exchange of information among nearby farmers about what they have learned. Learning groups help farmers to capture more value for themselves. The Center for Rural Affairs in Nebraska describes the value of small groups as: 'the best atmosphere. In a learning group, everyone's ideas, questions and experiences are valuable. The purpose of each meeting is to hear new ideas and help each other solve their problems. Everyone on the farm makes mistakes and has problems. Members of learning groups can have a great time learning from each other.'

Another successful institution is the Kansas Rural Center, which has supported family farming and the grassroots involvement of local people in farming and countryside matters in order to conserve natural resources. Their Heartland Sustainable Agriculture Network brings farmers together to enhance experimentation, exchange and education. The network has over 100 farmers in small clusters who work together on issues important to them. These include Covered Acres – farmers in central Kansas experimenting with legume cover crops; Smoking Hills – farmers working on grazing management in Saline County; Resourceful Farmers – crop, livestock and dairy farmers in south-central Kansas who give on-farm demonstrations of rotational grazing and clean water practices; and Quality Wheat – organic farmers in west Kansas seeking to improve soil fertility and to increase the protein content of wheat. The network is a clearing house for ideas on sustainable agriculture, helps to build support for new ideas, nurtures leadership, creates confidence amongst farmers to try something new, and works with conventional agricultural institutions to build support for rural regeneration through sustainable agriculture.

Practical Farmers of Iowa

One of the best examples of supporting farmer experimentation is Practical Farmers of Iowa (PFI, 1995; Hart et al, 1996). Like many other rural areas, Iowa is characterised by rural depopulation and a massive deterioration in the social fabric of rural communities. Modern agriculture was seen to exacerbate these problems, and so farmers felt that research and extension institutions were not addressing their needs. They wanted alternatives that reduced the impacts of agriculture on the environment but remained profitable. Established in 1985, PFI now has some 500 members who share practical, profitable and environmentally sound farming technologies and practices.

The cornerstone of PFI's work has been on-farm research run by farmers, combined with a wide range of sharing mechanisms to ensure that members and others learn about the innovations. Farmer cooperators run dozens of trials each to compare sustainable practices with the customary ones. There are field days and twilight tours, with more than 12,000 people attending field days since 1987; workshops and formal meetings; group get-togethers; and a regular newsletter. Various farmer and community group associations have emerged, some of which have started CSAs. But it has been the sheer breadth and depth of the 400-plus trials that has helped both to shift attitudes and practices and to add value to farming. There have been trials on nitrogen rates in maize, weed management, placement and timing of phosphorus and potassium, tillage, manure management, intercropping, cover crops, and management-intensive grazing. Sustainable agriculture now clearly means using fewer agrochemicals without losing on yields or net returns (see Table 5.1). The savings for farmers are significant. Each now saves at least US$80 per hectare per year in maize-based systems. This translates into a massive regional saving: the 500 member farmers are making savings of millions of dollars annually. This makes a big difference for local economies. More money in farmers' pockets means more is spent on local goods and services.

PFI has helped to build leadership amongst cooperating farmers who now speak to many other farmer groups, and has created a common language between farmers and scientists, which has, in turn, accelerated the diffusion of research findings. Farmers have changed too. Said one: 'we changed our complete way of farming after I got involved in PFI'. Another put it this way: 'I like the idea of small research on local farms – communicating with other farmers, accessibility to research. I've seen some good trials and changed some practices on my farm because of them.'

Labelling and Traceability

Ecolabelling, Assurance Schemes and Traceability

Although organic products have long been clearly labelled for consumers, it is only recently that there has been an expansion in the range of ecolabels. These have been given greater urgency by recent food scares about BSE, *E coli* and pesticide residues. The question consumers are increasingly asking is: can the food on the shelves be trusted? Ecolabels are important since they tell consumers something about the way that the

Table 5.1 *Selected findings from research run by farmers and their organisations with Practical Farmers of Iowa (PFI), US*

	Conventional farmers	Sustainable agriculture farmers in farmers' organisations	Sustainable as proportion of conventional (%)
Nitrogen management (74 trials)			
Nitrogen applied (kg/ha)	150	88	59%
Maize yields (t/ha)	8.20	8.08	99%
Weed management			
Maize yields (t/ha)	7.99	8.02	100%
Soybean yields (bu/ha)	106.7	104.7	98%
Costs ($/ha):			
Nitrogen	61	29.4	48%
Pesticides	66.8	24.1	36%
Drying	4.7	11.5	244%
Application operations	13.7	8.7	63%
Preplant tillage	37.7	18.6	49%
Other operations	34	57.7	170%
Total costs	230	150	65%

Note: for weed management, sustainable agriculture = zero herbicides.
Source: Practical Farmers of Iowa. 1995. *Shared Visions*. Report to the W K Kellogg Foundation by Gary Huber, PFI, Iowa; Hart A, Boddy P, Shequist K, Huber G and Exner D. 1996. 'Iowa, US: An effective partnership between the Practical Farmers of Iowa and Iowa State University'. In Thrupp L A (ed). *New Partnerships for Sustainable Agriculture*. WRI, Washington, DC

food was produced. They allow growers and processors to be rewarded for using environmentally friendly production processes. They also permit consumers to express their values while making purchases. And, if they work well, they help to push the food and agriculture industry towards more sustainable practices (Lang, 1995b; MacRae, 1997).

In Britain, the UK Register of Organic Food Standards sets strict standards for organic farmers; people buying products labelled with the Soil Association standard can be confident that they are getting what they expect. Recently, however, a range of 'assurance' schemes have emerged. These are not organic but do seek to assure consumers about the conditions under which livestock and crops are raised or grown. Interestingly, farmers' views of these schemes are changing from perceiving them as constraints to viewing them as vital in holding onto market share. One response has been to say that farmers should not subject themselves to tighter production rules than foreign competitors have, since they would lose market share. One farmer from Yorkshire, Peter Hepworth, put it like this in July 1997: 'the NFU wishes to saddle future farmer generations with restrictions and straitjackets under the banner of cereal assurance schemes... The idea is totally mad... Only when our main cereal competitors seek to impose such rubbish should we even consider following suit' (*Farmers' Weekly*, 4 July 1997).

However, most now see traceability as an increasingly vital part of the industry. The intention is for any food product to be traced back to the individual farmer. Not only will this provide a link when food scares do happen, but it is also hoped that consumers' confidence will rise as the information with which they are provided also increases. The Farm Assured British Beef and Lamb (FABBL) scheme was the first to be launched by farmers in 1992. By 1997, it had some 20,000 members. In 1996, Farm Assured British Pigs (FABPIG) was also launched. Like the beef scheme, it covers animals from birth to slaughter and includes strict standards of cleanliness and welfare for farmers, transporters and abattoirs. Quarterly farm inspections by local vets will provide independent auditing of farm practices. Other schemes include the Scottish Quality Beef and Lamb Assurance, with 8000 members by 1997; this scheme provides independent audits of all members, whether feed markets, farmers, auction markets, hauliers, meat processing plants or retail shops. The Assured Combinable Crops Scheme was launched at the end of 1996 to ensure that grain is produced with due regard to codes of practice and the quality of the environment.

Some of these new schemes have been criticised. One problem is that there are now so many that it is difficult to be sure what each guarantees. Some argue that all schemes need to be brought under one national umbrella that has strong teeth and is enforceable. Others say that they are not as well defined as organic standards. These have been tested and refined over many years and are widely accepted by growers, retailers and consumers alike. But some new labels give the impression that food is wholesome and produced in a sustainable fashion: these may simply be no more than hollow labels. 'Farm fresh' and 'environmentally friendly' may look good but have no effect whatsoever on food production methods. It has been alleged, for example, that intensively produced cattle in Britain are sometimes put out to graze on pastures for only very short periods, in order that the meat may receive a 'farm fresh' label. In Germany, 'Black Forest ham' is intended to conjure up images of environmentally sensitive pig rearing in a forest, perhaps where consumers once spent their holidays. Yet, according to Rainer Luick, '90 per cent of the pigs are fattened in pig units in The Netherlands, Denmark and Poland' (Luick, 1996). Nonetheless, new forms of labelling are vital since they let consumers know about the mode and place of production. It also lets them know more about the contribution that agriculture can make to social and natural capital. Real Foods of Ireland, for example, is a recent initiative intended to add value to food that is produced in an independent and responsible fashion. It has 45 members and, as Giana Ferguson of Cork put it, 'sees food culture at the centre of creating and building community' (Giana Ferguson, pers comm, 1997).

A good example of local labelling comes from Devon, where the Taw Torridge Estuary Project is bringing farmers together to make the best of distinctive local saltmarsh resources (Bell, 1996). Sheep grazed on saltmarshes can increase plant biodiversity if managed properly; they have fewer diseases; the lamb meat has a distinctive flavour – but management is more demanding owing to the uncertainties of tides and irregular inundation. Taking the success of the *pré salé* lamb raised on the saltmarshes of Normandy and Brittany as an example, Andy Bell and colleagues have been developing ways to market this lamb more effectively so as to increase returns to farmers.

A new Dutch green-label certification scheme is called *Milieukeur.* This has been established with the backing of the Ministry of Agriculture, Nature and Fisheries (de Vries, 1996). The *Milieukeur* label is applied to cereals, onions and field vegetables. This system is seen as a way of gradually raising standards without forcing farmers to change overnight, but it does rely on consumers being willing and able to pay premiums for the 'better quality' food. It contains tough conditions for pesticide use, including a commitment to reduce by 50 per cent the amounts applied compared with the 1980s' levels.

The potato label is the most advanced. In 1995, 100 growers produced 20,000 tonnes of *Milieukeur* potatoes. Some 3 to 4 per cent of apples and pears are also marketed this way. As *Milieukeur* producers meet the conditions set out in the government's Multi-Year Crop Protection Plan (which is seeking to reduce total pesticide use by 50 per cent by the year 2000), Pat Matteson and colleagues have predicted that *Milieukeur* will market, by 2000, some 60 per cent of all sugar beet, 30 per cent of apples and pears, 25 per cent of cooking apples and 10 per cent of arable, vegetables, pork and dairy products (Proost and Matteson, 1997a, b).

In Canada, many organisations are developing 'locally produced' symbols that allow consumers to identify foods from their own region. Ontario has four such labels. Says Rod MacRae of the Toronto Food Policy Council: 'such labels allow consumers to make purchases based on their feelings about a particular place'. But not all are in favour: 'for transnationals, this is a direct contradiction to the type of food system they are trying to develop – one where consumers have no allegiance to place, but only to price and perceived quality, as expressed through brand allegiance' (MacRae, 1997)

Labelling, therefore, exposes some vital contradictions. Foods must contain details of all the additives and preservatives incorporated during processing, but no such rules exist for details of pesticide residues. Some countries label foods that contain GMOs; others do not, giving consumers no choice. Many commentators are very concerned about how the international food standards bodies, the Codex Alimentarius Commission of the United Nations and the International Standards Organisation (responsible for the ISO 14000 environmental protocols), are interpreting the new rules of the WTO. The Codex, for example, has 105 countries participating in its committees that deliberate and arbitrate on a wide range of food issues, yet it also has representatives from 140 multinational companies. One study of the 1988–1991 committee cycle found that there were 26 representatives from public interest groups and 662 from industry (Avery et al, 1993).

Companies Ahead of Legislation?

Despite this confusion, some European and American companies are adopting rules for their growers that are stricter than required by legislation (see Box 5.5). Some are now encouraging all their growers to use ICM technologies, not in order to acquire a label that leads to increased food prices, but to guarantee a market share. Birds Eye Walls has stipulated that all contract growers of peas (on 50,000 hectares) will be required to register with the LEAF scheme in 1998.

In the US, the companies Gerber and Campbell have strict quality control systems on farms and in processing. Yet neither company markets their products in a differentiated way (North West Food Alliance, 1995). The companies say it would provide no obvious market advantage, but it may well be that they simply do not wish to draw the public's attention to the fact that pesticides were used in the past, even though amounts have been reduced now. Alistair Leake of Cooperative Wholesale Society (CWS) Agriculture in the UK recently put it this way: 'produce grown under an integrated regime will not, indeed should not, command a price premium. It is a standard to which every grower should aspire... The economic imperative will be the difference between selling and not selling their production, rather than any niche market "green" premiums' (Leake, 1996).

Other changes are being made ahead of legislation. In the US, the North West Food Alliance has brought together a wide range of stakeholder groups, including farmers, food producers, distributors,

Box 5.5 Examples of private-sector contributions to food quality ahead of legislation

- Sainsbury's introduced ICM systems for vegetable growers in 1992 – by the end of 1996, some 80 per cent of all fresh produce sold in 360 stores was grown under ICM.
- The Cooperative Society and its farming wing CWS Agriculture have adopted a right-to-know policy for consumers – letting consumers know about just how their agricultural products are grown.
- In Germany, Hipp, a baby food producer, has a 'no-pesticide' policy with its growers, whilst Nestlé Alete has a very restrictive approach.
- In the US, the Campbell Soup Company launched an effort to reduce pesticide applications by 50 per cent in 1992 – this has now been achieved with no sacrifice to yields or food quality.
- In the US, Gerber (a baby food company) requires all of its growers to use IPM and other environmentally sound production practices.
- Potato growers supplying Sainsbury's have to use ICM technologies and practices. This means 70 per cent of the annual throughput of 250,000 tonnes will be grown according to the company's ICM policy (which is in turn based on the protocols developed by the NFU).

environmentalists and consumer groups, to encourage the transition towards sustainable agriculture (North West Food Alliance, 1995–1996). The major concern is to support maximum farmer innovation in order to produce better quality food for consumers. The impact on farming practice has been significant: apple growers, for example, have found that by reducing nitrogen applications to 28 kilogrammes per hectare they can sustain yields since there is less aphid damage, earlier leaf drop and reduced pruning needs. Others have dramatically cut pesticide and fungicide use in vegetable cultivation.

This is surely good for consumers and for farmers; it widens choice and increases the number of farmers engaged in more sustainable practices. It does not, however, guarantee the complete transition towards sustainability.

Consumers Make a Difference with Fair Trade

Trade is enormously important to the global economy. Since 1970, it has more than tripled in value, to US$4,000 billion, and is now outpacing the expansion of the world economy by a factor of three (Scenario 2010 Working Group). The EU is now the world's largest trading bloc, making up 21.5 per cent of world trade, compared with 18 per cent for the US and 10 per cent for Japan.

Trade has become progressively more liberal since 1970, especially following the Uruguay Round of the GATT, completed in 1994. Now there are large trade agreements that seek to make trade 'free' of quotas, tariffs and other trade barriers. But there are many criticisms of the trend to globalisation and free trade. In particular, liberalisation is seen as encouraging companies to 'race to the bottom' in terms of social and environmental standards. It has also led to greater concentration: just five corporations control some 77 per cent of the cereal trade; three have 83 per cent of cocoa; three have 80 per cent of the banana trade; and three have 85 per cent of the tea trade. But only very small proportions of the price paid gets to farmers. For bananas, about 2 per cent of the price we pay in the shops goes to the fieldworker; 5 per cent to the farmer; and 88 per cent to the intermediary importers, wholesalers, retailers and freight companies (Lang and Hines, 1993; Madden, 1992; Paxton, 1994).

However, there are now alternatives in the form of 'fair trade' products. These try to guarantee a better deal for producers of the South and ensure that consumers receive good quality produce. Consumers can make a big difference to this market, as the EU imports three times as much food from developing countries than does the US, and twice as much as the other G8 partners put together. The Fairtrade Foundation is an example: it licenses and promotes the fair trade mark as an indicator to consumers that their products are giving a 'better deal to producers in developing countries' (Fairtrade Foundation, 1995–1996). The basic assumption is that the conditions of poverty in which people live are due, at least in part, to the manner by which industrialised countries trade with them.

The products most commonly subject to fair trade marketing are chocolate, coffee and tea. Producers are paid a higher than world price, and are guaranteed a market price for at least three years ahead. In 1995, it was estimated that £150,000 extra was paid to producers of Fairtrade marked products over and above what they would have received. The difference fair trade can make to local communities is enormous, as the case from Mexico illustrates (see Box 5.6). There are considerable social as well as economic benefits for local people. It is now estimated that there are some 100,000 Mexican farmers producing fair trade coffee with organic methods (Geier, 1996).

Other examples of successful fair trade products include organic cotton for textiles from Uganda, Senegal, Turkey and India; tagua nuts for buttons from Ecuador; bananas from the Caribbean; soft fruit from Chile; and tea

Box 5.6 The impact of fair trade in Mexico: the case of the Union of Indian Communities in the Isthmus Region

The Union of Indian Communities in the Isthmus region (UCIRI) was organised by farmers in remote areas in the state of Oaxaca in 1981. The landscape is mountainous and coffee is grown mainly on slopes, with maize and vegetables on the flatter lands. Coffee farmers were traditionally dependent on intermediaries, who controlled credit, bought the coffee, and supplied the basic necessities. Local people were poor and entirely dependent on outside groups and agencies. The basic family income was only US$250 per year.

In 1985, the union decided to move from traditional to organic agriculture. This was partly to reduce dependency on credit, but it was also hoped that yields would improve too. Contacts with fair trade organisations supported these goals since they offered a premium on the organic coffee. Organic coffee cultivation demands more active management and a higher labour input from farmers. Coffee is grown in the secondary forest, and farmers leave leguminous trees that are beneficial to the coffee. It is planted on the contour to stop soil erosion, and slashed weeds and pruned branches are laid on the contour too. Half-moon shaped terraces are constructed for each coffee tree. Formerly the coffee beans were depulped into waterways, causing significant water pollution. But now organic farmers return the pulp to the fields through composting. Other materials used for composting include animal manures, lime and green plant material. This improved system produces 600 to 1200 kilogrammes per hectare of coffee beans, an improvement of 30 to 50 per cent compared with earlier practices.

There are now 4800 members of UCIRI in 37 communities. The union has been able to build up its own infrastructure for the transport, storage, processing and export of its coffee. The premium received for the organic coffee from fair trade organisations is used for a range of social and economic purposes, including improving the educational systems. The union also runs a public transport system into the mountains and a medical insurance system, and owns several shops from which local people can buy basic necessities. The higher price paid by consumers for fair trade coffee has led to substantial improvements in welfare of whole communities. Today the average income is $480.

Sources: Pretty, 1995, *Regenerating Agriculture*; Bernward Geier, pers comm, 1997

from Sri Lanka (Robins and Roberts, 1997). All of these have resulted in greater returns to rural people, much of which goes towards improving the conditions of whole communities. Cafedirect is a brand of blended fair trade coffee now sold in 1700 supermarkets in the UK. Fourteen producer organisations in Mexico, Costa Rica, Peru, Nicaragua, the Dominican Republic, Uganda and Tanzania supply the coffee. In five years, it has captured 3 per cent of the share of the British market – which means that some 460,000 families in these countries are benefiting.

Community Food Security and the Foodshed

Foodshed Thinking

As described in this chapter, there is a range of ways for communities to take back more of the middle, preserving the value locally and improving local welfare and the environment. But, until recently, those concerned with poverty and hunger, particularly in urban areas, and those concerned with sustainable agriculture and conservation have not found a way to integrate their interests and activities. The emergence of community food security and the idea of the foodshed as integrating concepts have changed all this.

Jack Kloppenberg and colleagues at the University of Wisconsin–Madison have used 'foodshed' to give an area-based grounding to the production and consumption of food (Kloppenberg, 1991). The term foodshed was first used by Arthur Getz in Hawaii (Rod MacRae, pers comm). A feature of sustainable agriculture, as described in Chapter 3, is that it needs many farmers operating in a contiguous area to shift their practices before a significant difference can be made to natural and social capital. Foodsheds have been described as 'self-reliant, locally or regionally based food systems comprised of diversified farms using sustainable practices to supply fresher, more nutritious food stuffs to small-scale processors and consumers to whom producers are linked by the bonds of community as well as economy'. The foodshed, by bringing consumers and producers literally and figuratively closer, helps to regenerate and reinvigorate natural and social capital. As this book has made clear, few are saying that autarkic systems with no external linkages are best. Rather, it is a question of making the best of local capacities, resources and linkages before turning to externally sourced products.

Community food security has now become the way to implement this notion of the foodshed. It has received prominence in the US owing to the effectiveness of the Community Food Security Coalition, coordinated by Andy Fisher and colleagues from Venice, California, with a broad, nationally drawn steering committee. This is a diverse network of anti-hunger, sustainable agriculture, environmental, community development and other food-related organisations who, through good timing and good campaigning, persuaded politicians to incorporate community food security in the 1996 Farm Bill. The Community Food Security Act authorised the USDA for seven years to spend $2.5 million annually on local projects designed to: 'meet the food needs of low income people, increase the self-reliance of communities in providing for their own food needs, and promote comprehensive, inclusive and future-oriented solutions to local food, farm and nutrition problems' (Community Food Security News, 1996). This was a remarkable success, one of the few innovations in the Farm Bill, and the only one committing more financial resources.

Other important policy changes in the US include, at national level, the USDA's authorisation in 1992 that recipients of women, infants and children (WIC) assistance should be given fresh-produce vouchers to spend at farmers' markets. Over five years, 800,000 were issued in 24 states, and it is estimated that almost 90 per cent of the 6600 farmers participating in the programme have increased their sales (Cook and Rodgers, 1996). Before the programme, most WIC recipients had never visited a farmers' market; since then, participants have on average doubled their consumption of fresh fruit and vegetables. According to Marion Kalb, director of Southlands Farmers' Market Association in California, 60 per cent of women who initially shop at farmers' markets continue to use these markets after their coupons run out (quoted in Newman, 1997).

At the same time, a range of local food policy councils and systems have become increasingly effective, most notably in Toronto, Canada, in Hartford, Connecticut (see cases below), in Knoxville, Tennessee, in St Paul, Minnesota, and in Austin, Texas. Bringing together these different stakeholders is not just a nice idea. It works. It works for local people, it works for communities, it works for farmers, it works for the natural environment. Andy Fisher put it this way:

> *...the concept of community food security seems to strengthen the real connections between urban and rural populations, family farmers, farmers' markets and community gardening... The federal policies realise the need for grassroots economics initiatives, instead of welfare reforms that further separate low-income communities from sources of healthy food. The best guarantee of a sustainable food future is to empower local communities (Fisher, 1996).*

The Toronto Food Policy Council

The Toronto Food Policy Council (FPC) is an extended network of organisations concerned with food security, sustainable agriculture, public health and community development (Rod MacRae, pers comm, 1997; MacRae, 1994; Food Share, 1996; TFPC, 1994). The focus is on increasing low-income families' access to an affordable, nourishing diet and on fostering food micro-enterprises involving low-income people. It was set up in 1990 to bring together professionals and activists from a wide range of sectors: public health, farm and rural, food, labour, education, community, and hunger advocacy.

The FPC receives support from the city council, and is administered by the Department of Public Health. This gives it formal credibility while allowing it to work with many community groups. It has two key roles: the removal of public policy impediments which limit access to decent levels of nutrition; and the creation of progressive policies which promote community action on food issues. Why Toronto? Says Rod MacRae, the FPC coordinator:

> *...it's my favourite question, and I hear it almost every week. The short answer is: we all eat and Toronto has 650,000 mouths to feed. The long answer is that the food and agriculture system in Canada is not designed to provide optimal nourishment and, consequently, is contributing to a host of health problems for Toronto residents.*

At the time of the emergence of the FPC, Toronto was characterised by economic decline, a rapid shift from permanent to part-time and short-term jobs, increased dependence on social services, ingrained poverty in certain groups combined with increased poverty in lower-middle classes, and increased numbers of hungry people. During the 1980s, food banks emerged in response to the dismantling of social-security safety nets, growing to more than 400 in number by the 1990s. Now, 150,000 residents use food banks each year, and on average each receives US$10 worth of food every month. But food banks were also recognised as only a stop gap that focused on the symptoms of hunger rather than on the underlying causes.

The FPC set out to shift the food, welfare and public health systems from their emergency focus to give them a greater role in improving community self-reliance and social capital. The process has been complex. According to Rod MacRae, the fundamental basis was:

> *...a belief that as our environments become more turbulent and unpredictable, we need processes and forms of social organisation that mirror the complexity and diversity of the environments and ecosystems we are attempting to affect. In response, there is a growing demand for equally complex, multiorganisational and multisectoral responses.*

One example of the FPC programme is the Field to Table Programme, which developed from a partnership between community groups and

Ontario farmers. It has three modes of food delivery: under the Good Food Box (GFB), families can buy a box of fresh fruit and vegetables each month; from community markets, individuals can purchase fruit and vegetables and sell them locally; and in the Buying Clubs, community members can order fresh fruit and vegetables and have them delivered. The Good Food Box has been particularly effective. It targets people who wish to buy fresh foods, who are on low incomes, have disability or health problems, or who are senior citizens. Between 1500 and 2500 boxes per month were delivered in 1997, mostly to customers on low incomes, including a sizeable proportion of single parent mothers.

The impacts have been on many parts of the food system. Some 70 per cent of those buying food boxes now eat more vegetables; 21 per cent eat a greater variety; and 16 per cent now try new foods. More people also know about the recommendation that we should eat five or more servings of fruit or vegetables per day. At the start, a quarter of food in the food banks was sourced from Ontario farmers; by 1996 it had grown to 95 per cent. Even more interesting, a substantial number of GFB recipients said the scheme was affecting their social contacts. More than a fifth believed the scheme had made them more community-oriented. Other less tangible effects of the whole Field to Table scheme have been noted; for example, in schools, there is better attendance, less tardiness and better sociability in classrooms. The Field to Table Programme reaches 10,000 people.

Other benefits of the FPC's work include the rapid growth in community gardens in Toronto; the support now given by provincial government to school food programmes; and the wider impacts on policy processes. It has also been promoting local economic renewal through a wide range of institutions and sectors, including ecological tax reform, health care reform and agricultural policy reform. It is, therefore, engaged in large-scale social change. According to Rod MacRae, 'TFPC is a real experiment, trying to build cohesion out of diverse points of view'. It is slowly helping to make a shift away from welfare towards preventative health care.

The Hartford Food System, Connecticut

Hartford is a city of some 125,000 people situated in the fertile Connecticut River valley, on the eastern seaboard of the US. Hartford and its nearby towns of Bridgeport and New Haven are not special. They are like many other parts of the US, except that Hartford has a coordinated alliance of organisations concerned with food, poverty, hunger and community development. The Advisory Commission on Food Policy was set up to address the alarming inner-city poverty and hunger, leading to the establishment of the Hartford Food System (HFS), coordinated by Mark Winne (Mark Winne, pers comm, 1997; Advisory Commission on Food Policy, 1996).

Poverty and food insecurity are severe. Some four in ten children live in poverty, and nearly 80 per cent are eligible for free or reduced-price school meals – a family qualifies for free meals if it earns no more than 30 per cent above the poverty line (an income of US$15,166). Studies of low-income neighbourhoods have found that 25 to 41 per cent of residents experience hunger. At the same time, they spend up to 40 per cent of their annual income on food, whereas those on higher incomes spend only 9 to 15 per cent. Over the years, farming and food retailing have become more centralised and large scale. In the 1940s, there were 22,240 farms in Connecticut; now there are just 3500. In the late 1960s, there were 13 supermarket chains in Hartford; now there are only two. As a result, poor families without access to transport pay more – food prices are up to 35 per cent higher in local food stores than in out of town supermarkets, and they have to rely on public transport to get them there. Fresh foods are also scarce in local shops.

The general approach by agencies dealing with poverty and hunger was of an emergency nature. It concentrated on welfare, rather than being concerned with the more difficult task of building self-reliance over the long term. Some $30 million is spent annually on food assistance programmes aimed at women and children. Each month, more than 41,000 people collect food stamps, and 160 to 190,000 emergency meals are served. But the HFS has started to amend all this. It is helping to promote a wide range of changes in consumer behaviour, in urban agriculture and in local farmers' markets. The HFS has had a number of impacts. It encourages better food education and collective food consumption in schools. New awards have been created, such as the Golden Muffin Awards, to raise the profile of the National School Breakfast programme. Over three years there has been a 35 per cent increase in the number of children eating breakfast at school. The farm to school education programme exposes children to fruit and vegetables and local agriculture in their classrooms, and links farms with schools to provide fresh fruit and vegetables in cafeterias.

The Food Policy Commission conducts and publishes regular supermarket surveys to monitor prices. It publishes the *Hunger Report* which tracks public assistance programmes, emergency meals served and other welfare programmes. These are vitally important, as they get information to people about short-term changes in the local food system. Through links to the Knox Parks Foundation, the HFS promotes urban agriculture in 16 community gardens, with 200 families now growing affordable and nutritious food. It also runs Holcombe farm, a CSA farm which grows 50 varieties of fruit, vegetables and herbs. This CSA is unusual in that half of the produce is made available to low-income families. The 6.5-hectare farm currently supplies 150 family members of the CSA, each of whom receives about 180 kilogrammes per season, and ten community organisations, each of whom receives 1800 kilogrammes, which is then passed on to some 1000 low-income families.

To broaden the links between farmers and consumers, HFS also encourages farmers' markets in Hartford. There are now five local farmers' markets and 50 in Connecticut altogether. The prices are lower than in supermarkets, and the quality of food is said by many to be superior. There is also a coupon programme, with low-income families receiving $10 coupons to spend at farmers' markets. In 1996, $130,000 worth of coupons were distributed, bringing benefits both to farmers and these families. State-wide, some 60,000 families have benefited from the farmers' markets nutrition programme. As a result, four-fifths of recipients of the WIC coupon programme report eating more fruit and vegetables (Nugent, 1997).

Summary

Half a century ago, at least half of the pound, franc, mark or dollar spent on food found its way back to the farmer and the rural community. The rest was spread amongst suppliers of various inputs (feeds, pesticides, fertilisers, seeds, machinery, labour, and so on) and manufacturers, processors and retailers. Since then, the balance of power has shifted increasingly away from the middle, with value captured on the input side by agrochemical, feed and seed companies, and on the output side by those who move, transform and sell the food. Sustainable agriculture can mean a cut in some production costs, as knowledge and information substitute for external inputs. In addition, rural communities need to find new ways of adding value to the goods and services derived from available natural capital. This is called taking back the middle.

The first option for taking back more of the middle is for farmers to find ways of selling their produce directly to consumers. Farmers' shops and pick-your-own operations have long been an important means for farmers to sell directly to consumers. When successful, they can make a very significant contribution to individual farm incomes. But they can be costly to establish and run and rely on customers driving to farms to make purchases.

Produce markets, in which growers sell directly to customers, are another option. Produce markets offer the opportunity for consumers to buy organic or sustainably produced food without paying a premium. There is a long history of markets in rural towns and centres. Weekly or twice weekly markets are a part of the vitality of many small towns in Europe. These markets are almost entirely dominated by traders selling on wholesale produce. Very few farmers are directly involved, and so there are few direct links with the public.

In the US, the situation is rather different, as farmers' markets have emerged on a huge scale in recent years. Held on a weekly or twice weekly basis, farmers and consumer groups have established new market sites to foster direct selling to the local public. There are at least 2400 farmers' markets, involving more than 20,000 farmers as vendors, one third of whom use them as their sole outlet. Each week, about one million people visit these farmers' markets, 90 per cent of whom live within 11 kilometres of the market. The annual national turnover is about US$1 billion. The benefits these markets bring are substantial. They improve access to local food; they improve returns to farmers; they also contribute to community life and social capital, bringing large numbers of people together on a regular basis. Farmers' markets also recycle resources into other important community functions, contributing particularly to social capital.

Community-supported agriculture (CSA) is a partnership arrangement between producers and consumers. The basic model is simple: consumers provide support for growers by agreeing to pay for a share of the total produce, and growers provide a weekly share of food of a guaranteed quality and quantity. CSAs help to reconnect people to farming. Members know where their food has come from, and farmers receive payment at the beginning of the season rather than when the harvest is in. In this way, a community is sharing the risks and responsibilities of farming.

Organic farmers and growers in the UK have recently made great strides in the alternative marketing and sale of their produce directly to consumers. There has been spectacular growth in box schemes. These schemes also emphasise that payment is not just for the food, but for support of the farm as a whole. It is the link between farmer and consumer that guarantees the quality of the food. This encourages social responsibility, increases the understanding of farming issues amongst consumers, and increases the diversity of crops grown and available over time.

Food cooperatives are an important way to get good food to urban groups with no direct access to farms and the countryside. Direct links between consumers and farmers have had spectacular success in Japan, with the rapid growth of the consumer cooperatives, *sanchoku* groups (direct from the place of production) and *teikei* schemes (tie-up or mutual compromise between consumers and producers). This extraordinary movement has been driven by consumers rather than farmers, and mainly by women. There are now some 800 to 1000 groups in Japan, with a total membership of 11 million people and an annual turnover of more than US$15 billion. These consumer–producer groups are based on relations of trust and put a high value on face-to-face contact. Some of these have had a remarkable effect on farming as well as on other environmental matters.

Community gardens have a great impact on social capital. In the UK, there are now several hundred city farms or community gardens which cultivate formerly derelict land. These provide food, especially vegetables and fruit, for poorer urban groups, and other natural products – wood, flowers and herbs. They add some local value to produce before sale. They help to reduce vandalism through the involvement of schoolchildren. They are important for providing local children with an educational opportunity to learn about farming and animals. They also provide an opportunity for mental health patients to engage in work that builds self-esteem and confidence. Case studies in this chapter are drawn from Newcastle, Oxford, and Bradford in England, and New York in the US.

Another way that farmers can create extra local value is to work together in groups. A major challenge for sustainable agriculture is to create and support processes that foster learning. When farmers and other rural people are well linked and trust each other, then it is possible for learning mechanisms to be established. Sustainability needs perpetual novelty and adaptive performance. The more that information and knowledge is locally generated and locally applied, the more that value is taken back into local systems. Farmers' organisations can help research institutions to become more responsive to local needs, if scientists are willing to relinquish some of their control over the research process. But this implies new roles for both farmers and scientists, and it takes a deliberate effort to create the conditions for such research-oriented local groups. Nonetheless, there have been successes in both industrialised and developing countries.

Another option is to enhance labelling and traceability of foodstuffs in order to increase consumer confidence about both the source and quality of foods. This can be done through ecolabelling and other assurance schemes, organic standards, and fair trade schemes. Labelling is important, as it tells consumers something about the way that the food was produced. Although organic products have long been clearly labelled for consumers, it is only recently that there has been an expansion in other labelling schemes. Nonetheless, labelling policy is still full of contradictions: there is full disclosure for additives but not for pesticides.

Fair trade is now emerging as another type of labelling scheme. Fair trade schemes try to guarantee a better deal for producers of the South as well as ensuring that consumers receive good quality produce. The agricultural products most commonly subject to fair trade production and marketing are chocolate, coffee and tea. Producers are paid a higher than world price and are guaranteed a market price for at least three years ahead. There are considerable social as well as economic benefits for local people. It has been estimated that there are some 100,000 Mexican farmers now producing fair trade coffee.

The emergence of community food security and the idea of the foodshed as integrating concepts help communities to take back more of the middle. The foodshed, by linking consumers and producers, helps to regenerate and reinvigorate natural and social capital. As this book has made clear, few are saying that autarkic systems with no external linkages are best. Rather, it is a question of making the best of local capacities and resources before turning to externally sourced products. Community food security has now become the way to implement this notion of the foodshed.

Case studies of the Toronto Food Policy Council in Canada and the Hartford Food System in Connecticut show just how system-wide changes are possible. Fundamental changes to the agriculture and food systems are two parts of the living land triumvirate. The third is sustainable rural development, and the enhancing of rural social capital.

Part III

Towards Sustainable Rural Communities

Chapter 6

Dying Rural Communities: the Social Costs of Countryside Modernisation

'I thought it would last my time –
The sense that, beyond the town
There would always be fields and farms,
Where the village louts could climb
Such trees as were not cut down;
I knew there'd be false alarms...

And that will be England gone
The shadows, the meadows, the lanes,
The guildhalls, the curved choirs.
There'll be books; it will linger on
In galleries; but all that remains
For us will be concrete and tyres.

Most things are never meant.
This won't be. Most likely...
I just think it will happen soon.'

Philip Larkin (1922–1985)
from *Going Going*

Change in Agricultural and Rural Communities

The Decline of Social Capital

The 20th century's period of remarkably successful agricultural growth brought great social change in rural areas throughout Europe (Newby, 1980; Derounian, 1993). In the quest for greater food production, landscapes, rural livelihoods and farming systems have all been progressively simplified. Where there were diverse and integrated farms employing local people, there are now operations specialising in one or two enterprises that largely rely on farm or contractor labour only. Where processing operations were local, now they are centralised and remote from many rural communities.

The result is that few people who live in rural areas now have a direct link to the process of farming. Fewer people make a living from the land and so, of course, they understand it less. The lack of employment has also coincided with the steady decline in rural services, including schools, shops, doctors and public transport. The social costs of modern agriculture and rural development have been as significant and far-reaching as the environmental costs detailed in Chapter 2. The social capital of rural areas has also declined. Horizontal networks within communities have diminished, often replaced by vertical links to distant organisations. Farmers have fewer direct contacts with local people, and as a result trust and confidence have declined. Opportunities for informal and formal horizontal exchanges have fallen, and so norms of cohesive rural societies gradually have eroded.

Changing Farm Size

In every European country, farms have become both fewer in number and larger during this century. When the CAP was established in 1957, the six member countries (France, Germany, Italy, The Netherlands, Belgium, and Luxembourg) had 22 million farmers. Today that number has fallen to about seven million. In England and Wales, the number of farms has fallen from 363,000 to 184,000 since the mid 1940s, while the amount of agricultural land has remained roughly at 19 million hectares. In just the last 30 years, the number of cereal holdings fell by 56 per cent to 75,500; potato holdings by 80 per cent to 22,100; and dairy holdings by 70 per cent to 41,500 (Body, 1996; Office of Science and Technology, 1995).

For most countries, these declines have been a continuation of long trends. In France, there were nine million farmers in 1880; in 1970 just 2.75 million; and by the early 1990s only 1.5 million. During the later 1990s, another half a million more farmers will have left the land. In the US, there were some 5.5 million farms in 1950 with an average size of 80 hectares; in 1996, there were 2.1 million, averaging 200 hectares each (USDA, 1997). In Japan, the number of farmers fell from more than six million in 1950 to four million in 1990. These changes have left very different landholding structures in each European country (see Table 6.1). The average farm size across the EU is about 15 hectares. Only 6 per cent of farms are more than 50 hectares in size; and 60 per cent are smaller than five hectares. The UK has the largest farm size structure. At an average farm size of 67 hectares, it is almost double that of the next highest – which is Denmark at 34 hectares. A third of UK farms are greater than 50 hectares and, put together, the UK has one in six of all the farms in Europe that are greater than 50 hectares in size. Greece,

Table 6.1 *Land holding structure across 12 European countries, 1994 (thousands of farms in each category)*

	Less than 5 ha (000 farms)	*5–50 ha (000 farms)*	*More than 50 ha (000 farms)*	*Average farm size (ha)*
Belgium	32	48	5	15.8
Denmark	2	63	16	34.3
Germany	219	378	56	26.1
Greece	719	202	3	3.9
Spain	971	535	86	15.4
France	249	495	179	29.5
Ireland	19	131	19	25.9
Italy	2100	526	38	5.6
Luxembourg	1	2	1	32.0
Netherlands	40	78	6	16.0
Portugal	492	97	9	6.7
UK	34	128	81	66.7
EU – 12	4877	2683	499	14.8

Note: comparable data for the three most recent EU members, Austria, Finland and Sweden, are not yet available.
Source: Eurostat. 1995. *Agriculture Statistical Yearbook*. Brussels

Italy and Portugal together have 3.3 million farmers with less than five hectares each, and average farm sizes do not exceed seven hectares.

Many argue that progressively larger farms are an economic necessity. They permit economies of scale to be made. They mean that more efficient producers can take over or absorb the operations of smaller and more inefficient producers. These individual farmers benefit, as does the economy as a whole. The National Farmers Union (NFU) of England made these points in 1997: large farms are better for Britain, better for the environment and better for farmers themselves. However, many people contest these views on the grounds that small farms provide other benefits to society. They employ more people than larger farms do, probably because access to labour is easier than to capital. In the UK, farms under 40 hectares provide five times the per-hectare employment than those over 200 hectares. Small farms contribute to both rural social capital and natural capital. Their greater on-farm diversity maintains both plant and animal biodiversity; they are more efficient users of energy; they are better at preserving and enhancing landscape and wildlife; and they tend to have a better record with animal welfare (Raven and Brownbridge, 1996; Rawles and Holland, 1996).

Nevertheless, small family farms have became increasingly vulnerable. Only 89,000 of the holdings in England are considered full time, yet these account for 98 per cent of agricultural output (DoE/MAFF, 1995). Small farmers rely more on diverse sources of off-farm income and so are dependent upon the wider success of the rural economy. When small farms are given up, they tend to be amalgamated into ever larger holdings, with a resulting radical change in the landscape structure (Lobley, 1996; Moss, 1996; Munton and Marsden, 1991). Many such amalgamations lead to intensified land use and the removal of woods and hedges. One consequence of these changes is that the mixed farm is more of a rarity. There are three to four times as many mixed farms in the Mediterranean region than in the north of Europe. These mixed farms are smaller and use less than half the amount of fertiliser per hectare (CLM, 1994). In the UK, farms are now broadly divided into an eastern and southern section devoted to intensive arable production, and a western and northern half devoted to livestock rearing. Large farms produce a greater proportion of agricultural output. They also receive most from public subsidies.

Livestock rearing is very concentrated and much now takes place indoors. Of the 34,000 egg producers in the UK, just 300, each with flocks of more than 20,000 birds, produce three-quarters of all eggs. Flock size in the biggest broiler operations is five times larger, with just 200 of all 2000 producers accounting for 55 per cent of production. A tenth of all 12,000 pig producers account for 59 per cent of production, with herds of over 1000 animals in size (MAFF, 1997). In the EU as a whole, 6 per cent of the farmers produce 60 per cent of the cereals, and 1 per cent rear 40 per cent of all farm animals.

Effects on Social Cohesion

A good example of how farm size affects whole communities comes from crofting communities in Scotland. Frequently labelled as agriculturally inefficient, primitive and backward, they have tended to maintain more cohesive rural communities compared to areas with only large and more 'modern' farms. The islands of Coll and Tiree are situated in the Inner Hebrides, some ten to twenty kilometres west of Mull. They are geographically adjacent, roughly the same area and physically similar. But Tiree is a crofting community. It has 275 crofts and a population in excess of 780. Coll, by contrast, has ten farms and eight crofts, and just 153 residents (see Table 6.2). Even though the islands are naturally similar, they are socially very different. As Frank Rennie of Lewis Castle College on the Isle of Lewis has put it: 'Tiree has managed to retain a much more vibrant rural community'.

There have also been huge changes in rural culture in the US this century. Since 1900, the proportion of the national population who are

Table 6.2 *Comparison between two adjacent islands in the Inner Hebrides, Scotland*

Attribute	*Tiree*	*Coll*
Land area (km^2)	79	68
Number of crofts	275	8
Number of farms	4	10
Population	783	153
Number of beef cows	1250	313
Number of breeding sheep	4850	3990

Note: the peak population for Tiree was 4453 and for Coll 1442 (in 1830–1840).
Source: Rennie F. 1996. 'Sustainable rural development'. In: Mitchell K (ed). *The Common Agricultural Policy and Environmental Practices.* Proc. of a seminar organised by European Forum on Nature Conservation and Pastoralism, 24 Jan 1996, Brussels, EFNCP and WWF, Brussels

farm residents has fallen from 40 per cent to just 1.9 per cent, or 4.6 million people (*Alternative Agriculture News*, 1993). Family farms have been consolidated into larger farms; labour opportunities have fallen; and farm enterprises have been concentrated in fewer hands. This modernisation has been most visible in the declining number of farms and the replacement of family farming by modern large-scale farming. But it has also had significant impacts on social systems.

A classic study conducted in 1946 by Walter Goldschmidt showed what happens when the social structure in the countryside changes during modernisation (Goldschmidt, 1978). He studied the two rural Californian communities of Arvin and Dinuba in the San Joaquin Valley. These were matched for climate, value of agricultural sales, enterprises, reliance on irrigation, and distance from urban areas. The differences were in farm scale: Dinuba was characterised by small family farms, and Arvin by large, commercialised farms. There were striking differences between the two communities. In Dinuba, there was a better quality of life, superior public services and facilities, more parks, more shops and retail trade, more diverse businesses, twice the number of organisations for civic improvement and social recreation, and better participation by the public. The small farm community was a better place to live, 'perhaps because the small farm offered the opportunity for "attachment" to local culture and care for the surrounding land' (Perelman, 1976). A study of the same communities in the late 1970s reaffirmed these findings: social capital – in the form of connectedness, reciprocity and trust – was greater where farm scale was smaller (Small Farm Viability Project, 1977).*

* For a review of the pros and cons of the Goldschmidt hypothesis, see Lobao et al, 1993.

However, the financial squeeze on family farmers has continued and many thousands have lost their businesses. Some do not see this as a problem, but as a desirable means to national agricultural prosperity. Michael Perelman quotes a Bank of America official who said in 1969: 'what is needed is a program that will enable the small and uneconomic farmer – the one who is unwilling or unable to bring his farm to the commercial level by expansion or merger – to take his land out of production with dignity' (Perelman, 1976).

Small farmers are widely viewed as economically inefficient. But their disappearance has been a severe loss to rural society. Linda Lobao's 1990 study of rural inequality shows the importance of the locality that both Rennie and Goldschmidt's research have illustrated (Lobao, 1990). The changing structure of farming has brought a decline in rural population, has increased poverty and income inequality, has lowered numbers of community services, has diminished democratic participation, has decreased retail trade, has increased environmental pollution, and has led to more unemployment. The decline of family farming does not just harm farmers. It hurts the quality of life in the whole of society. Corporate farms are good for productivity, but not much else. Says Lobao: 'this type of farming is very limited in what it can do for a community...we need farms that will be viable in the future, correspond to local needs and remain wedded to the community'.

Some of these kinds of social benefits can be provided by active management of the countryside. This option is open to private owners of large estates – and the effect on social structure can be very beneficial. Large estates may not appear to fit in the sustainable agriculture and community development paradigm. But the Duke of Westminster's estate management at Eaton and Abbeystead shows what can be done for local social capital. A range of policies help to ensure that farming stays closely linked to community welfare. These include the preferential selection of younger families as tenants (in Abbeystead, the average age of tenants was 55 in 1980; now it is 32), including the acceptance of low tenancy bids; the direct support for schools and shops; encouragement to small businesses to set up in villages; conditions on leases to encourage farms and businesses to buy and sell through local shops; and support to youth and sports clubs. Some would argue that this all costs more, and only large operations can afford to be benign. But the Duke of Westminster says: 'we're investing in people, and this makes people feel good. Our wages are higher, and our farm profitability is in the top three in the country' (the Duke of Westminster, pers comm, 1997).

Land Abandonment and Population Decline

A different but equally damaging problem is that of land abandonment. The regions most prone to population decline since the 1950s have been rural Ireland, parts of Scotland and Wales, central France, most of Italy and Greece, and Spain and Portugal. Generally there has been very little population decine in England, Germany, Belgium and The Netherlands (Clout, 1984). As described in Chapter 3, there are at least 56 million hectares of low-intensity farmland of a high-biodiversity nature conservation value in Europe. Their value is maintained by the practice of farming – and when farmers leave, the value declines significantly (Bignall and McCracken, 1996).

On the west coast of Ireland, County Galway has seen 224 villages abandoned in the past 65 years. In the one year of 1986, the three counties of Galway, Mayo and Roscommon saw 10,000 people emigrate (Douthwaite, 1996). In Ireland as a whole, 670,000 people gained their livelihood from the land in the mid 1920s; by the mid 1990s, there were just 150,000, some 14 per cent of the national workforce.

Land abandonment has led to a dramatic decline in farmed areas in many rural parts of Europe as largely self-sufficient rural communities have collapsed or contracted. But there is little data on just how bad the problem is. Portugal records the abandonment of 245,000 hectares of farmland in recent years, and it is known that similar problems are occurring in large parts of southern and upland France, northern Greece, Spain, and Italy (IEEP and WWF, 1994). The impact on the countryside and rural economy can be as significant as if farms were amalgamated. Farmland is no longer positively managed. In Spain this means fewer birds on the *dehesas*; in southern France it means encroachment of scrub and trees, and a much greater incidence of summer fires – in Provence and the Côte d'Azur, some 15,000 hectares burn annually; in the uplands it means terraces slip and are breached, leading to greater soil erosion.

In the Apennines in Italy, 30 to 50 per cent of the population was lost between 1950 and 1980. Massimo Bianchi (1996), of Italy's Forest and Range Management Research Institute, recently described the consequences:

> *...the abandonment of mountain fields, pastures and forests is slowly changing the landscape we know today and which was created over centuries. Wild nature is beginning to prevail... at least in the short term, it results in a loss of biodiversity... as regards tourism and the recreational exploitation of the mountains, the cultivated landscape is much more attractive than the wild landscape.*

Abandonment is also happening in parts of the US. In his description of how the first CSA farms emerged, the community economic development specialist Richard Douthwaite, described what was occurring in Massachusetts:

> *...dairying in Massachusetts, which in the 1930s supplied a large part of New York's requirements by rail, was in rapid decline because it had become so cheap to transport milk 2400 kilometres by road from Wisconsin...Half of Massachusetts' milk producers went out of business between 1980 and 1993... In many cases their land went unsold and is now reverting to forest, while their houses are either occupied as holiday homes for a few weeks a year, or collapsing from neglect and decay.*

The farm population now accounts for only 1.9 per cent of the population as a whole, and only about 7 per cent of the rural population. The USDA classifies rural counties into five categories – farming, mining, manufacturing, persistent poverty and retirement. Only one in seven of the rural population lives in farming counties, while more than half live in retirement or mining counties. The pattern, as with large parts of continental Europe, is one of depopulation in some areas (the young are leaving and farming is losing out), with growth in others (the old are coming in). As a result, the age structure of rural areas is changing substantially. Rural population growth is occurring near metropolitan areas. In the 1980s, 83 per cent of the growth was in counties next to urban areas; most of this growth occurred in retirement counties (Unnevehr, 1993). This means that rural areas are seeing depopulation in purely farming areas and increasing suburbanisation and retirement. Now, almost half of the population lives in the sprawling suburbs. Many are becoming 'edge cities' – homogenous, modern spaces for living with little or no distinctive character (Garreau, 1992).

The Collapse in Farm Jobs

Changing farm size and abandonment have also brought a dramatic decline in the numbers of people employed in agriculture across Europe. Most countries now have unemployment rates higher in rural than in urban regions (Bollman and Bryden, 1997). For the OECD as a whole, there were some 25 million people officially unemployed in 1995, a doubling over the previous two decades (von Meyer, 1997). In the EU–15, the 1997 total was 18 million. Average unemployment rates for whole economies vary greatly – from a low of 3 per cent in Japan to a high of 23 per cent for Spain in 1995. On average, 11 per cent of

Europe's total workforce cannot find a job – for the 15- to 24-year-olds it is about 20 per cent. Within countries, there is also great variation. According to OECD measures, rural communities are defined as those with a population density of less than 150 people per square kilometres. This means that only one third of people live in rural areas which cover 90 per cent of OECD territory.

National shares vary from a low of less than 5 per cent in The Netherlands and Belgium to a high of greater than 50 per cent in Ireland, Greece, Finland, Sweden and Norway. The predominantly rural and often very remote regions tend to have higher rates of unemployment than do urban regions. All rural areas are also characterised by a low representation of rural women in the labour force. The gap between men and women is least pronounced in Sweden and Canada. Most rural areas have also been affected by a steady increase in the age of working people, combined with a decline in opportunities for young people to find work.

What is also true is that farming, once the backbone of rural economies, is no longer the provider of large numbers of jobs. The number of people employed on farms has fallen steadily all century. During the 1980s, there was a 10 per cent fall in total agricultural labour force across the EU, accounting for more than 1.93 million jobs (see Table 6.3). The greatest proportional falls were in Denmark, France and Luxembourg, while Italy saw no overall change and Greece had an increase in employment. But the greatest absolute fall in numbers occurred in France and Spain, which lost some 600,000 and 800,000 jobs respectively between 1980 and 1990. Over the same period, every country, except for Portugal, saw a small increase in non-agricultural employment (such as tourism or landscape services) in rural areas (Post and Terluin, 1997). In many countries, particularly in remoter regions, tourism and related activities account for more jobs than does agriculture. Between 10 and 19 per cent of rural jobs are tourism-related in Austria, Finland, the US and Switzerland, and 5 to 10 per cent in Germany, Canada, France and the UK (Bontron and Lasnier, 1997).

In the UK, there was a fall of some 112,000 in the total labour force from 1981 to 1995, with regular hired workers hit the hardest (MAFF, 1997; Pretty and Howes, 1993; see Table 6.4). The total number of people directly engaged in agriculture is now only about 2 per cent of the total workforce. Some say that these falls will continue, perhaps to as low as 480,000 by 2005 (Baldock and Bishop, 1996).

These differing rates of change in employment and farm structures have left very different densities of labour use per hectare of land. The UK has the second lowest labour force per hectare of agricultural land in the EU – just 2.9 employees per 100 hectares of agricultural land. This contrasts with highs of 21 for Greece and Portugal; 11 to 12 for

Table 6.3 *Changes in total agricultural labour force and numbers of land managers in 12 European countries, 1980–1990 (in thousands)*

	Total labour force 1980	*Total labour force 1990*	*% change between 1980 and 1990*	*Land managers 1980*	*Land managers 1990*	*% change between 1980 and 1990*
Belgium	186	141	–14%	114	84	–26%
Denmark	234	123	–47%	120	80	–23%
Germany	1983	1776	–10%	828	625	–25%
Greece	1841	2022	+ 10%	997	921	–8%
Spain	3436	2839	–17%	1793	1432	–20%
France	2659	1859	–30%	1210	908	–25%
Ireland	468	312	–23%	214	165	–23%
Italy	5300	5287	no change	2760	2574	–7%
Luxembourg	12	9	–25%	4.8	3.6	–25%
Netherlands	302	289	–4%	146	119	–18%
Portugal	1666	1561	–6%	619	569	–8%
UK	723	659	–9%	237	200	–16%
EU–12	18,810	16,878	–11%	9043	7681	–15%

Notes: Spain and Portugal data are for 1987 and 1990; Germany is counted as constituted after 3 October 1990; comparable data for the three most recent EU members, Austria, Finland and Sweden, are not yet available.
Source: Eurostat. 1995. *Agriculture Statistical Yearbook*, Brussels

The Netherlands and Italy; between five and seven for Ireland, Germany and Belgium; four to five for France, Spain and Luxembourg; and the lowest of all for Denmark, with just two jobs per 100 hectares (Eurostat, 1995).

What is now quite clear is that agricultural support through the CAP has not maintained rural employment and farm businesses in many parts of Europe. Unless there are big changes in the way farming operates, any further loss of farms though amalgamation or abandonment would only reduce social and natural capital. Some say that the complete removal of agricultural support would only slightly increase agricultural unemployment (MAFF, 1995). But such a change is not desirable, as it would also reduce the number of agricultural holdings. Farms would go on getting bigger and fewer in number, with an even worse effect on rural landscapes and communities.

Table 6.4 *Changes in labour force on agricultural holdings in the UK, 1981–1995*

Class of worker engaged in agriculture	*1981 number (thousands)*	*1995 number (thousands)*
Total farmers, partners, directors and salaried managers	301.5	289.3
Spouses of farmers, partners and directors doing farm work	76.9	74.9
Regular hired whole and part-time workers	182.2	103.2
Regular family workers	61.9	56.2
Seasonal or casual workers	97.0	84.0
Total labour force (in thousands)	719.5	607.7

Source: MAFF Agricultural Statistics, from www.maff.gov.uk

Growing Rural Deprivation

Rural Poverty and Declining Services

Fewer farms, fewer jobs and larger-scale farming operations have also played a role in the rise of rural poverty and lack of services. A range of UK national inquiries conducted in the 1990s, chaired by the Duke of Westminster (DoW), or led by the House of Lords Select Committee on the European Communities, the Archbishops' Commission on Rural Areas (ACORA), the National Association of Citizens Advice Bureaux (NACAB), and the Rural Development Commission (RDC), have shown that the incidence of rural poverty is considerably greater than previously supposed (DoW, 1992; HL, 1990; ACORA, 1990; NACAB, 1995; RDC, 1994). According to both NACAB and an unpublished study by the then Department of the Environment, about a quarter of rural households are living on or below the official poverty line (DoW, 1992; NACAB, 1995). A study for the Rural Development Commission of 3000 households in 12 areas of England found up to 40 per cent of households falling below the poverty line. These were in central Nottinghamshire, central Devonshire, and north Essex (RDC, 1994). In some low-income areas, more than half of the households had gross salaries of less than £8000 per year.

Often this poverty is hidden. Wiltshire has a successful and wealthy high-technology industry area. Yet income disparities can be huge. According to Action with Communities in Rural England (ACRE), some 30 per cent of rural people in the county were on some form of benefit in 1996, even though unemployment levels are very much lower than the national average (ACRE, 1995). In Devizes, a small market town, just under half of the population was on benefits to top-up income. In one small village of 200 people outside Swindon, a third of households annually earn less than £6000, while another third earn more than £40,000 (Community Council for Wiltshire, 1996). Poverty is now endemic – the number of families under the poverty line in Britain (measured as less than half the average income for the country as a whole) increased from 14.3 per cent in 1983 to 17.2 per cent in 1993. Only four countries (Spain, Portugal, Ireland and Greece) had a higher proportion than Britain. By comparison, only 5 per cent of households were under the poverty line in the Benelux countries.

The gap between the rich and the poor has, however, widened more rapidly in Britain than anywhere else in the EU. In 1979, the poorest tenth of the population earned 4.3 per cent of the wealth; by 1996 this had fallen to just 2.9 per cent. By contrast, the share of the richest tenth rose from 20.6 per cent to more than 26 per cent. Over the same period, the proportion of families with no full-time adult worker rose from 29 per cent to 37 per cent; in 1979 it took an average of one and a half years for someone without work to find a job. In 1993, it took four and a half years. The problem is that much of this poverty and deprivation is missed. Although the 1995 *Rural White Paper* did much to identify common concerns in the countryside, it is worrying to note that much of the evidence to the House of Commons Environment Committee's deliberations following the issue of the *White Paper* focused on rural poverty and low pay, since these matters had been given too little attention in the *White Paper.*

What does this kind of poverty mean for people in rural areas? ACRE put it this way in their 1996 evidence to the House of Commons Select Committee on the Environment:

> *Rural disadvantage, poverty and inequality have become heightened in recent years... In rural England, it is easy to fail to recognise levels of poverty which exist in rural communities. As rural communities are mixed, poverty cannot be identified in specific places... but tends to be hidden both by the apparent prosperity of some counties, and by the process of social exclusion deriving from increased polarisation of income and attitude (ACRE, 1996).*

Endemic poverty means a lack of food for some, particularly amongst older people. It means a lack of affordable housing, few job opportunities and inadequate childcare facilities. Many people are active in the informal economy. Many people cannot travel, lacking access to both public transport and cars. Certain groups, especially women, are much less likely to have access to a car during the daytime. Many people feel left out and marginalised in their own community. Yet all this deprivation tends to be invisible – the poverty is 'thatched' and hidden behind the 'rose cottage' (Derounian, 1993).

Declining Services

The decline in rural services has also been marked. The poorest and most deprived live in areas where key services, such as schools, doctors, pharmacies and shops, are concentrated in the larger towns, while people in outlying areas have to travel for these services (see Box 6.1). Four out of ten parishes in rural England have no shop or post office; six out of ten no primary school; and three-quarters no bus service or GP practice.

Village post offices are important to local people, yet they are widely under threat. Six out of seven people on pensions or benefits collect them from their local post office. More than half found the post office vital for keeping in touch with village life. Many people see the post office as a focal point of village cohesion. Yet since 1979, some 2000 have closed and a further 1000 are said to be under threat of closure (Liberal Democrats, 1994). Poor public transport and tight weekly budgets also put banks and regular travel to the nearest town out of reach for most people. Many rural households do not have a car. More than half of women living in rural areas do not have driving licences, rising to three-quarters for women aged 50 and over. There is, however, great regional variation: only 54 per cent of households in south Wales have a private car, yet this rises to 74 per cent in south-east England – and many of those with cars have two or more. For many in the countryside, the car is no longer a luxury that can be given up when times are hard. Over two-thirds of rural journeys are by car, compared with a national average of about a half (WACAB, 1995; National Federation of Women's Institutes, 1996; Rural Viewpoint, 1993; DoE/MAFF, 1995).

But not everyone thinks access to these services is important. A decade and a half ago, *The Economist* accused those who point to low incomes and sparse services in rural areas of indulging in 'the usual bleating of the rural lobby which seems to think country folk should have all the advantages of rural splendour and city convenience' (*The Economist*, 1982). Nevertheless, the political attitude as to whether these things are important does make a difference. In the UK and Belgium, the closure of rural schools has long been seen as inevitable. Small schools were believed to deliver a poor education and to be more costly. In southern Belgium, some 1800 schools were closed in the 1970s alone (Crout, 1984). In

Box 6.1 Services in rural parishes and status of households in England

Of rural parishes:

- 13% had no bus service whatsoever*;
- 30% had no village hall, community centre, public house or daily bus service;
- 40–45% had no permanent shop or post office;
- 50–60% had no school** or church or chapel with resident minister;
- 83% had no GP;
- 98% had no permanently staffed police station.

Of rural households:

- 8% had no telephone;
- 12% had no car;
- 16% of houses have no central heating;
- 20% of families experience some conflict with neighbours, farmers or other users of the countryside;
- 25% were below the poverty line, rising to 40% in some areas;
- 75% do their shopping at a supermarket or shopping centre in a distant town.

Source: Rural Development Commission survey, reported in DoE/MAFF. 1995. *Rural White Paper: Rural England*. HMSO, London
Notes: * NACAB put the figure at 73 per cent with no bus service.
** 350 rural schools closed between 1983–1995, even though small primary schools in rural areas with fewer than 100 pupils achieve higher standards than pupils in large schools – DOE/MAFF, 1995.

England, schools with fewer than 30 pupils have been found to have 20 to 90 per cent higher unit costs, prompting authorities to conclude that net savings would be made by closing them (Derounian, 1993). But in France, rural planners rejected this approach, recognising that village schools often act as a focus for local people in thinly populated areas, whether they had children attending or not: 'once the school had been closed, that critical point of identity was gone'. Their approach was to value the contribution schools made to social capital, not just base value on economic expenditure.

The Loss of Social Cohesion

The result of this poverty, deprivation and declining services is the gradual unravelling of communities. Although there has been an increase in material affluence, this is not linked to social, cultural and spiritual strength. Since 1950, GNP in Britain has steadily increased year on year to stand at two and a half times the 1950 value in real terms. By contrast, the alternative Index of Sustainable Economic Welfare, which incorporates a wider range of measures into national welfare than just GDP, rose only to the mid 1970s and has since fallen back to the 1950 value (Jackson and Marks, 1994). We may have progressed in monetary terms, but we have lost a huge amount in the process.

Throughout the world, external agencies have routinely undermined social capital in order to encourage economic 'development'. It may not be intentional, but the effect of doing things for people, rather than encouraging or motivating them to act as much as possible for themselves, is substantial. Local management and skills have been lost, leading to increased dependence of local people on formal state and private institutions; local information networks have been replaced by external sources; banks and cooperatives have substituted for local credit arrangements; cooperatives and marketing boards have been replaced by input and product markets; and supermarkets have replaced local shops.

All of this has led to a decreased capacity amongst local people to cope with environmental and economic change. As once thriving communities are plunged into dependency, so people have increasingly lost a sense of belonging to, and commitment for, a particular place. This spiral of decline in local communities has diminished social capital. This is lowest when there is little investment in an area, when there is little participation in community affairs, and when there is high mistrust of external public and private bodies. ACRE have identified 14 key indicators of decline (see Box 6.2).

One recent study of two very different rural communities in the US has illustrated how communities become divided because of changing economic circumstances. In a coal-dependent Appalachian community, Cynthia Duncan and Nita Lamborghini found that the limited job opportunities and volatile industrial sector had significantly contributed to a divided community (Duncan and Lamborghini, 1994). People turned against their neighbours as stress increased. But in a northern New England community, where there were better opportunities for work, the social context was more inclusive. Importantly, local people invested more in public goods, doing things for the benefit of the community rather than just for themselves. The researchers suggested that there was a 'tipping point' for social capital – a threshold at which a community slides away from more supportive and inclusive relations and institutions. They concluded that 'policies and programmes that encourage cooperation and inclusive community institutions may be of growing importance to prevent decline in the social fabric that plagues inner cities'.

Box 6.2 Indicators of rural deprivation

- lack of local service and facilities, eg medical, telephones, banks, post offices, job centres and pubs;
- high cost of living, eg petrol, food prices in shops;
- lack of public transport for access to employment, shops;
- lack of low-cost housing;
- distance from sources of information, eg advice and counselling centres;
- low income and often seasonal jobs;
- lack of jobs, especially for women without transport;
- limited adult education and vocational training;
- inadequate social facilities, eg sports, meeting places;
- lack of services for certain groups, eg disabled, elderly;
- lack of political influence, eg small voter base;
- lack of control over resources;
- stigma attached to certain groups, eg single parent families, unemployed;
- lack of anonymity when visiting personal services, eg GPs.

Source: ACRE, in Derounian J G. 1993. *Another Country: Life Beyond the Rose Cottage.* NCVO Publications, London

Community Continuity and the Younger Generation

What does this mean for the continuity of communities? Continuity of individual farms is held dear by many farm families. Most wish to see their children as successors. Yet, the evidence suggests that few will see this happen. Since the late 1960s, the proportion of British farmers planning to pass their businesses to their heirs has fallen from about three-quarters to less than a half (Ward, 1996; Pugh, 1996). Succession is least likely for farmers in the less prosperous areas. Farming's declining economic fortunes seem to have eroded the commitment of successors in family farming; the prospect of a farming career appears to have become less attractive to farm children.

In the EU as a whole, there are more than four million farmers over the age of 55 – more than half of the total (see Table 6.5). A quarter are now past retirement age, and only 8 per cent are under 35 years of age. The problem is equally acute elsewhere in the industrialised world. In the US, half of farmland is operated by farmers within ten years of retirement. In Japan, 43 per cent of farmers were over 65 years of age in 1995. These changes are alarming: a recent Iowa State University survey found that farmers younger than 40 are much more likely to experiment with sustainable agriculture than are those over 60 (Vorley and Keeney, 1998).

Table 6.5 *Age structure of the land manager population in Europe, 1990 (in thousands)*

Land managers	Less than 35 years (000)	35–54 years (000)	55–64 years (000)	More than 65 years (000)	% over 55 years of age
All European Union (12)	620	2930	2270	1870	54%
France	120	387	276	125	44%
Germany	99	311	174	41	34%
Spain	106	545	457	323	55%
Ireland	22	70	37	36	44%
Italy	134	875	748	817	61%
Netherlands	11	57	33	17	42%
UK	15	89	53	43	48%

Source: Eurostat. 1995. *Agriculture Statistical Yearbook.* Brussels

There is also a growing dislocation between generations, with the young perhaps never knowing or appreciating what a sense of community may feel like. Almost every rural area in Europe is getting older. The young have fewer opportunities, and so get less involved in community life. This is part of a wider trend in all age groups. Many now feel that there has been a significant downward spiral in social trust and civic engagement in virtually all areas, whether cities, suburbs, small towns or the countryside.

Time budget studies of average Americans show that since the mid 1960s, there has been a 25 per cent fall in the time spent socialising and visiting, and a 50 per cent fall in the time devoted to local clubs and organisations. People are spending less and less time with their neighbours and friends in their community. Membership records seem to confirm these trends, with a 25 to 50 per cent fall in membership of parent–teacher associations, trade unions, bowling leagues and so on over the past two to three decades (Putnam, 1995). Age is a key predictor of civic engagement. As Robert Putnam put it: 'older people are consistently more engaged and trusting than younger people, yet we do not become more engaged and trusting as we age'. There is an important cohort of people born between 1910 and 1940 who vote more, join more, read newspapers more, and trust more. There are also important gender differences: seven out of ten volunteers in UK rural communities are women (Derounian, 1993).

In some regions, the exodus of rural youth has virtually become an epidemic. In Canada, the traditional values of small communities can seem backward to young people. Big cities appear more exciting. David

Hajesz and Shirley Dawe (1997), community economic-development specialists from Ottawa, report on a 1994 *New York Times* article about young people leaving Great Plains communities. Following a school graduation dinner in a small town, the 13 graduates were asked to stand and share their plans for the future: 'Twelve proudly described their plans to leave and attend college. One young man stared at the floor and apologised that he would "just be staying here to farm with (his) dad"'. Similar changes are occurring in Canada, though Hajesz and Dawe found that young people in small rural and coastal communities in Newfoundland put a high value on their communities and natural environment. They said they liked the 'small-town feeling'; it was 'close to everything'; and 'everyone was friendly'. Many wanted to set up their own businesses, saying: 'it would be good for the town', and 'I want to help out'. Despite these values put on social capital, they would still have to leave if job opportunities did not emerge.

Geoff Mulgan and Helen Wilkinson's (1995) study for Demos identified a growing band of 'underwolves' amongst young people in Britain aged 18 to 24. These people had been given 'unprecedented freedoms but [are] living in an increasingly unstable environment, with chronic relationship breakdown, uncertainty about jobs and profound disconnection from the political process'. Action is clearly needed to build new rules of commitment for young people at home, in the workplace – if they have one – in the school, and in civil society at large.

Young people feel real pressures in rural communities. James Garo Derounian (1993) indicates that 'the goldfish bowl of small village life makes everything so public; friendship patterns are so narrow and unorthodox dress behaviour tends to be frowned upon'. He quotes a community development worker from Norfolk, Jerry Smith, who said: 'boredom is a significant factor for rural youth...in many villages, there is just nothing for young people to do – and how do you participate, with nothing to participate in?'

Stress in the Countryside

Another result of the decline of social capital arising from changes in farming is a marked increase in mental distress amongst farmers. In Britain, the Office of Population and Census Surveys has shown that farmers and farmworkers are two and a half times more likely to commit suicide than is the rest of the population, and that suicide is the second most common form of death for farmers and agricultural workers aged 15 to 44 years. During the ten years spanning the end of the 1980s and early 1990s, some 600 farmers and farmworkers took their own lives.

The stress and deteriorating confidence are associated with the growing loneliness of the occupation and the social isolation felt by farmers. The highest rates of suicide occur in the most remote areas, such as in central Wales. The Duke of Westminster's 1992 report on rural poverty described these problems in this way:

> *Hidden in the rural landscape, which the British so much love, people are suffering poverty, housing problems, unemployment, deprivation of various kinds, and misery. Traditional patterns of rural life are changing fast, causing worry, shame and distress. Those most affected are often angry and bitter but feel they have little chance of being heard. The suicide rate is very high. Neither the public nor the private sector is showing any signs of caring very much about all this.*

In France, suicide rates amongst both men and women are highest in rural communities, and fall steadily with increasing size of village and then town (Philippe, 1974). This is the exact opposite of the reported situation at the turn of the century – with the highest rates recorded in the cities. In the mid-west of the US, suicide rates amongst male farmers are twice the national average. Some 913 took their own lives between 1980 and 1988, producing annual rates higher than for any other documented occupation.

Research by sociologists from Iowa State University in the US has shown that there are greater levels of stress amongst people of both poorer and depopulated areas (Hoyt et al, 1995). People are more likely to suffer stress if they have low levels of social support from their own community. According to Danny Hoyt and colleagues: 'regional economic and social trends have important mental health consequences for rural communities'. As communities become economically depressed and people emigrate, so those who remain become more prone to mental ill-health.

In the mid-west farming country, community attachment to, and social integration into, local activities were identified as vital in another study by researchers at the University of Missouri. When people felt little attachment, mental health problems and depression were more likely (O'Brien et al, 1994). Local people know that social capital makes a difference to their lives. As David O'Brien and colleagues put it: 'rural residents' concerns with preserving local community are not merely a reflection of romantic visions, but rather are rooted in measurable effects of community attachment on personal well-being'. This is also true in France and New Zealand. Dorothy Fairburn's recent study of rural stress found that farmers suffering the least stress were those with the most connectivity with the local community. In New Zealand, 'those who are the most healthy are those who get off the farm and become involved in community activities'. This was most often in the form of relations with the village school and fundraising for the community. In France, the strong connections with extended families were vital, along with civic institutions and farmers' cooperatives.

The greater responsibility passed down to communes for managing public services helps to involve more people in community life. France has 36,000 maires (mayors), of whom 32,000 preside over 'communes' of fewer than 2000 people (Derounian, 1993). They are complemented by 150,000 deputies and 500,000 councillors, producing a very strong network of civic engagement. This contrasts with 8159 parish councils in England with 70,000 elected representatives in all. Responsibility in France is greater too. James Garo Derounian quotes the case of the small community of Ambierle near St Etienne: it maintains 150 kilometres of roads, public buildings such as the church, school and village hall, and is responsible for water supply and assisting with local education. In UK terms, the financial clout of these small communities is extraordinary: in the early 1990s, Ambierle operated on an annual budget of £505,000.

Whose Rural Idyll?

Urban and Suburban Desires

Despite all these economic and social problems in rural areas, large numbers of people want to live there. Surveys in the UK by the Henley Centre for Forecasting and by Mintel indicate that nearly half of those who live in cities want to live in a village or country town; about a quarter expect to move out in the next decade. Of those who wanted to move, 45 per cent mentioned the appeal of open spaces; 50 per cent that cities were noisy and dirty; and 20 per cent that rural life would be less stressful (Mintel, 1992). Some 60 per cent of people say the countryside is a vital element in their quality of life, and many visit for day trips or longer holidays. They make a very important contribution to the rural economy. Urban people make about 660 million day visits to the countryside each year, each spending some £4 to £8 per trip (Countryside Commission, 1994). This exceeds the amount of public money paid to farmers under the CAP.

The number of people in rural areas is increasing, though it tends to be younger people who are emigrating and being replaced by older, particularly retired, new entrants (this is quite the opposite elsewhere in Europe – in many areas of France and Spain, the whole fabric of rural life is at threat). Between 1981 and 1991, the population of England and Wales largely remained the same, but it fell in urban areas (by 5 per cent in London, by 7.4 per cent in the larger cities, by 3.6 per cent in smaller cities) and rose in rural areas (by 6.4 per cent in remote rural areas, by 4.1 per cent in accessible rural areas, and by 2.9 per cent in the south-east). This growth in rural areas will continue – forecasts suggest a further growth of 12 per cent by 2025 in all rural areas (DoE/MAFF, 1995). Most will be in search of the rural idyll.

The key questions are: where will all these people go and what will they do? Some will continue to work in cities, commuting by car or public transport. This may turn yet more communities into ghost towns or dormitory only villages, where social capital is by definition low. Many will need new houses. In rural areas, there are already some 377,000 people in need of low-cost housing, 10,000 of whom are homeless – 12 per cent of the national total (CLA, 1995). Shifting family sizes as well as urban to country migration has led the government to estimate that another three to four million houses will be needed in the next 20 years. Much of these will be built on previously developed land, but this will still leave a large number allocated to 'green field' sites. Most of these people will be looking for a 'traditional' country life in some form or other. Incomers, once they are there, are more likely to object to new developments which are intended to supply housing or contribute to jobs. They show not only Nimby attitudes (not in my back yard), but also Nodam (no development after mine) (Derounian, 1998). They are also less likely to understand farming and its positive contribution to the landscape and to natural capital. As one farmer, Sir Julian Rose, put it: 'they will not expect to find a rural and cultural desert dominated by theme parks, golf courses and hypermarkets. But that is exactly what they will find if agriculture continues to be regarded only as a tool for constantly increasing the "efficient" production of food, in which human input is regarded as an undesirable expense' (Rose, 1996). Today few rural communities are bonded by a common understanding and economic interest in the land.

Edge Cities – Places with No History

The extraordinary phenomenon of Edge Cities in the US has brought into sharp relief many of these issues relating to social cohesion in rural areas. Described in great detail by Joel Garreau in his book *Edge Cities*, they began to emerge in the 1980s as places designed to provide modern working, living and shopping space (Garreau, 1992). They represent the third phase in population movements in and around cities. People first moved out to live in the suburbs. They were then followed by shopping and leisure facilities, which set up larger and more concentrated operations on green field sites. And, finally, they have been joined by the means for creating wealth. Now, two-thirds of all office space in the US is in Edge Cities. People no longer go on the old-fashioned commute from the suburbs to downtown. They travel by car, and skirt around the old centres.

Edge Cities are designed to be very efficient. They accommodate different groups, provide child care, offer safety and have a great diversity of shops and corporate opportunities. They are characterised by high-rise buildings and shopping plazas or malls. The new monument – no longer a cathedral – is the atrium, a climate-controlled environment for the shopping experience. They are tied together by roads and satellite dishes. But the worlds created are 'in-between'. Many have no name. There are 20 major Edge Cities around New York alone, each of which is bigger than Memphis. In New Jersey, half-way between the ocean and the Pennsylvania border, is '287 and 78', a city named after two roads (or highways). There are no welcoming signs, as the cities have no beginning and no end.

But these new worlds created by 'master planners' have neither community nor history. Indeed, Francis Fukuyama would describe them as emerging at the 'end of history'. They are not places that result from the interactions between different groups and institutions. They are designed from above without people being involved. Garreau paraphrased the approach like this: 'we're going to build this thing that is perfect for you. We haven't met you but we know what you're like and we know you're going to like it here.' People do find Edge Cities convenient and easy to work in. But they do not like them. There is no civic structure or sense of community. Joel Garreau describes how he asked people at Tyson's Corner in Virginia, an Edge City that recently sprung up out of farmland, about what they thought of their 'brave new world': 'the words I recorded were searing. They described the area as plastic, a hodgepodge, Disneyland (used as a pejorative) and sterile. They said it lacked livability, civilisation, community, neighbourhood and even a soul. These responses are frightening.'

Southern California, in the 100-kilometre radius around Los Angeles, would be the eleventh richest country in the world if it were a separate state. It has 26 fully developed Edge Cities in five counties. Half a million new jobs were created in the 1980s, and only 4 per cent of the region's jobs are now in downtown Los Angeles. There are five major airports and 133 colleges or universities. Irvine was originally a Spanish ranch of 250 square kilometres in Orange County. It now has three Edge Cities, with a total population of about 200,000. The tourist guides describe Orange County as 'a theme park – and the theme is you can have anything you want. It's the most Californian-looking of all California: the most like the movies, the most like the stories, the most like the dreams.' The danger of the dream, however, is that a place that reminds people of Eden can also taunt them. The homes in Irvine are more repetitive than in the old city suburbs. They are so perfect, why should anyone want to individualise them? Irvine authorities have deed restrictions that forbid people from customising their homes. They have become Le Corbusier's machine for modern living, once again repeating the modernist approach that objected to any personalising of houses. As David Harvey put it: 'public housing tenants were not allowed to modify their environments to meet personal needs, and the students living in Le Corbusier's Pavillon Suisse had to fry every summer because the architect refused, for aesthetic reasons, to let blinds be installed' (Harvey, 1989).

Nevertheless, the developers still sell this as a community, a place where there is a 'lively interplay between commerce and the arts, between nature and technology, between work and leisure... Here is a place where individuals are free to shape their own future, as a sculptor shapes a city.' It is the individuals who are seen to be the shapers – not civil society in all its diversity. There are no mayors, no city councils and very little public participation. Compare this with Walter Goldschmidt's 1946 study of two communities in the San Joaquin valley, and how social capital was created. There is some hope, however, where people are beginning to create their own local associations and groups, as they would anywhere else. But one thing they cannot do is to recreate history. Clearly, huge Edge Cities will never emerge in the much denser countryside of the UK and continental Europe. But edge villages and edge estates already exist.

Policies for Rural Development

The Dominant Exogenous Model

There are two competing schools of thought about the best way for authorities to promote development in rural areas. One seeks to build on and develop existing resources and capacity – so-called 'endogenous' development (see Chapter 7). The other relies on external investments and capital to help modernise the countryside and therefore implies competing with other recipients for mobile investment and professionals. The exogenous model is currently predominant. It says that we, in industrialised countries, must look to other sectors of the economy to provide for new jobs and wealth creation. Farming is seen, inevitably, to lose labour as countries get richer. It also contributes proportionally less and less over time to GDP. Farming has become more mechanised and modern at the same time as countries have become richer, and the way for them to continue to get richer is to encourage the same process and trends. This school of thought implies that agriculture is best left to its own devices. Agriculture is for producing food. We are very good at it now – but it clearly has no major role in rural economic development. Where inward investment is to be encouraged, it is best targeted at alternatives to agriculture and natural resource-based activities. Golf courses, high-technology industry, craft-based rural industry, telecottaging, and industrial relocation can all bring jobs; but few have much to do with agriculture.

Will any of these options make a difference? Some believe not. Even if all the demand for golf courses were met, this would only mean reallocating some 30,000 hectares of agricultural land. And there is only a limited amount of growth in rural areas that mobile capital can create. There cannot, effectively, be more than one Silicon Valley for Britain. There are a limited number of urban-based manufacturing and service businesses that can or wish to relocate to rural communities.

Alan Whitehead and Judith Smyth (1996) call this exogenous approach 'civic boosterism'. Authorities offer tax incentives to employers and investors, provide leisure facilities for mobile, high-earning professionals, and support prestige projects. Yet, as they point out, while some communities can do well, all communities in a country cannot follow this approach. The first movers benefit, and this simply inflates the price that the most fashionable investors can demand for relocation. It also 'assumes that citizens have no loyalty or commitment to their city... Social capital is what makes a city livable, lived-in and lively.' Nonetheless, many local authorities and development agencies offer attractive incentives to encourage businesses to set up in their areas. This is what Americans call 'smokestack chasing'. At first this can look good, as extra jobs are created for local people when new factories or offices begin to operate. But all too often, the net effect is simply to move jobs around the country – from areas where businesses do not receive subsidies to those where they will. Take the case of a new factory announced in South Wales in 1996. The company was offered inducements by the Welsh Development Agency since it would bring many new jobs. So it simply relocated its operations from Hull and Peterborough and closed factories. The net result was little change in jobs since they were simply transferred from one part of the country to another – except now they are publicly subsidised.

Another problem is that many of these new operations are creating a new low-wage culture. The Rural Development Commission has estimated that 30 per cent of people in rural areas earn less than £180 per week (Temperley, 1996). Workers in Wales employed by South Korean companies are now paid less than workers in South Korea (*Independent on Sunday*, 1996). A 1996 study by the Low Pay Unit highlighted the low pay culture prevalent in South Wales: some 30 per cent of men and 60 per cent of women earn less than the Council of Europe decency threshold of £6.03 per hour (a sum calculated to provide a decent standard of living). Rates of pay of £2.20 to £2.70 per hour were common. In recent years, an increasing number of firms from Taiwan, South Korea and Japan have relocated to Wales, producing several thousand jobs. Thirty years ago, these countries had the reputation for being low-pay and long-hour cultures. Now the situation has reversed.

The need for rural alternatives was clearly identified by the Intergovernmental Consultation on Sustainable Mountain Development in Aviemore, Scotland. European mountain communities are characterised by major demographic, economic and social instability, with a crisis of unemployment, disrupted communities and growing uncertainties over traditional achievements. To prevent emigration, there is a need for creating employment and livelihood opportunities 'from within', making the best use of locally managed and owned resources.

National and EU Programmes for Rural Development

Most rural development policies have adopted the exogenous model, leaving farming to agricultural policies and ministries. Throughout Europe, economic development support is heavily biased towards urban and industrial regeneration.* Of the EC structural funds allocated for 1994–1999 (some £117 billion), only a tenth are for agricultural and rural programmes. In the UK, urban development corporations provided £364.2 million of support in towns and cities in 1992–1994, but only about £24 million went to the Rural Development Commission in the same period (DoE, 1994).

The Rural Development Commission targets expenditure on designated areas of rural England with the most deprivation. There are also

* Five priority areas were defined in EC Regulation 2052/88, amended by 2081/93: *Objective 1:* economic adaptation of the most marginalised rural areas in the EU, characterised by rural decline and depopulation, limited potential for economic diversification, and the high cost of developing basic infrastructure. Objective 1 regions are the whole of Greece, Ireland and Portugal; many parts of Spain; the Italian Mezzogiorno; Corsica; the French overseas departments; the districts of Douai, Valenciennes and Avesnes in France; Northern Ireland; Merseyside in England; the Highlands and Islands of Scotland; Hainault in Belgium; Flevoland in The Netherlands; and the five new German Länder;
Objective 2: economic reconversion of regions affected by industrial decline;
Objective 3: combating long-term unemployment;
Objective 4: adaptation of workers to industrial changes and preventative anti-unemployment measures;
Objective 5a: adjustment of processing and marketing for agriculture and fisheries;
Objective 5b: economic diversification of rural areas.

rural community councils, one for each county and supported by the RDC. These play a vital role in facilitating local development, linking different agencies and institutions within and between villages, and running projects of their own.

The Single Regeneration Budget (SRB) is the UK government's main instrument for encouraging local regeneration, though again mostly in urban areas. In the first round (starting in April 1995), some 40 schemes with a rural component were approved. The SRB combines 20 formerly separate programmes from five government departments into a single fund to provide for integrated and local regeneration. The 1995–1996 budget of £1.33 billion covers four areas: funding for the Urban Development Corporations, English Partnerships, and Housing Action Trusts; SRB bids from local partnerships; support for the Estate Action, City Challenge, Urban Programme and Inner City Task Forces; and grants for Business Start-up, local initiatives funds and regional enterprise funds. Resources are also available for rural areas from EC structural funds. Under the Objective 5b schemes, specific rural areas receive help to improve development and structural adjustment. Partnership is said to be a key feature, encouraging links between government agencies, training and enterprise councils (TECs), local authorities, environment bodies, and the private and voluntary sectors. For eligibility, rural areas must have a low level of agricultural income, low population density or significant depopulation, a high percentage of employees in the agricultural sector, and a low GDP per person. Approved areas receive partial funding (not more than 50 per cent) for agreed rural development programmes, and governments must match this with their own funds.* These are complemented with LEADER I and II support, which focuses on innovative structures and building the skills of local groups.

The persistent problem with these national and EU schemes for rural development is that they are still piecemeal. Not every part of rural Europe is covered, and in some cases there is considerable overlap – such as between Objective 5b areas and Rural Development Areas. Other criticisms are that bureaucracies are deliberately slow to agree to projects since this means that they have to allocate matching funds. Worse still, many programmes have failed to deliver. The 1996 Social Justice Commission chaired by Lord Borrie put it this way:

* In 1988–1993, only Devon and Cornwall were in a scheme, receiving £66 million. In 1994–1999, six areas (Devon and Cornwall, parts of East Anglia, Wales, Lincolnshire and the Fens, the Midland Uplands, English Northern Uplands, Dumfries and Galloway, Borders and the Scottish Southern Uplands) with a population of 1.7 million people, were allocated £410 million.

> *One reason for the repeated disappointment of policy initiatives over the last 15 years is that they have failed to understand the importance of social capital. As a government-sponsored report found out, the £10 billion spent on property-led regeneration in inner cities during the 1980s has largely been wasted. Instead the aim must be to build amongst local people the capacities and institutions which enable them to take more responsibility for shaping their own futures (Social Justice Report, quoted in Conaty, 1997).*

Summary

The 20th century's period of remarkably successful agricultural growth brought great social change in rural areas throughout Europe. In the quest for greater food production, landscapes, rural livelihoods and farming systems have all been progressively simplified. In every European country, farms have become both fewer in number and larger during this century. When the CAP was established in 1957, the six member countries had 22 million farmers. Today that number has fallen to about seven million.

These changes have left very different landholding structures in each European country. The average farm size across the EU is about 15 hectares. Only 6 per cent of farms are more than 50 hectares in size; and 60 per cent are smaller than five hectares. Many argue that progressively larger farms are an economic necessity. They permit economies of scale to be made and they enable more efficient producers to take over or absorb the operations of smaller and more 'inefficient' producers. But small farms provide other benefits to society. They employ more people than do larger farms. Small farms contribute to both rural social capital and natural capital. Their greater on-farm diversity maintains both plant and animal biodiversity; they are more efficient users of energy; they are better at preserving and enhancing landscape and wildlife; and they tend to have a better record with animal welfare. But large farms receive most from public subsidies.

A different but equally damaging problem is that of land abandonment. The regions most prone to population decline have been rural Ireland, parts of Scotland and Wales, central France, most of Italy and Greece, and Spain and Portugal. Generally there has been very little population decline in England, Germany, Belgium and The Netherlands. There are at least 56 million hectares of low-intensity farmland of high biodiversity value in Europe. Their value is maintained by the practice of farming – and when farmers leave, the value declines significantly.

Changing farm size and abandonment have also brought a dramatic decline in the numbers of people employed in agriculture. Most countries now have unemployment rates higher in rural than in urban regions. For the OECD as a whole, there were some 25 million people officially unemployed in 1995, a doubling over the previous two decades. Farming, once the backbone of rural economies, is no longer the provider of large numbers of jobs. During the 1980s, there was a 10 per cent fall in total agricultural labour force across the EU, accounting for more than 1.93 million jobs. In many countries, particularly in remoter regions; tourism and related activities account for more jobs than does agriculture. Between 10 and 19 per cent of rural jobs are tourism-related in Austria, Finland, the US and Switzerland, and 5 to 10 per cent in Germany, Canada, France and the UK.

Fewer farms, fewer jobs and larger-scale farming operations have also played a role in the rise of rural poverty and the lack of services. A quarter of rural households in the UK are living on or below the official poverty line. Endemic poverty means a lack of food for some, particularly amongst older people. It means a lack of affordable housing, few job opportunities and inadequate childcare facilities. Many people are active in the informal economy. Many people cannot travel, lacking access to both public transport and cars. Certain groups, especially women, are much less likely to have access to a car during the daytime. Yet all this deprivation tends to be invisible – the poverty is 'thatched' and hidden behind the 'rose cottage'.

The decline in rural services has also been marked. The poorest and most deprived live in areas where key services, such as schools, doctors, pharmacies and shops, are concentrated in the larger towns, while people in outlying areas have to travel for these services. Four out of ten parishes in rural England have no shop or post office; six out of ten no primary school; and three-quarters no GP practice. Yet the political attitude as to whether these things are important does make a difference. In the UK and Belgium, the closure of rural schools has long been seen as inevitable. But in France, rural planners rejected this approach, recognising that village schools often act as a focus for local people in thinly populated areas, whether they had children attending or not. Their approach was to value the contribution schools make to social capital, not just base value on economic expenditure. The result of this poverty, deprivation and declining services is the gradual unravelling of communities. Although there has been an increase in material affluence, this is not linked to social, cultural and spiritual strength. Throughout the world, external agencies have routinely undermined social capital in order to encourage economic 'development'. It may not be intentional, but the effect of doing things for people, rather than encouraging or motivating them to act as much as possible for themselves, is substantial.

All of this has led to a decreased capacity amongst local people to cope with environmental and economic change. As once thriving communities are plunged into dependency, so people have increasingly lost a sense of belonging to, and commitment for, a particular place. This spiral of decline in local communities has diminished social capital. Continuity of individual farms is held dear by many farm families. Most wish to see their children as successors. Yet, in the EU as a whole, there are more than four million farmers over the age of 55. A quarter are now past retirement age, and only 8 per cent are under 35 years of age. There is also a growing dislocation between generations, with the young perhaps never appreciating what a sense of community may feel like. Almost every rural area in Europe is getting older. The young have fewer opportunities and so get less involved in community life. In some regions, the exodus of rural youth has virtually become an epidemic.

Another result of the decline of social capital arising from changes in farming is a marked increase in mental distress amongst farmers. In Britain, farmers and farmworkers are two and a half times more likely to commit suicide than is the rest of the population. In France, suicide rates amongst both men and women are highest in rural communities and fall steadily with increasing size of village and then town. This is the exact opposite of the reported situation at the turn of the century – with the highest rates recorded in the cities. According to research in the US, as communities become economically depressed and people emigrate, so those who remain become more prone to mental ill-health.

Despite all of these economic and social problems in rural areas, large numbers of people want to live there. Surveys in the UK indicate that nearly half of those who live in cities want to live in a village or country town; and about a quarter expect to move out in the next decade. Most of these people will be looking for a 'traditional' country life in some form of other. Incomers, once they are there, are more likely to object to new developments which are intended to supply housing or contribute to jobs. They are also less likely to understand farming and its positive contribution to the landscape and to natural capital.

There are two competing schools of thought about the best way for authorities to promote development in rural areas. One seeks to build on and develop existing resources and capacity – so-called 'endogenous' development. The other relies on external investments and capital to help 'modernise' the countryside. The exogenous model is currently predominant. It says that we, in industrialised countries, must look to other sectors of the economy to provide for new jobs and wealth creation. Farming is seen, inevitably, to lose labour as countries get richer.

Nonetheless, many local authorities and development agencies offer attractive incentives to encourage businesses to set up in their areas. At first this can look good, as extra jobs are created for local people when new factories or offices begin to operate. But, all too often, the net effect is simply to move jobs around the country – from areas where businesses do not receive subsidies to those where they will. Most rural development policies have adopted the exogenous model, leaving farming to agricultural policies and ministries. If we are to see sustainable rural development, there will be a need for agencies and policies to become much more oriented towards endogenous approaches and processes that build on, and make the best of, the existing social resources in rural areas.

Chapter 7

Participation and Partnerships for Community Regeneration

Many of the answers must lie in the hands of local people seeking local solutions to local problems.

Rural England (1995)
DoE/MAFF Government White Paper

Endogenous Development Patterns for Europe

Why Endogenous?

The dominant pattern of rural development in both the South and North has been 'exogenous': the aim is to attract external capital, technologies or institutions into rural areas in order to promote change. An alternative school of thought focuses on 'endogenous' patterns of development, which implies 'growing or originating from within' (Oxford English Dictionary, 1976). The priority is to look, first, at what natural and social resources are available in rural areas – agriculture, people, natural resources and wildlife – and then to ask: can anything be done differently that results in the more productive use of these available resources without causing damage to natural and social capital?

Such a change in emphasis will be difficult since it means operating in a very different way. The current biases of institutions, policies and funding are against 'endogenous' development. This is true for local people too. When asked what they need, a common response is to turn to external solutions: 'we are waiting for the government to solve our problems'; 'we need a change in exchange or interest rates to give us more money'; 'we need more tractors and fertilisers'; 'we need the local council to fix the potholes in our road'; 'we need a new factory to relocate to bring us new jobs'. In the US, this external dependency is often called 'smokestack chasing'. When consulted, communities at first see manufacturing industry as the only means for local economic growth. But not all areas can attract factories and, even if they are successful, they are just as likely to lose them again to another location when conditions change.

These attitudes foster the dependency deadlock. Local people become entirely dependent on external agencies and actors to provide solutions to local problems. Again, this may appear reasonable. If you pay your taxes, why should you not expect the local council to fix the road or to supply other services? But there are always things that local people can do better – they know local conditions. They will be living there long after external bodies have departed; they just may not realise their own capabilities.

Community Regeneration in 'Underdeveloped' Europe

Most remote and peripheral parts of Europe, such as the Western Isles of Scotland, or the mountains of Spain, Portugal and Italy, or the wild landscapes of northern Greece, or the forests and tundra of the Saami Laplanders in northern Sweden and Finland, are subject to severe economic and social stresses. These areas have suffered chronic loss of population this century, with those who remain steadily ageing. Young people are no longer interested in farming or making a living in rural areas and are losing touch with local resources and traditions. Communities tend to be economically poor, fragile and increasingly threatened by distant development plans and patterns.

All are characterised by the endogenous development myth, which suggests that the best way to deal the problems of poverty, environmental degradation and emigration is to turn to external solutions. This means that either local people move away or they hope that external technologies, investments and infrastructure will somehow miraculously appear. Both are largely misguided and inefficient. Many remote and culturally diverse areas are poor precisely because external solutions have failed to work. They are unlikely to suddenly start working in the future. What is needed are processes to raise the awareness and motivation of local people, so that they can do something different to lift themselves out of poverty while preserving their locally unique aspects of environment and culture. The key elements of this process are participation, working together, education and awareness raising. The following four cases of community regeneration show just how endogenous development can work, despite all of the adversity.

Husa community, Jämtland Province, Sweden

Husa is a remote village in the central province of Jämtland, Sweden. It is an example of a community that refused to die when the external solutions finally stopped (Pearce, 1996). Success came through self-help and mobilisation over two decades and has led to both economic growth and an increase in the number of families making a local living. The contribution of the process to social capital is very significant.

In the 1700 and 1800s, Husa was the centre of a copper mining industry and was the largest town in the province. Since the 1880s, Husa has been in steady decline, from 900 people at its peak to just 90 in 1979. The village shop closed and the school was under threat. In 1979, a local artist helped the community to write and present a play about the village. This was the spark that ignited people's motivation to tackle their own problems.

Since 1979, 15 village associations have been formed, as well as new businesses for tourism and skiing. Holiday homes and local housing have been built by two cooperatives; another has established and run a sawmill; yet another has converted the old mine-owner's house into a museum and restaurant; and another group was formed for roads maintenance in the winter. There are many other individual enterprises, including goat farming, a local bakery, pony-trekking, and holiday chalets. In 1990, an umbrella village association emerged as local people came to appreciate the need for greater coordination and strategic planning. By the mid 1990s, Husa had a population of 160, of whom 40 are children of nursery and primary school age.

Cilento, near Naples, Italy

Olive trees have been cultivated for at least two millennia on the Mediterranean coast, contributing to local livelihoods and producing rich and varied habitats for wildlife. But over recent decades, Cilento has suffered from mass emigration, with the young no longer wanting to be olive farmers. Combined with competition from Tunisia, where production costs are lower, pressures on the system are growing.

The CADISPA project, coordinated by Geoff Fagan at the University of Strathclyde, Glascow, works closely with local groups in a range of countries to support local regeneration. In Cilento, CADISPA–Italy began working with a local olive oil cooperative, Nuovo Cilento, to introduce organic farming and new marketing methods. Now 130 farmers located in the national park of Cilento are fully organic, using a wide range of resource-conserving practices to minimise input use and to recycle valuable products, such as using olive husks for fertilisers. They now produce Cilento *verde*, an extra virgin organic oil of high value.

The president of Nuovo Cilento, Peppino Cilento, says: 'once again, olive trees are important in the culture and history of the community. We hope to rescue our land; we don't want it to be forgotten.' Since the successful regeneration of olive production, cooperative businesses have been set up for wild chestnut flour production and for ecotourism. These new ventures are largely run by young people who are increasingly opting to stay in Cilento and use their abilities and skills to develop high-quality local goods and services.

The Aragón Region of Pyrénées, Spain

For 500 years, Aragón was a rich sheep, wool and weaving region. But local crafts could not compete with mass production, and the last weaving mill closed in 1950. The future of the local Pyrennean culture is now at great risk. People prefer to leave in search of employment, and traditions are being lost. As a result, the old communal economy came close to breaking down forever. A local resident of Biescas, Maximo Palacio Allue, says:

> *...the future of Pyrennean culture is lost if we cannot lead a modern life and conserve traditions at the same time. We must educate people and raise their awareness, and do it quickly... Yet people want to be employed, be paid... this means people from outside come in and invest in the area. And we have become used to asking the authorities for everything, and they only make short-term plans without affecting the root of the problem (CADISPA, 1996).*

Various initiatives, including *Por un Pireneo Vivo* (For a Living Pyrénées) and *L'home i la Muntaya* (Man and the Mountain), are producing educational material for children, taking them on field trips to learn about their mountain home, running round table meetings, and organising travelling exhibitions. Another initiative is supporting local weaving using traditional looms and designs; another is renovating housing. Through all these, local people are relearning their traditions and history, with the hope that this will enable them to find new and sustainable livelihood opportunities based on local resources.

Highlands and Islands, Scotland

Highland and Island communities in the west of Scotland have three common characteristics. They are far from centres of commercial or industrial activity, suffer from relatively high unemployment and outward migration, and traditional economic activities no longer provide adequate employment opportunities. CADISPA works with local communities to create economic ventures which use and preserve the special economic, cultural and environmental characteristics of their islands. Small-scale ventures are seen to be best as they are less dependent on external resources and therefore are more likely to persist. On the island of Tiree, a range of community groups, including farming, school, village hall, dance and business, have come together to develop a common Tiree Development Plan, which they hope will more effectively draw in European money from Objective 1 funds for local projects. These include plans for a dance studio, lecture hall, swimming pool and the local school, and redevelopment of an old mill into a Gaelic arts centre. As shown in Table 6.2, Tiree is fortunate to have a small farm and community structure that provides the basic social capital for community regeneration.

Another example of the success of local organisation is the Assynt Crofters Trust. This was formed by 120 crofters in 13 townships after recognising that land was being heavily overgrazed when under private, especially company, ownership. The trust bought 8500 hectares of north-west Scotland and is working together as a whole group on regenerating natural resources and strengthening local communities. It is hoping to access grants usually only available to large landowners. It is engaged in tree planting, working on a croft entrance scheme for young families, and finding properties for retired people.

The Importance of Local Participation in Rural Development

Development Without Participation

Rural communities have seen many revolutions throughout history, from the advent of agriculture some 10,000 years ago to the far-reaching 17th- and 19th-century agricultural revolution in Europe, and the massive rural transformations of the later 20th century. Two guiding themes have dominated this century's worldwide transformations (Pretty, 1995a; see Chapter 2 for a review of transformations in Africa, Asia and Latin America). One has been the need for increased food production to meet the requirements of growing numbers of people. The other has been the desire to prevent the degradation of natural resources, perceived to be largely caused by growing numbers of people and their bad practices.

As a result, food production and the amount of land conserved have increased dramatically. But both have been achieved within the framework of modernisation, which is firmly rooted in and driven by the enlightenment tradition of positivist science (Habermas, 1987; Harvey, 1989; Kurokawa, 1991). Scientists and planners identify the problem that needs solving. Rational solutions are proposed, and technologies known to work in a research station or other controlled environments are passed on to rural people and farmers. Central to modernisation is the assumption that technologies are universal and therefore are independent of social context. New technologies are assumed to be better than those from the past and so represent 'progress'. The new and modern displace the old and traditional. This iconography is powerful and usually implies that what worked before is not as good as what we have now. This notion is not new. Modernity has sought to sweep away the confusion of diverse local practices and pluralistic functions accumulated over the ages so as to establish a new order. This order is supposed to bring freedom from the constraints of history, and liberty in the form of new technologies and practices. This is captured in one of the slogans of the modernist architect Le Corbusier, who said 'by order, bring about freedom'. Regrettably, many institutions which encourage conservation in agriculture have had all the makings of modernity. Farmers have been first encouraged, then later coerced, into adopting technologies that are known to work. When these farmers fail to maintain or adopt these measures, then interventions remould local social and economic environments to suit the technologies (Pretty and Shah, 1997).

The contrast with what is required to sustainably manage natural resources is crucial. Called, by some, post-modernism (coming after or contrasting with modernism), it favours heterogeneity, difference and human capacity as liberating forces. What post-modern traditions have in common is the rejection of 'meta-narratives', or large-scale plans, technologies or theoretical interpretations that purport to have universal application. The new theme is that different stakeholder groups have a right to speak and act for themselves and their communities, in their own voices, and to have their voice accepted as authentic and legitimate.

Induced Social Disruption

Anyone who has worked in countries of the South will be familiar with the failures of development – tractors inoperable for the lack of spare parts; exotic trees which suck up groundwater; farmers' groups imposed and unwanted; pesticides which damage farmers' health; abandoned project buildings with no current use. The list is almost endless. But it is one thing to have a technology fail. It is quite another to induce new forms of social disruption and discontent. Yet this is exactly what many conservation-oriented programmes have managed to do. In the long term, this can lead to a reduction in natural capital.

The failure to involve people in designing soil conservation measures has had a massive social impact in many countries. This is true of the US as well as in developing countries. The hundreds of thousands of hectares of terracing constructed by the US Soil Conservation Service in the 1930s and 1940s were later found to be rarely maintained by farmers. In Kenya, the enforced terracing and destocking of livestock in the 1950s, coupled with the use of soil conservation as a punishment for those supporting the campaign for independence, helped to focus opposition against both authority and soil conservation. This led, after independence, to the deliberate destruction of many conservation structures because of their association with the colonial administration (Pretty and Shah, 1997).

In Rwanda, a massive terracing programme using forced labour prior to 1960 created such negative feelings towards soil conservation that no further activities were possible until the late 1970s (Musema-Uwimana, 1983). In the Uluguru Mountains of Tanzania, where ladder and step terraces were common, a scheme that introduced compulsory bench terracing in the 1950s had to be abandoned after serious riots by local people (IFAD, 1992). Elsewhere in Tanzania, the HADO project completely removed livestock from whole communities, with tens of thousands of animals taken away from individual districts. Such a policy was only possible 'after mustering the cooperation of the ruling party and government machinery at village, district, regional and national levels. Inevitably some of the actions necessary to reverse soil degradation processes are a bitter pill to swallow' (Mndeme, 1992).

Social conflicts have also become increasingly common in and around national parks and protected areas. There are now close to 8500 major protected areas throughout the world, covering about 7.7 million square kilometres or some 5.2 per cent of the world's land area. Traditional conservationists, however, tend to see the biological value of, say, a rainforest, but are blind to the people. Indeed, local people are actively excluded, leaving only the visitors and tourists. In Africa, some 134 million hectares are now 'protected' in over 700 sites, of which 67 per cent permit no use of wild resources by local communities. This neglect of resident people strongly persists today. Until quite recently, many plans for protected area management made no mention of the people living inside forests, coastal strips, wetlands and other biodiversity-rich areas earmarked for conservation. In India, a 1985 study of 171 national parks and sanctuaries found that 1.6 million people were living in 118 of the parks. Yet by 1993, protected areas in India had already displaced some 600,000 tribal people (Ghimire and Pimbert, 1997; Gómez-Pompa and Kaus, 1992; Kothari et al, 1989). As a result of this type of displacement, tribal people, poor farmers, fishermen and pastoralists have seen their livelihoods undermined in their new, more risk prone, environments. Lack of livelihood security ultimately undermines conservation objectives as poverty and rates of environmental degradation intensify in areas surrounding parks and natural reserves. Furthermore, there is a risk that growing conflict induced by such management schemes actually destroys what has been protected. Open protest and rallies against protected areas, attacks on park guards, poisoning of animals and the deliberate burning of forests have become common in many countries (Pimbert and Pretty, 1995).

In India, resentment of national parks legislation and of enforcement agencies has led to acts of sabotage and civil disobedience. Villagers have set fire to large areas of parks, such as in the Kanha National Park in Madhya Pradesh. Others, such as those displaced from the Manas Tiger Reserve in Assam and the Kutru Tiger and Buffalo Reserve, have joined insurgents. In Africa, active and passive resistance against the impositions of protected areas is common. Created on lands traditionally used by Masai pastoralists, the Amboseli National Park in Kenya denied the Masai access to dry-season grazing lands and watering points. The Masai expressed their resentment towards the park by spearing lions, rhinos and other wildlife.

A Brief History of Participation

Fortunately we do have somewhere to turn. There has been a revolution in the past ten years in the methodologies for creating social capital. Emerging from a range of different traditions and disciplines, participatory methods have expanded in use and efficacy during the 1980s and 1990s. The greatest expansion has occurred in the developing world context, where participatory approaches are now being used in almost every country. Recent years have also seen an expansion in their use in Europe.

As a result, the terms 'people's participation' and 'popular participation' are now part of the normal language of many development agencies, including NGOs, government departments and banks (Adnan et al, 1992; World Bank, 1994). This has created many paradoxes. The term participation, for example, has been used to justify the control of the state as well as to build local capacity and self-reliance; it has been used to justify external decisions as well as to devolve power and decision-making away from external agencies; it has been used for data collection as well as for interactive dialogue.

In conventional rural development, participation has commonly centred on encouraging local people to sell their labour in return for food, cash or materials. Yet these material incentives distort perceptions, create dependencies, and give the misleading impression that local people support externally driven initiatives. Few have commented so unequivocally as has Roland Bunch, a sustainable agriculture specialist based in central America, on the destructive process of giving things away to people, or doing things for them (Bunch, 1983). But, as he says, 'obviously, though, programmes must do something for the people. Were they able and willing to solve all their own problems, they would have done so long ago... It should be emphasised that anything we do that people can do for themselves is paternalistic'. When little effort is made to build local skills, interests and capacity, then local people have no stake in maintaining structures or practices once the flow of incentives stops.

In one sustainable agriculture project in northern Thailand, farmers were given cash incentives and free fertilisers and pesticides to encourage them to adopt sustainable agriculture technologies. Many did so and transformed the landscape. But when the project decided to adopt participatory approaches and drop the direct incentives, farmers voted with their feet and stopped using the technologies that had appeared so successful. It took five years for the project to reach these levels again through careful building of local interests, capabilities and skills. This time, farmers were engaged in sustainable agriculture because it paid, not because someone else was paying.

In Britain, the meaning and use of the term participation has long been interpreted in very different ways. At the end of the 1960s, for example, the Skeffington Report indicated that planning could only be legitimate if it had an input of responses from the public. Yet in the early 1970s, local government reorganisation made larger authorities less accessible. The Development Commission encouraged its field officers to 'put their energy into making the new system work' by educating people into what was expected of them rather than involving them in the system of planning. Several rural officers, however, took a more active role in developing local participation (Wright, 1992). Some saw village appraisals as a way of developing a collective view of local needs, generating new social and political relations in villages and strengthening local capacity to deal with authorities. But this self-help ethic was later challenged in the face of cuts to services in the 1980s. Some said that encouraging local people to take on the role of the state simply legitimised public spending cuts.

Types of Participation

The many ways that organisations interpret and use the term participation can be summarised in seven distinct types. These range from manipulative and passive participation, where people are told what is to happen and act out predetermined roles, to self-mobilisation, where people take initiatives largely independent of external institutions (see Table 7.1). The problem with participation as used in types one to four is that any achievements are likely to have no positive lasting effect on people's lives. The term participation can be used, knowing it will not lead to action.

Table 7.1 *A typology of participation*

Typology	*Characteristics of Each Type*
1 Manipulative participation	Participation is simply a pretence.
2 Passive participation	People participate by being told what has been decided or has already happened. Information being shared belongs only to external professionals.
3 Participation by consultation	People participate by being consulted or by answering questions. Process does not concede any share in decision-making, and professionals are under no obligation to take on board people's views.
4 Bought participation	People participate in return for food, cash or other material incentives. Local people have no stake in prolonging technologies or practices when the incentives end.
5 Functional participation	Participation seen by external agencies as a means to achieve project goals, especially reduced costs. People may participate by forming groups to meet predetermined objectives related to the project.
6 Interactive participation	People participate in joint analysis, development of action plans and formation or strengthening of local groups or institutions. Learning methodologies used to seek multiple perspectives, and groups determine how available resources are used.
7 Self-mobilisation and Self-reliance	People participate by taking initiatives independently of external institutions to change systems. They develop contacts with external institutions for resources and technical advice they need, but retain control over how resources are used.

Source: Pretty J N. 1995a. *Regenerating Agriculture.* Earthscan Publications Ltd, London.

What this typology implies is that the participation should not be accepted without careful clarification (Hart, 1992; Rahnema, 1992).

The World Bank's internal Learning Group on Participatory Development, in seeking to clarify the benefits and costs of participation in the mid 1990s, saw differences in the way their institution used the term: 'many Bank activities which are termed "participatory" do not conform to [our] definition, because they provide stakeholders with little or no influence, such as when [they] are involved simply as passive recipients, informants or labourers in a development effort' (World Bank, 1994).

It is clear that interactive participation pays. A study of 121 rural water supply projects in 49 countries of Africa, Asia and Latin America by Deepa Narayan found that participation was the most significant factor contributing to project effectiveness and the maintenance of water systems (Narayan, 1993). Most of the projects said that community participation was important, but only a fifth scored high on interactive participation. Clearly, intentions did not translate into practice. When people were involved in decision-making during all stages of the project, from design to maintenance, then the best results occurred. If they were just involved in information sharing and consultations, then results were much weaker.

Great care, therefore, must be taken over using and interpreting the term participation. It should always be qualified by reference to the type of participation, as most types will threaten rather than support the goals of community regeneration. What will be important is for institutions and individuals to define better ways of shifting from the more common passive, consultative and incentive-driven participation towards the interactive end of the spectrum.

Participatory Learning Is More than Teaching

The central feature of sustainability is that we need new ways of learning about natural and social systems. Such learning should not be confused with 'teaching'. Teaching implies the transfer of knowledge from someone who knows to someone who does not know and is the normal mode of educational curricula and extension systems.

A move from a teaching to a learning style has profound implications for all agricultural and rural development institutions. The focus should be less on what stakeholders learn and more on how they learn and with whom. This implies new roles for professionals, with new concepts, values, methods and behaviours. These new professionals will have to make explicit their underlying values, will have to select methodologies to suit needs, will have to work closely with other disciplines, and should not be intimidated by the complexities and uncertainties of dialogue and action with a wide range of stakeholders (Pretty and Chambers, 1993). Many existing agricultural professionals will resist such paradigmatic changes; they will see this as reducing the value of research. But Roger Hart, specialist on children's participation, has put it differently: 'I see it as a "re-professionalisation", with new roles for the researcher as a democratic participant' (Hart, 1992).

A major challenge is to institutionalise these participatory approaches and structures that encourage learning. Most organisations have mechanisms for identifying departures from normal operating procedures. This is what the organisational specialist Chris Argyris calls single-loop learning (Argyris et al, 1985). But most institutions are very resistant to double-loop learning, as this involves the questioning of, and possible changes to, the wider values and procedures under which they operate. For associations to become learning organisations, they must ensure that people become aware of the way they learn, both from mistakes and from successes.

Institutions can improve learning by encouraging systems that develop a better awareness of information. The best way to do this is to be in close touch with external environments and to have a genuine commitment to participative decision-making, combined with participatory analysis of performance. Learning organisations also have to be more decentralised, with greater multidisciplinarity and with diverse outputs responding to the needs of local people. These multiple realities and complexities have to be understood through networks and alliances, with regular participation between professional and public actors. It is only when some of these new professional norms and practices are in place that widespread improvements to natural and social capital are likely to be achieved.

Peter Senge defines a learning organisation as 'an organisation that is continually expanding its capacity to create its future' (Senge, 1990). Earlier chapters have shown how learning communities and networks have emerged amongst farmers and rural people. For real change to rural social capital, it will be necessary to use innovative participatory approaches and new partnerships to bring together different stakeholders so that they can start to shape their own futures.

Participatory Methodologies in Use in Britain

Background

In recent years there have been worldwide and rapid innovations in participatory methods and approaches for stakeholder learning and interaction in the context of community development. There are now more than 50 different terms for these systems of learning and action, some more widely used than others.* This diversity is a good thing. It is a sign that many different groups are adapting participatory methods to their own needs.

---

* A selection of terms for systems of participatory learning in common use in the mid to late 1990s include: Action Planning, Agroecosystems Analysis (AEA), Beneficiary Assessment, Citizens' Juries, Community Audits, Community Profiles, Community Visions, Development Education Leadership Teams (DELTA), Diagnostico Rurale Participativo (DRP), Evaluacion Rural Participativa (ERP), Farmer Participatory

Some of them have spread much more rapidly than have others. (The informal journal *PLA Notes* was first published by IIED in 1988, and 28 issues were published to mid 1997 containing 400 articles from more than 50 countries.) There appears to be a direct relationship between flexibility, openness and honesty and the rate of spread. It works this way: if practitioners find that aspects of a highly designed and fixed methodology do not work for them, then they have only one option. They reject the whole methodology. But if they find that part of a flexible methodology does not work, then they drop that part and invent something new. This is one of the reasons why Participatory Rural Appraisal has spread so rapidly in recent years. Some methodologies, however, have been registered or trade-marked to prevent their use by anyone else. This may be good for these individuals, but experience suggests that it is counterproductive if widespread adoption is desired.

What is important though, is that this huge diversity is largely built upon common principles of co-learning, involvement and action. When these principles are followed, then the participation is likely to be interactive and ultimately successful for the stakeholders involved. One thing all these methodologies do is to emphasise people's capacity to change, while grounding this in a realistic understanding of what is possible. There are always things that local people can do themselves with no outside help or money. There are also things local people can do with some money. And, importantly, there are things local people cannot do, but external agencies can.

Research, Farming Systems Research, Future Search, Groupe de Recherche et d'Appui pour l'Auto-Promotion Paysanne (GRAAP), Méthode Active de Recherche et de Planification Participative (MARP), Open Space Technology, Parish Appraisals, Participatory Appraisal (PA), Participatory Analysis and Learning Methods (PALM), Participatory Action Research (PAR), Participatory Forest Resource Assessment (PFRA), Participatory Monitoring and Evaluation (PME), Participatory Poverty Assessment (PPA), Participatory Research Methodology (PRM), Participatory Rural Appraisal (PRA), Participatory Rural Appraisal and Planning (PRAP), Participatory Technology Development (PTD), Participatory Urban Appraisal (PUA), Planning for Real, Post-Occupancy Evaluation (POE), Process Documentation, Rapid Appraisal (RA), Rapid Assessment of Agricultural Knowledge Systems (RAAKS), Rapid Assessment Procedures (RAP), Rapid Assessment Techniques (RAT), Rapid Catchment Analysis (RCA), Rapid Ethnographic Assessment (REA), Rapid Food Security Assessment (RFSA), Rapid Multi-perspective Appraisal (RMA), Rapid Organisational Assessment (ROA), Rapid Rural Appraisal (RRA), Real Time Strategic Change (RTSC), Regenerated Freiréan Literacy through Empowering Community Techniques (REFLECT), Samuhik Brahman (Joint Trek), Soft Systems Methodology (SSM), Theatre for Development, Training for Transformation, Village Action Plans, Village Appraisals, and Visualisation in Participatory Programmes (VIPP).

It is important to bear in mind, however, that participation alone is not the route to system-wide change. There are significant risks. A consensus approach can appear to smooth over deep-seated social and political differences and inequity. Participation may not affect decision-making processes, rendering public involvement eventually worthless. Participation may be used as an input to achieve public support for decisions that are already made. Participation may also give the impression that the whole system can be changed, raising expectations unfairly.

Village and Parish Appraisals

Village appraisals (VAs), also known as parish and community appraisals, were first developed at the end of the 1970s to help communities, in villages or parishes, develop their own plans for development based on their own priorities. Philip Allies and James Garo Derounian (1997) of Cheltenham and Gloucester College define VAs as a process in which: 'a survey is carried by and for the local community. This aims to identify local characteristics, problems, needs, threats and opportunities. It is a means of taking stock of the community and of creating a sound foundation of awareness and understanding on which to base future community action'.

The questions are developed by a group of local people and then self-administered. Responses are gathered together and a report is published summarising key priorities and areas for action. Some are very large, involving several hundred households, others considerably smaller. Average response rates are about 60 per cent, though in some communities they may be in the high 90s. People can participate in the appraisal process on several levels: as part of the steering group; designing or analysing the questionnaire; completing the field survey or taking photographs; responding to the questionnaire; and getting involved in new projects. Very few communities have completed whole appraisals in less than a year – two years is typical.

By late 1997, some 2000 appraisals had been completed. About one million households have provided information on their circumstances, with some 10 to 15,000 actively involved in appraisal processes (Moseley et al, 1996). As there are some 8600 parishes in England and Wales, this represents good coverage so far (note that some communities have conducted several village appraisals). The impacts of successful VA processes include:

- increased confidence and cohesion amongst local people as they come to realise that new things can be achieved;

- increased motivation amongst parish councils to act, with more people turning up to parish council meetings; the emergence of new leadership, leading to a revitalisation of existing structures;
- the emergence of new groups, especially for young people, such as a Youth Council established in Tenterden in Kent as a parallel structure to the parish council;
- increased likelihood of external agencies acting on local plans – some county councils have now accepted VAs as legal components of structural development plans, such as in North Kesteven where the structural plan has been changed to take account of local needs;
- benefits for external organisations, including better information to target resources; lower expenditure derived from the input of free time by local people; and legitimisation through the involvement of large numbers of local people in development activities.

Malcolm Moseley and colleagues (1996) recently studied 44 parish appraisals conducted in 1990–1993 in Gloucestershire and Oxfordshire. Some 40,000 households were involved, coming up with 422 recommended action points (RAPs) (see Table 7.2). What was particularly interesting is what happened after the parish appraisals. There were two key changes – physical improvements, and enhancements to local skills, awareness and confidence. It is difficult to measure the improvements to social capital, though typical comments from people involved included: 'a new community identity has been created', and 'the report... led to action groups being set up in a number of areas'.

However, despite the good intentions, slightly less than a third of the RAPs were fully implemented. Another third were partially implemented, but slightly more than a third not at all. The top ten action points most likely to be implemented were information (newsletters or boards); footpaths; good neighbour schemes; playgrounds; neighbourhood watch schemes; recycling; street furniture; local environmental improvements; control of traffic speeding; and litter control. The least likely ten were youth facilities; education; health; street lighting; policing; traffic control; shops; road conditions; low-cost housing; and employment schemes.

Interestingly, all of the top ten (except for speeding) could be fully implemented by local people alone. None of the bottom ten fall into this category. They all require significant action by external agencies (save for street lighting). Despite the participatory process, it still is very hard for local communities to get any change to education, jobs, policing, healthcare and low-cost housing, even though they have identified these as important.

Table 7.2 *Selection of recommended action points from village appraisals conducted in 44 parishes of Gloucestershire and Oxfordshire*

More than half of parishes wanted action to:	• reduce traffic problems (introduce and enforce speed limits; reduce traffic); • develop housing, especially low cost housing for local people, but also curtail speculative housing.
Between a quarter and a half of parishes wanted action to:	• improve local environment (plant trees; street lighting; safeguard and improve footpaths; reduce litter; local recycling schemes); • improve quality of community life (more social events; better parish newsletters and information boards); • reduce crime and fear of crime (create neighbourhood watch; retain/increase police presence); • retain and restore village services (build/restore village hall; retain/improve bus service; retain/improve schools and shops).
Up to a quarter of parishes wanted action to:	• reduce dog fouling; • increase local job opportunities; • retain/ improve health care; • improve roads.

Source: Moseley M, Derounian J G and Allies P J. 1996. 'Parish appraisals – a spur to local action?' *TPR* 67 (3), 309–329

Village and parish appraisals work best when there is a dynamic facilitator or community development worker who helps to motivate local people. They are also more likely to succeed when there is attention to follow-up; enthusiastic endorsement by the parish council; and legitimisation through local consultation and consensus building. But appraisals still tend not to make sufficiently good links to external agencies, which in turn still tend to be unwilling or unable to respond to bottom-up needs. Many agencies and companies have expressed an interest in social innovations since they provide information on local needs, resources and priorities; stimulate self-help; and legitimate policy decisions. But it remains to be seen whether they themselves can make the necessary changes to respond to local needs.

Participatory Appraisal (also known as PRA or PLA)

Participatory Appraisal (PA) is a structured process of learning in which people participate in joint analysis, develop action plans and formulate or strengthen local groups or institutions. Learning methodologies are used to seek multiple perspectives, and groups determine how available resources are used. Originating from Rapid Rural Appraisal and several methodologies for anthropology and sociology in the 1980s, PA was developed in a rural developing world context. It is now used by governments and non-government organisations, by research and extension bodies, and by community groups and planners. The methods are now successfully employed by groups in more than 150 countries.

There are six basic principles:

1. PA is a methodology for collective and cumulative learning by all actors – the processes are structured but rarely used as a blueprint; methodologies are context-specific and so there are many variants; and the methods encourage interaction, more than just consultation.
2. PA is user-friendly and quick – the visual and dialogue methods are simple and widely applicable; processes are group-based and interactive, with people from different disciplines, sectors and mixes of professionals and non-professionals; the process creates enthusiasm.
3. Diversity is represented so as to give multiple-perspectives; individuals and groups evaluate situations differently, and this leads to different actions.
4. External actors help to play a key role – they facilitate learning and are concerned with transformations that people in the situation regard as improvements; new attitudes and values amongst professionals are crucial, with listening and facilitating more important than teaching; external professionals also contribute technical knowledge and support.
5. The process emphasises self-assessments – people in their situation carry out their own study and so achieve something; the skills and knowledge of different stakeholders are put at the centre of the process of mediation, dialogue, learning and action.
6. PAs should lead to an enhanced capacity for action – the learning process should be the basis for lasting change and the development of both individual and organisational capacity and social capital; action plans identify responsibilities for action and potential sources of funding, and analysis and debate about change lead to an increased readiness to contemplate action; the motivation to act increases as people find they can achieve goals.

A distinctive feature is that the methodology does not depend on people coming together at special events or meetings. PA can be used with groups and individuals wherever they happen to be – such as in pubs, clubs, mother and toddler groups, schools, street corners, shops and kitchens.

There is a strong emphasis on visualisation to encourage interaction. PA involves a shift from verbally oriented methods (formal interviews and written assessments) to visually oriented ones (participatory diagrams and visualisations). Everyone has an inherent ability for visual literacy, and the impact of visual methods on communication and analysis can be profound. Diagrams and visualisations seem to work because they:

- provide a focus during discussions by stimulating people's memories about their past and present situations;
- represent complex issues or processes simply;
- are sequentially modified and extended;
- stimulate exchanges from both non-literate and literate people;
- reinforce the written or spoken word.

There are many different diagrams and visualisations. These include resource maps (agro-ecological zones, land tenure, land use), social maps (health, wealth and well-being), matrix scorings, preference rankings, mobility maps, transects, time lines and historical profiles, wealth rankings, seasonal calendars, daily routines, systems and flow diagrams, impact diagrams, crop and animal biographies, pie diagrams and venn diagrams. These are supplemented by a range of interviewing and dialogue methods: semi-structured interviewing and conversation; direct observation; group meetings (casual, focus, structured and feedback); key informants; ethnohistories and biographies; oral histories; questionnaires and local stories; portraits and case studies. Triangulation is the use of different sources and methods to cross-check findings, and to assess the coherence and trustworthiness of diverse information.

PA is suitable in a wide range of circumstances, including agriculture, forestry, fisheries and rural community development; health and nutrition; urban environmental improvements; and technology testing and the development of local plans. In all of the contexts where PA has been used interactively, local people have benefited substantially. PA has been used in a wide range of contexts in Britain (for natural resource and fisheries management in Scotland; for community planning in Berkshire; for health in Derbyshire and Hull; for the elderly in South Armagh; and for housing needs in Scotland).

Future Search

Future Searches are structured visioning events held over two and a half days and involving 64 people organised into eight stakeholder groups of equal size (Weisbord and Janoff, 1995; Centre for Community Visions,

1997). The emphasis is on the quality and representativeness of participation, rather than on involving all members of a community. Legitimacy arises from the diversity of groups within the community. Participants include those with information to share, those with the authority and resources to act, and those affected by what could happen. First developed by Marvin Weisbord in the US, it has now been used in a variety of contexts in Britain.

The process has four stages. The first involves participatory documentation and visualising a common history at three levels: the individual, the community or town, and the global level. This is non-controversial and helps to bring all the participants together. The second involves reviewing the present, so as to understand what is happening and why. The third stage involves developing a common vision for the future, incorporating both the normative (what is the desired future; what do we want to become?), and the exploratory (what are all the possible futures, whether desirable or not?). The final stage is action planning. Participants know where they want to go, and they now establish the means and mechanisms to achieve this vision.

The good thing about Future Searches is that they create great energy and motivation to achieve change. The event focuses attention on what they want, on what others want, and on what can be achieved. But these events do take a long time to organise, need large space, and require follow-up to guarantee that achievements will be sustained. One of the most successful uses of visioning events like Future Search is the Chattanooga process, where large numbers of citizens have been involved in turning a polluted, low-job, low-pride city into a clean, employment-rich and successful city (see Box 7.1). In Britain, the Centre for Community Visions has been facilitating Future Search processes in a number of communities, with the most successful project occurring in Gloucestershire. There, the Vision 21 process has involved several hundreds of people in the county in dozens of meetings and workshops spread over four years. It has mobilised people from many different groups, and could become a model for the way that local authorities should interact with the local citizens (Vision 21, 1997).

Community Audits

Community Audits are a participatory and visioning methodology developed by James Kelly and colleagues for the district of North Kesteven in Lincolnshire (Kelly, 1997). This is a sparsely populated rural area of heathlands and fens, with some 85,000 people spread over 118,000 hectares. The district has acute development needs. As a result of agricultural change, defence cuts, and an ageing population, traditions are dying

Box 7.1 The Chattanooga Vision Process

Chattanooga is a city of some 150,000 people situated on the banks of the Tennessee River in southern US. In 1969, it was designated the most polluted city in the country. But heavy manufacturing industry declined rapidly in the 1970s, producing huge lay-offs. By the 1980s, social tension was high and civic pride at a low. As a result of these increasing threats to their city, and following a visit to Indianapolis to see community participation first hand, civic leaders launched Vision 2000 in 1984. A wide range of different groups were brought together in an open and diverse process. One participant, Countess Bernie Jenkins of Bessie Smith Hall, described what happened: 'People had been angry and cynical because they thought someone else was making all the decisions... But when we all got together in the same room, we realised that no one was making all the decisions. No one was creating a vision for the city.'

The process produced 40 jointly agreed goals for the year 2000. Vision 2000 was immensely successful. By 1993, it had been responsible for 223 projects and programmes; 1381 new jobs and 7300 temporary construction jobs; $793 million of investment; 800 homes renovated or financed for low-income families; and the manufacture of electric buses, which now ferry one million passengers. In 1990, Chattanooga met all six of the National Ambient Air Quality Standards, one of the few south-eastern cities to do so. By 1993, the police department credited a 20 per cent fall in crime rates to active neighbourhood associations.

Citizens were so pleased that they held a follow-up ReVision 2000 series of events during 1993 and 1994 to assess progress and set new goals. A process of neighbourhood, workplace, government and school meetings produced another 2559 ideas. These were presented back at an open Vision Fair, where people voted for what they would like to work upon, and task forces were formed for each concern. The family violence group, for example, soon set up a new shelter – all members had wanted it to happen, but had never been able to form the necessary critical mass. Other major initiatives include Chattanooga Neighbourhood Enterprise to eliminate substandard housing, Chattanooga Venture to provide shared facilities, and the Recycle Network. Some 2600 people participated in ReVision 2000, a third of whom were under 25, and 85 per cent of whom had not taken part in the original exercise.

Source: Centre for Community Visions. 1997. *Chattanooga. Case Study No 4.* CCV, London

and there is a steady erosion of social capital and local people's sense of community. The district council adopted a range of community development activities as part of its Agenda 21 process, from public meetings to decentralised planning committees and citizens' juries.

Community audits were developed to help organise and motivate local people while making the most of local distinctiveness. After a community approaches the district council, usually via its parish council, a Rapid Rural Appraisal is conducted to organise existing information. This is followed by stakeholder meetings with local people, whole community questionnaires, and 'visioning' meetings with selected groups. The resulting proposals form the basis for an action plan, which is presented back to the whole community in an exhibition. This allows for further cross-checking and feedback. A five-year action plan is then finalised, and a community contract is signed by all relevant agencies and the community. An implementation group is then set up to oversee action and performance.

Sixteen Community Audits were carried out during 1995 to 1997, with a further four to be completed in 1997. The principal impacts have occurred in communities and in external organisations. There have been more actions at community level, including the emergence of new leadership and a readiness to work together. And there have been changes in council officer attitudes and much less conflict between local people and the council. James Kelly put it this way: 'officers are much more responsive... and there has been a profound impact on officers' attitudes'. Successful community audits bring forward many new challenges and potential conflicts, particularly over how the district council can work better with other agencies; over the allocation of scarce resources; and over the time that needs to be allocated to the whole process.

Parish Maps

The Parish Maps project promoted by Sue Clifford and colleagues of the charity Common Ground encourages people to come together to explore and express what they value about their local place. It centres on the production of a map that reflects as many points of view as possible, and which identifies common actions. The visual principle is important, and those maps with the most profound impact are initiated by local people themselves. The process varies from place to place, but generally involves the formation of a core group, followed by group meetings, walking and talking, and thematic studies of local archaeology, field names or wildlife.

Hundreds of communities have now made their own parish maps, some drawn and painted, others embroidered or designed from a mix of materials. The findings are often surprising. The wildlife-focused map of Norton Radstock in Avon was coordinated by the Cam Valley Wildlife Group. Says Steve Preddy: 'it forms an historic snapshot, so that we can compare today's wildlife with that of the past and of the future. The results already showed us that 17 of our 26 ponds have disappeared through natural deterioration or removal; and 4.5 miles of hedgerow have been removed... in the last 40 years' (Common Ground, 1996). The maps are good for functional planning and for developing common histories; they are also vital in helping to regenerate social capital.

Through joint mapping, communities learn to work together and this helps to develop their common sense of community. As one villager from Barrow on Soar, Helen Sudler, put it: 'Our map created a little hiatus into which a sense of pride and purpose can creep before we spill over the edge of the next millennium. It certainly proved to me that community spirit is not dead; it simply has to be approached in the right way' (in Clifford, 1997). The maps give people not normally engaged in social groups extra meaning and some empowerment. Sue Clifford has seen the effect on local women in particular: 'many maps are driven by women, who find all kinds of skills and courage in the process'.

Action Planning

Action Planning is primarily an urban-based design process involving multidisciplinary teams of specialists working with local people (Wates, 1996). Commonly lasting five days, it is highly structured, visual, dynamic and transparent to all involved. The process involves an outside team going on reconnaissance transects and walkabouts, followed by the event in a large common space. Small groups work on what they would like to see changed and what can be done to achieve it. Design days follow with information being restructured and organised; a physical report is produced with clear proposals; and this is followed by a public presentation.

Action Planning works because of the high profile given to the wide range of community needs and problems. Professionals are present to learn and listen, as well as to stimulate the interest of the local people present. The high publicity of an action planning event raises the profile of the process, helping to focus attention on development issues that are not usually addressed. It also motivates local people. At the Poundbury Action Planning five-day event in Dorchester, some 2000 people attended the weekend to explore the implications of building a new town on Duchy of Cornwall property. Three-quarters felt it was worthwhile to them, and 90 per cent wanted continued involvement as the project progressed. Nick Wates's handbook on Action Planning documents the testimony of many participants at these design events. One woman from West Silvertown in London put it like this:

> *...I only went to be nosy. I just went to see what was going on and to know what had happened; I was in the thick of it.... I thought it was brilliant. I really enjoyed it. It took me a week to sleep properly afterwards; all these items were springing back into my head... Having everybody in one room together slogging it out got a lot of ideas out.*

The contribution to social capital can also be important, with participants telling of 'new bonds being created' (Berlin), of the 'sheer amount of creative energy' (Manchester), of 'the enthusiasm of councillors and officers being fired anew' (Haringey, London), and of 'personal batteries being recharged' (Glasgow). Once again, the key elements of long-term success rest on whether existing or new institutions can take on proposals and implement them. The chairman of the Urban Design Group, Jon Rowland, said in 1995: 'you shouldn't do these things unless you are able to follow up for two to three years at least. Action planning events must be the beginning of the process, not isolated events.'

Planning for Real

First developed by Tony Gibson of the Neighbourhood Initiatives Foundation, Planning for Real is used for both community development programmes and as a way of consulting with local people about key policies or local issues (Gibson, 1997; 1993). A central feature of the methodology is a physical model. At normal public meetings and consultations, local planners and other outsiders sit on a platform, behind a table, maintaining their superiority. When only a few people turn up, and only a few of them speak, the planners blame local indifference. Planning for Real attempts to bridge this gap by focusing on a model of the neighbourhood. Unlike an architect's model, this is intended to be interactive. After a neighbourhood model is constructed, using houses and apartment blocks made from card and paper on a polystyrene base, it then goes into the community, to the launderette, the local school, and the fish and chip shop so that people see it and get to hear about the next stages of participatory planning.

At later meetings the aim is to find out whether the planners have got it right. There is no room for passivity; there are not many chairs and no platform, with the model in the middle of the room. People spot the landmarks, discuss and identify problems and develop solutions. They can put more than one solution on the same place, allowing for conflicts to surface. Often people who put down an idea wait for others to talk first about it. The process permits people to have first, second and third thoughts – they can change their minds. The model allows people to address conflicts without needing to identify themselves.

The professionals are important too. These local planners, engineers, transport officials, police and social workers wear a badge identifying themselves, but can only talk when they are spoken to. The result is that they are drawn in and begin to like this new role. The 'us and them' barriers begin to break down. Priorities are assessed, and local people become more committed to the follow-up stages which can involve local skills surveys, group formation, and partnership with external agencies.

Planning for Real has been used mainly in urban contexts, often in poor and marginalised estates. The impacts have been mostly positive, with greater local organisation leading to improvements in the ability to make change. People change during the participatory process. As one resident of Collingwood in the north of England put it:

> *It's getting exciting now. The nitty-gritty we started with, there was a lot of suspicions... But when I've gone home after a meeting and I've had time to sit and think, there are a lot of things to mull over...We are all so excited about the things that are going on and how far we've come. We're a determined group and we know we've got to do this (in DoE, 1991).*

Citizens' Juries

Citizens' Juries are small groups of 12 to 25 people who are brought together for three to five days to consider an issue of public policy or a difficult decision before a public body. The individuals are chosen to represent a cross-section of society or are drawn randomly from the electoral register. They hear presentations from expert witnesses who give different sides of the argument. They cross-examine the witnesses, discuss the matter amongst themselves and, with a moderator to ensure fair play, draw conclusions. A report is presented to the commissioning body, who is then expected to act upon the recommendations. Because the jury contains a range of views that are representative of those in wider society, the Citizens' Jury process helps to enrich democratic processes and improves the quality of governance (Stewart et al, 1993; Ward, 1997; Coote and Lenaghan, 1997; Kings Fund, 1997).

Citizens' Juries first emerged during the 1970s in Germany and the US, developed by Peter Dienel of the University of Wupperthal, where they are known as *plannungszelle* or planning cells, and Ned Crosby of the Jefferson Center in Minneapolis (the term citizens' jury has been patented to ensure that no one can apply it in practice without acknowledging the Jefferson Center; *plannungszelle* have not been patented). Citizens' Juries were first introduced in the UK in the mid 1990s and have been used in a wide range of contexts. Typically, these have been sponsored by local health care trusts or by local authorities to address the tough decisions needed over allocating scarce monetary resources. So far, the role has been primarily to advise public bodies rather than to determine outcomes. Some important environmental issues have also been considered, including biotechnology, clean-up of pollution and charging of car drivers for road use. The face-to-face nature of the debate is a way for people to be drawn into complex decisions, enabling them to act as citizens for the common good of society (Ward, 1997).

Perhaps the most vital issue raised by Citizens' Juries and *plannungszelle* is how participation and the learning process changes the attitudes of jurors. Jurors almost always change their minds during the sessions as they learn more about the issues and become better informed. In a US 1993 Citizens' Jury on the federal budget, half of the jurors at first favoured cutting taxes. By the end, they all favoured raising taxes by US$70 billion: a quarter on alcohol and tobacco, a half on high-income earners, and a quarter on energy and fossil fuels. *Plannungszelle* in Germany permit jurors to add their own ideas to those originally established in the terms of reference – in the city of Heidelberg they came up with new ideas for traffic management; in Koln, they did the same for the redevelopment of the city hall. Interestingly, conclusions reached by citizens' juries lean neither to the right nor to the left of the political spectrum. As John Stewart, Anna Coote and colleagues put it in their their IPPR report: 'not only do voting arrangements allow for a variety of views to be expressed, but it is common for the opinions formed by juries to confound predictions and to defy traditional political categorisation'. It is for these reasons that citizens' juries and the learning process they embody represent an important option for greater public participation in difficult decisions over public resources and goods.

However, Citizens' Juries can be time-consuming and expensive. Jurors and witnesses need to be paid for their time, and appropriate facilities hired. However, the cost of a typical jury may still be less than paying for a consultancy team of 'experts' to address the matter. The jury, by contrast, is able to claim some added legitimacy through its representative nature.

New Partnerships for Rural Europe

Why Partnerships between Farmers and Rural People?

As documented in Chapter 2, policies are increasingly encouraging farmers to adopt environmentally sensitive practices. Most of these, however, have a critical failing: they do not seek changes within a contiguous area. The focus is commonly on individual farmers who have a prior interest. Where there are defined areas, such as in ESAs, the level of uptake by farmers is not good – about 28 per cent for all the designated land in Britain.

The need for co-managing of natural resources is both a technical and social issue. It is important for pest and predator management, nutrient management, controlling the contamination of aquifers and surface watercourses, maintaining landscape value, conserving soil and water resources, and sustaining access to the countryside. The problem is that if one farmer does something positive, this may be undermined by neighbours or others in a community who act differently. For example, one farmer who encourages predators through beetle banks and conservation headlands could have neighbours who use non-selective pesticides which kill predators, preventing regional predator populations from reaching a viable size. Someone who adopts practices to reduce nitrates and pesticides from leaching into groundwater could have little impact on water quality if other farmers on land overlying the same aquifer continue to apply large amounts of nitrogen, manures or pesticides, or use practices which permit leaching.

There are also good social reasons for working together in partnerships. Regular exchanges and reciprocity increase trust and confidence, and lubricate cooperation. There are great assets already within communities. It is just that local people often do not realise they have them. The existing assets are primarily in social capital: everyone knows each other, and they tend to help each other when in need. They bring good expertise; they experience problems at first hand; they are aware of local networks; and they are much more likely to sustain initiatives if they feel ownership. In general, group-based collaborative action by farmers and communities in Britain is unusual. This was not always the case. In the 18th- and 19th-century agricultural revolution, it was partnerships and exchanges between farmers that enhanced the spread of new technologies (Pretty, 1991). Some of these survive, such as in the close-knit crofting communities of Scotland. Others have emerged, in the form of machinery rings, research groups and farm tourism cooperatives. But these are very much the exception rather than the rule.

It is now increasingly recognised that there is a need to enhance community participation in rural areas, and ACRE have suggested that there is a real opportunity to increase support for community groups and infrastructure (ACRE, 1995). Rural partnerships which have emerged in recent years are beginning to show what is possible (Pretty and Raven, 1994; Oma, 1995). There are no plans, as yet, for spreading these to a national scale – even though the benefits for jobs, communities and the environment are amply demonstrated by relatively small-scale schemes in a number of places in Europe and North America.

Landcare Groups in Australia

Sustainable agriculture and the conservation of natural capital is not just about adopting sustainable technologies on individual farms. It is about the complex business of creating individual and institutional linkages that encourage coordinated local action. One of the best examples of rural partnerships comes from Australia, where a remarkable national social experiment has been underway since the 1980s. Landcare encourages groups of farmers to work together with government and rural communities to solve a wide range of rural environmental and social problems (Campbell, 1994). By the end of 1997, there were 4300 active local groups, comprising more than one half of all Australian farm families. For a country where individual farmers have prided themselves so long on their frontier spirit and their capacity to cope alone with problems, this is an extraordinary society-wide recognition that some problems can only be dealt with by working together.

Landcare groups have emerged to deal with many different local problems that affect the whole community. Groups deal with pest, weed and rabbit problems; with tree decline; with dune regeneration; with conservation farming; with soil salinity; with wildlife conservation; and with farm profitability and business management. The huge diversity of the 4300 groups is quite impossible to represent here, save to say that the most successful have been able to transform their communities and environments.

One example is the Morbinning Catchment group from the wheat belt of Western Australia. The Morbinning Catchment consists of 20 families on 25,000 hectares of farmland. They formed the group in 1989, united by their common problems of increasing soil salinity, poor drainage and the effects of periodical flooding. These problems could only be dealt with by planning and cooperating across farm boundaries. Over eight years, the group has revegetated 300 hectares of creeklines; treated 550 hectares of saltland; planted 440,000 trees, including 91 kilometres of windbreaks and 90 hectares of fodder trees; erected 249 kilometres of fencing to protect natural bush; planted 460 hectares of alley farming systems and 80 hectares of permanent pastures; and installed 145 piezometers in order to measure regularly the watertable depth. The group has also been at the forefront of local farm improvements, including in oil seeds, reduced tillage, alternative fertilisers, soil aeration, floriculture, sandalwood planting, and school visit programmes.

But it is not just environmental and farm benefits that the group has seen. Bob Hall, president of the group in 1997, put it this way: 'Before the group, farms were amalgamating, young people moving away, the community was falling apart. But now we meet six to eight times per year, with the regular involvement of member families. As trust increased, so the opportunities to learn from each other also increased, and this has brought the community together' (Bob Hall, pers comm, 1997). The Morbinning group won the National Landcare Award for catchment groups in 1995.

Landcare has brought major individual and institutional changes. Many of the people involved are learning more about their own land, about the land in their district, and about issues they may have rarely considered in the past. Extension staff have also changed, becoming more than providers of information. They are evolving into facilitators of learning. They are being trained to work with groups and to help groups become self-reliant. Landcare, by involving committed people closest to the land, has the potential to be the first step in evolving new land-use systems and new relationships between people and the land, building upon human resources instead of discounting them or seeing them as part of the problem.

Prespa National Park, Greece

The Prespa National Park is close to the borders of Albania and Macedonia. It comprises a montane valley with two lakes and their surrounding floodplain. It is home to the largest nesting colony of Dalmatian pelicans (*Pelecanus crispus*) in the world. The area is very remote, with the main economic activity of the 12 villages consisting of farming beans, with some livestock and fishing to supplement incomes. The traditional land management system has been important for natural capital, as livestock graze the wet meadows and keep down the reeds, creating valuable habitats for birds and fish. However, the adoption of intensive cultivation methods for the beans led to the conversion of some meadows to arable, and a big increase in fertiliser and pesticide use. Both of these have had a significant impact on aquatic resources, and the consequent loss of spoonbills and glossy ibis. In 1993, various organisations, including the Aristotle University of Thessaloniki, CADISPA and the Society for the Protection of Prespa, began to promote organic bean cultivation, the diversification of agriculture, and the development of the wildlife tourism potential of the park (Cuff and Rayment, 1997; CADISPA, 1996; Geoff Fagan, pers comm, 1997; Kalburtji, 1996).

After an initially difficult conversion period, farmers are now getting higher bean yields as well as premium prices. This encourages more farmers to adopt sustainable practices. With the focus on ecotourism, the number of visitors to the park has increased from 5300 in 1993 to 13,000 in 1995. These visitors are better spread throughout the year. Young people from the communities have been trained in environmental management, and two tourism centres have opened. These have helped to change local attitudes to conservation as well as those of visitors. The growth in ecotourism has prompted the establishment of two guest houses run by local women, and several restaurants and tavernas have benefited from increased spending by visitors. Some 50 to 60 people are now employed in the ecotourism sector. The government has also helped by investing in infrastructure for ecotourism. But this is only just tackling the deep problems. The population is still ageing; and agri-environmental support is not sufficient enough to encourage all farmers to convert to sustainable practices. Although some young people have benefited, new ways are still needed to help young people to develop interesting local livelihood opportunities.

Green Enterprises in Willapa Watershed of the Pacific North-West, US

In recent years, the protection of either local jobs or the environment has been on a collision course in the Pacific north-west. Bitter disputes have erupted over the spotted owl; after it was declared endangered in 1990, the volume of timber harvested in Oregon and Washington fell by a half. But now, new partnerships between formerly hostile groups have emerged, showing just how much the sustainable management of natural resources can contribute to local economic growth (Maughan, 1995).

The Willapa watershed comprises 275,000 hectares on the coast of Washington state. It is rich in natural resources, including oysters, clams, crabs, sturgeon, salmon and dense forests. But the four counties that comprise the watershed are extremely poor and are listed as 'economically distressed' by the state. In addition, natural resources have become diminished: salmon runs have dwindled, sturgeon have almost disappeared, oyster size has fallen, and old growth forests have been replaced by conifer plantations. There is a close connection between the state of resources and local poverty. As Janet Maughan put it: 'resources were harvested and shipped out, with few jobs and little income created en route'. The challenge was to create businesses and products that made sustainable use of natural resources and also added value to them. Ecotrust, an environmental group based in Portland, helped to form a new partnership of farmers, oyster growers, fishermen, small businesses, native American groups and others. This Willapa Alliance commissioned studies on resource use and assets, and developed a joint management plan. It was clear that many business ideas existed, but skills and access to markets and credit were in short supply. Ecotrust then contacted a well-established community bank in Chicago, the South Shore Bank (SSB), which had invested US$345 million in low-income neighbourhoods for community regeneration since the 1970s.

With the support of the Ford Foundation and the SSB, the Willapa Alliance market tested development banking for the watershed. Help has now been given to a range of local businesses that add value to natural resources, including:

- Willapa oysters that are now marketed locally rather than shipped out wholesale;
- cranberry growers who now produce a wide range of products – all cranberries used to be transported out;
- mushrooms which are collected from the forests for sale;
- alder which is harvested from secondary forests for high-quality wood products;
- fish and crab which is marketed with the north-west image of wholesome foods;
- interweaving of *Spartina* grasses with denim to make women's summer shoes (*Spartina* is a weed of shellfish beds and otherwise would be controlled with herbicides).

Many of these have helped local people to make a new livelihood, including the Shoalwater Bay Tribe, who now have formed an oyster company and harvest from 300 hectares of tidal beds. After several years of growing success, a new platform has emerged. The Willapa Economic Development Task Force has designed an ambitious regional development plan based on the sustainable use of natural resources. The process has been neither quick nor easy. Alana Probst of Ecotrust put it this way: 'the hardest part is having the patience to make this work. When people are unemployed, they want things to happen quickly.'

The Red Kite Project, Wales

The Red Kite project is an example of how wildlife tourism and appropriate farming can stimulate a rural economy (Cuff and Rayment, 1997; Rayment, 1997). Mid Wales, comprising north and west Brecknock, north Dinefwr, Ceredigion, south Montgomeryshire and west

Radnorshire, is an area of low wages, declining employment in agriculture and economic stagnation. It is also an area that supports remnants of the native population of red kite, a rare bird of prey. Agriculture here is heavily dependent on livestock farming. The population density is low, and the rural population is ageing.

The Kite Country Project was launched as a partnership between county and district councils, the RSPB, the CCW, the Wales Tourism Board, the Development Board for Rural Wales, and the Mid-Wales Tourism and Forest Enterprise in 1994. It is supported by Objective 5b funds, the Welsh Office and private sector sources. The aim is to increase tourism to the region, to promote wildlife and the environment, introduce a wider public to the area, and to encourage visitors to stay longer and so to spend more in the local economy. The project has set up six visitor centres; promoted public transport, cycling and walking; installed remote video technology and special feeding stations so that birds can be observed without disturbance; set up interpretation panels; developed community partnerships for green tourism; and promoted a stay-on-a-farm scheme with 130 farms as participants.

The impacts have been substantial. In 1995 and 1996, there were 148,000 visitors to the Kite Country centres. They spent £5.4 million in the mid Wales economy during these visits, about half of which has been attributed to the presence of the kites (Griffiths, 1996). A third of the visits are during winter, formerly a very low season for tourism. People who come tend to stay longer and come back more often. The project has created and safeguarded 114 full-time equivalent jobs in the local community, and created a further 14 directly through the employment of staff and contractors. 1995 was also the best breeding year seen for the red kite, with 120 pairs fledging 112 young.

Can Sustainable Agriculture Lead to More Rural Jobs?

Taking Back Some of the Middle

A healthy and vibrant agricultural sector is vital for the well-being of economies and environments. Sustainable agriculture is more knowledge, management and labour intensive than is conventional agriculture, and so can help to make the link between people and natural resources. Instead of relying first on external inputs, it seeks to make the best use of locally available natural and social resources.

Agriculture and natural resources are still the main source of income and livelihoods for some three-quarters of people in developing countries.

When it is sustainable, agriculture contributes to economic growth and development; to food security and poverty alleviation; and to the conservation of the environment. Few countries have achieved rapid economic growth without a correspondingly fast growth in the agricultural sector. Agricultural growth is also an important factor for the expansion of non-farm rural income and employment.

In industrialised countries, agriculture is no longer a significant contributor to jobs and rural livelihoods. Yet sustainable agriculture's need for more management skills, knowledge and labour represents a huge and emerging opportunity. How can farming be successful by making use of more labour? One of the ways that farming has become more efficient was to reduce its reliance on wage labour, losing millions of jobs across Europe this century. At the same time, farm output has greatly increased. However, increasing the number of jobs associated with farming should be good for local economies and the people who live there. As we have seen, rural areas have high numbers of poor families; economic growth is generally weak; and opportunities appear only to exist through 'exogenous' patterns of growth based on mobile capital. There are three opportunities for taking back the middle with a greater emphasis on sustainable agriculture and food systems. These are:

- increasing upstream close-to-farm information, management skills and services, thereby transferring some of the value from fossil-fuel-based products to people-based; this includes environmental services – hedgerows, drystone walls and woodlands all need care and management to ensure their complementarity with the food production components of farms; integrated pest management, which needs specific knowledge of new measures and technologies such as beetle banks; and integrated nutrient management.
- increasing on-farm labour, such as hand or mechanical weeding; field scouting and walking for pests and predators; establishing and maintaining soil conservation technologies and practices; and the high-technology, computer-aided field-mapping technologies;
- increasing downstream close-to-farm processing and marketing, adding more value to produce before it leaves rural communities, including farm shops and farmers' markets, direct marketing, box schemes, and mailing for direct delivery.

Sustainable agriculture is able to make contributions in all three of these areas. It does not, contrary to some opinion, represent a backward step to traditional agriculture, in which agriculture relied on many low-grade jobs, such as weeding or stone picking. Much of what is required needs higher levels of skill in the workforce.

Evidence for Local Job Opportunities

What is becoming increasingly clear is that sustainable agriculture farmers do three things. They spend more money locally on knowledge-based goods and services; they employ slightly more people in the business of farming; and they add value where they can through processing and direct marketing. However, it is not yet clear which of these represents the greatest potential job dividend.

It is also well established that more labour is needed on organic farms compared with conventional ones. Studies from a range of different European countries put the extra at between 20 per cent and 100 per cent, depending on whether crops and livestock are part of the enterprises, and whether on-farm marketing and processing are included. In Austria, Denmark and Germany, where there were 17,000 organic farms in the later 1990s, it has been estimated that additional labour needs are 10 to 50 per cent higher (Lampkin and Padel, 1994). The SAFE Alliance's 1997 study of jobs that have been created by organic farming in Britain has shown that an extra job per farm was created after conversion (Hird, 1997). On the 47 farms studied, there were some 90 family, full-time, part-time and casual jobs, rising to 179 after conversion. About half of these jobs were for marketing, processing and packaging. Tim Finney, farm manager of Eastbrook Farm in Wiltshire, emphasised the impact of this type of farming on the local community: 'The school and village pub would have disappeared by now if we had employed at conventional farming levels' (see also Box 5.1).

Important evidence has also emerged from recent studies of the Countryside Stewardship Scheme in England and the Tir Cymen scheme in Wales. Both of these schemes have given support to farmers to encourage more conservation-oriented activities on their farms. Most of the activities involve 'greening the edge' – regenerating wildlife-rich habitats and landscape features such as drystone walls. Fewer activities incorporate 'greening the middle' – encouraging the transition to more sustainable farming. Nonetheless, recent studies have shown that the 5000 farmers participating in Countryside Stewardship, and some 900 in Tir Cymen, have contributed significantly to local economies.

The recent Countryside Stewardship (CS) evaluation of the 3900 participants in the scheme at the time showed that these farmers were making a significant contribution to both natural and social capital. Their farms were more sustainable: they were using fewer fertilisers, pesticides and animal medicines; they were spending more on on-farm labour and local contractors; they were spending more locally; and importantly the majority had increased their farm income by 10 to 15 per cent (Table 7.3). This study indicates that the Countryside Stewardship scheme had

Table 7.3 *The impacts of the Countryside Stewardship scheme*

Farming performance	
Farm income	up 10–15% for 60% of CS farms; neutral or down for the other 40%
Fertiliser use	down £704 per farm
Pesticide use	down £389 per farm
Veterinary medicine use	down £119 per farm
Machinery	up £1349 per farm
Feedstuffs	up £1141 per farm
Fencing	up £1935 per farm
Net expenditure	up £152 per farm
Labour and businesses	
Farms with increased on-farm labour	10% farms with increase = 0.013 jobs per farm
Jobs created in local communities	220 new jobs created = 0.056 jobs per farm
Local expenditure	
Proportion of inputs and services purchased from small settlements	
Before CS	71%
After CS	80%
Proportion of inputs and services purchased within 15 kilometres of farm	
Before CS	43%
After CS	59%
Proportion of farms reporting more visitors and greater spend during the year	85%

Source: Harrison-Mayfield L, Dwyer J and Brooks G. 1996. *The Socio-Economic Effects of the Countryside Stewardship Scheme.* Countryside Commission, London

created some 0.013 on-farm jobs per farm, and an additional 0.056 local contractors' jobs per farm, bringing the total to 0.089 jobs per farm. This puts the total job premium at some 16,000 if all farms were to adopt the CS approach.

Some of these increases in local spending, particularly on on-farm labour and local contractors (such as for hedge-laying, orchard management and farm advice) are offset by losses elsewhere in the food system. The net effect of a shift to more sustainable agriculture is an increase in rural jobs, particularly in rural villages and small towns. But there are also losses in urban areas, especially from the pesticide and fertiliser industry. From the sample of 2000 farms reporting changes, there were estimated losses of 41.8 jobs from the fertiliser industry, a further 13.5 from the pesticide industry, and 3.7 from veterinary medicines. If these farms are representative of all 184,000 British farms, this would mean a loss from these input industries of some 5428 jobs, not all, of course, being lost in Britain.

The ADAS evaluation of the Welsh Tir Cymen scheme (see Chapter 8 for more details of the scheme), found both job creation and a substantial multiplier effect in local economies (ADAS, 1996; CCW, 1996). For the 718 farm agreements to the end of 1995–1996, it was estimated that 263 jobs had been created (equivalent to 0.37 jobs per farm). Interestingly, this multiplier is greater than for the CS scheme, though lower than for organic farming. This would appear to be because Tir Cymen emphasises whole farm agreements and a closer partnership between farmers and external authorities.

As a result, some greening of the middle is occurring as well as greening the edge. The ADAS evaluation found that Tir Cymen benefited farming, wildlife, and local economies. Some 29 out of 35 local businesses reported an increase in demand for services and materials; and 16 new jobs were created. Importantly, farmers' expenditures on services and materials in local communities (a total of £448,000 for 131 farms) were 23 per cent greater than the public support they received.

Other important evidence comes from the US, where the North-West Area Foundation's study of 2800 sustainable agriculture farmers found that these farms need more labour and management skills than do comparable conventional farms. Some 19 to 70 hours per week per farm were needed, equivalent to 0.4 to 1.4 extra jobs per farm (based on a 50-hour week) (NAF, 1994). The sustainable agriculture farmers also spend more money on local goods and services. In Iowa, each sustainable farm contributed £85.2 per hectare to its local economy, which is £33.4 per hectare more than conventional farmers. This amounts to some £13,500 for a farm size of about 150 hectares. If these were conventional farms, much of this would have gone to distant firms.

These studies of organic agriculture, Countryside Stewardship, Tir Cymen and NAF put the job multiplier at between 0.09 to 1.4 per farm. This fifteen-fold variation is partly accounted for by farm sizes and partly by the type of sustainable agriculture adopted. It is clear that the more organic the farm, the greater is the job dividend (Table 7.4). It is difficult to generalise from this patchy evidence. Clearly more sustainable agriculture can create more rural jobs. It will take some away from the agrochemical industry, but the net effect is positive. But from such limited evidence, it is difficult to say exactly what the overall effect on British or European farming as a whole would be if all farms were engaged in more sustainable agriculture. Labour intensity varies by farm type (livestock operations need more than arable) and by farm size (smaller farms need more labour per hectare than larger ones).

Table 7.4 *Effect of different multipliers for types of sustainable agriculture on jobs in British and EU farms*

	Low job multiplier (x 0.01)	*Mid range job multiplier (x 0.40)*	*High job multiplier (x 1)*
Example schemes	Countryside Stewardship (CS)	Tir Cymen and integrated crop management; sustainable agriculture in the US	Organic agriculture
Farms in UK 184,000	1840	73,600	184,000
Farms in EU–12 8.06 million	80,600	3.22 million	8.06 million

Jobs from Woodland Management and Energy Farming

There are also opportunities to increase employment in rural communities in the forestry sector, both in woodlands and in the field. Coppicing of woodland has long been important in rural Britain. It involves cutting trees down to the stump, allowing regeneration over an eight- to ten-year period, and then harvesting the abundance of new growth which can be converted into a variety of products.

During the last century, some 38,000 people worked in the coppicing trade, but this fell to less than a hundred in the 1980s (Rayment, 1997). Recent interest in coppiced woodlands has brought the employment up to 230 in the mid 1990s. According to the Hampshire County Council, the county with 60 per cent of Britain's working coppice, good-quality harvested coppice supports 80 FTE jobs per 1000 hectares – about ten times as much as a similar area of modern conventional forestry. The Wessex Coppice Group was established in 1995 to help the expansion of coppicing. The UK currently consumes 60,000 tonnes of charcoal each year, yet all but 2 per cent of this is imported. If this were all replaced with local charcoal, then some 2000 new jobs would be created. A further 5000 jobs would be created from other coppice products.

Recent years have also seen renewed interest in energy farming, which involves the cultivation of renewable energy crops, such as willow and poplar trees, grasses and oilseed rape. The trees are managed on a short rotation coppice (SRC) system. One hectare can produce ten to 15 tonnes of dry matter every year, which can then produce 200 to 300 gigajoules (GJ) of energy (equivalent to six to seven tonnes of coal).

SRC is very energy efficient, producing as much as 30 times more energy than is consumed in its production (DTI, 1994). SRC also contributes to wildlife biodiversity (particularly songbirds), is a low-input crop, and can be used for the absorption of sewage wastes. It has been estimated that just less than three million hectares of SRC could supply a quarter of Britain's domestic energy needs. This is unlikely to happen quickly, but a tenth of this, some 250,000 hectares which represent just 5 per cent of arable land, could supply farmers with 50 to 75 million gigajoules of energy if they could find a way of burning it locally for on-farm heat and electricity generation. This looks like an increasingly viable option, with cheaper and smaller electricity generation units being developed. Unfortunately, set-aside rules have not permitted the cultivation of energy crops, and so land that could be productively used to supply energy and to sequester atmospheric carbon remains unused.

There are emerging opportunities for secondary jobs through rural energy industry using coppicing and farm-wastes on farms. Yorkshire Environmental's new ten-megawatt plant at Eggborough near Selby is set to provide sufficient energy for 18,000 people derived from 2000 hectares of short rotation coppice after it opens in 1999. Two commercial power stations fuelled by chicken litter in Norfolk and Humberside recently opened and now supply electricity to the national grid. Each employs about 20 staff and provides employment to a further 15 to 20 people locally (DoE/MAFF, 1995).

Towards a National Scheme for Rural Partnerships

We know that community partnerships work. They can be good for social capital. They can be good for the environment. They can mean more local jobs. But we also know that they are given only patchy coverage and small amounts of resources – only a very small number of people are currently benefiting. The key constraints include a lack of skilled facilitators and community development workers. Rural participation takes time and money. Greater investment is needed early in the process, though empirical evidence does show that most participatory processes pay off handsomely in the end. What is needed are clear policies to establish the value of working with local communities; and to allocate extra financial resources. A national scheme with supportive policies could help the very rapid growth of this sector.

A national scheme for rural partnerships should have six elements. It should create a national identity for rural community partnerships. This would indicate its availability to all farmers, but would not necessarily mean establishing a new institution in competition with existing organisations. It should also aim to build up and strengthen existing institutions, structures and processes, such as Rural Community Councils, which already promote and support innovative approaches to economic and community development; and parish councils, which were identified in the 1995 *Rural White Paper* as having the potential for taking on new roles and responsibilities.

Support for the formation of new community institutions and supporting institutions, such as friendly societies, credit unions, farmers' and community groups, is also vital. There is much to learn about practical ways to get resources to local people. This in turn would help to strengthen social capital. We should continue to strengthen and nurture emerging initiatives, such as Local Exchange Trading Systems (LETS), food cooperatives, community transport, and community-supported agriculture. Taken together, they can build robust networks for local economic interaction.

At a higher level, new links need to be made. Support is required so that partnership arrangements between public, private and independent sectors, and local and central government, can emerge and be tied together at specific localities. Rural partnerships can only happen if an individual or institution takes the lead and also does not claim sole ownership of new initiatives. There may, however, still be conflicts between existing institutional roles. Using innovative participatory methodologies to mobilise both local people and external professionals is essential. A more systematic application of existing methodologies that are known to work should be a central part of a national scheme.

Summary

Endogenous development focuses on growing or originating from within. The priority is to look, first, at what resources are available in rural areas, primarily agriculture, people, natural resources and wildlife. Then ask: can anything be done differently to improve the use of these already available resources? Can it be done without incurring environmental and social costs? This is difficult, however, as the current patterns of institutions, policies, funding and intellectual thought all discriminate against endogenous development. When asked what they need, most people turn to external solutions.

These attitudes foster the dependency deadlock. Local people become entirely dependent on external agencies and actors to provide solutions to local problems. Again, this may appear reasonable. If you pay your taxes, why should you not expect the local council to fix the road or supply other services? But there are always things that local people can do better – they know local conditions. They will be living there long after external bodies have departed; they just may not realise their own capabilities. Four cases of community regeneration in Sweden, Italy, Spain and Scotland show just how endogenous development can work, despite all the adversity.

New participatory processes are needed to bring together different stakeholders in the renewal of the countryside. History tells us that coercion does not work. The terms 'people's participation' and 'popular participation' are now part of the normal language of most development agencies, including NGOs, government departments and banks. This has created many paradoxes. The term participation has been used to justify the control of the state as well as to build local capacity and self-reliance; it has been used to justify external decisions as well as to devolve power and decision-making away from external agencies.

The many ways that organisations interpret and use the term participation can be resolved into seven distinct types. These range from manipulative and passive participation, where people are told what is to happen and act out predetermined roles, to self-mobilization, where people take initiatives largely independently of external institutions. What this typology implies is that the term participation should not be accepted without careful clarification.

A major challenge is to institutionalise participatory approaches that encourage learning. Most organisations have mechanisms for identifying departures from normal operating procedures. This is single-loop learning. However, institutions are very resistant to double-loop learning, as this involves the questioning of, and possible changes to, the wider values and procedures under which they operate. For organisations to become learning organisations, they must ensure that people are aware of the way they learn, both from mistakes and from successes.

Recently, there has been rapid innovation in new participatory methods and approaches for stakeholder learning and interaction in the context of community development. There are now more than 50 different terms to describe systems of social learning. This diversity is a good thing. It is a sign that many different groups are adapting participatory methods to their own needs. These have common principles. There is a defined and organised methodology for cumulative learning by all actors. The inquiry and learning processes are user-friendly, with a particular emphasis on simple visual and verbal methods. Diversity and inclusion are emphasised throughout in order to reveal multiple perspectives. External actors facilitate learning and are concerned with transformations which people regard as improvements. These self-assessments lead to new visions for the future. And the analysis and debate about change leads to an increased motivation to act.

Detailed case studies of eight methodologies increasingly used in Europe show what kind of impacts are being achieved. These are Village or Parish Appraisals, Participatory Appraisal (or PRA/PLA), Future Search, Community Audits, Parish Maps, Action Planning, Planning for Real, and Citizens' Juries. New rural partnerships are needed for Europe. There are good social reasons for working together in partnerships. Regular exchanges and reciprocity increase trust and confidence, and lubricate cooperation. There are great assets within communities. The existing assets are primarily in social capital: everyone knows each other, and they tend to help each other when in need. They bring good expertise; they experience problems at first hand; they are aware of local networks; and they are much more likely to sustain initiatives if they feel ownership. Case studies from Australia, Greece, the US and Wales illustrate what has been achieved.

In industrialised countries, agriculture is no longer a significant contributor to rural jobs and livelihoods. Yet sustainable agriculture's need for more management skills, knowledge and labour represents a huge and emerging opportunity. What is becoming increasingly clear is that sustainable agriculture farmers do three things. They spend more money locally on knowledge-based goods and services; they employ slightly more people in the business of farming; and they add value where they can through processing and direct marketing. The jobs dividend could be very significant – in the range of 0.4 to one job per farm. There are also opportunities to increase employment in rural communities from the forestry sector, both in woodlands and in the field.

We know that community partnerships work. They can be good for social capital. They can be good for the environment. They can mean more local jobs. But we also know that they are currently only given small amounts of resources – only a very small number of people are currently benefiting. There is a pressing need for the right policy environment to nurture these local processes and help them spread to a wider scale.

Part IV

Making a Difference

Chapter 8

Financial Support and New Policies for a Living Land

Land of the hedgerow and the village spire,
Land of thatched cottages and murmuring bees,
And wayward inns where one may take one's ease, ...
Listen, the skylark singing overhead
That's the old country, that's the old home!
You never forget it whenever you roam.

E V Lucas (1868–1938),
from *The Old Country*

Spreading the Successes

The dividend derived from moving towards sustainability would make a significant impact on rural people's welfare and livelihoods. As has been illustrated throughout this book, sustainable agriculture, localised food systems and rural community partnerships can contribute significantly to natural and social capital. But we will need coordinated action to encourage and nurture the transition from modernised systems towards these more sustainable alternatives. Without appropriate policy support at a range of levels, these improvements will remain, at best, localised in extent or, at worst, will wither away.

Another recurring theme of this book is that much can be done with existing resources. At present, most value is captured by the few. If benefits were spread more evenly amongst stakeholders, then new resources would not be needed. A living land will not bankrupt the treasury. It will not undermine the effectiveness of other sectors of our economies. It will not, however, happen without some help and money. There are always costs associated with shifting from one way of doing things to another – the costs of learning new knowledge, the costs of developing new, or adapting old, technologies, the costs of learning to work together, the costs of institutions having to break free from existing patterns of thought and practice. It will also cost time and money to rebuild depleted natural and social capital.

Most of the improvements reported in this book have arisen despite existing national and institutional policies. These will need major reforms. Policies devised to deliver increased food production will have to be changed if they are to deliver environmental and social benefits too. Food policies framed to help deliver cheap and abundant food regardless of quality also will have to change. And rural development policies and institutions focusing on 'exogenous' solutions to the economic and social problems of rural communities are often ill-suited to the needs of community-based and participatory development.

But there are very real constraints to overcome. Vested interests in maintaining the status quo will make any change difficult. Why should fertiliser companies support a transition to legume-based farming when this could mean a loss of several hundred million pounds each year in Britain? Why should a pesticide company be balanced in its presentation of different types of farming, when it knows some types of sustainable agriculture mean that little or none of its products will be used? Why should a multiple retailer support the presence of networks of small shops that provide so much in the way of social capital?

These are difficult questions to answer. What we do know, however, is that both financial and policy support will be vital to help make the transition towards a more sustainable future for rural Europe. Which of the current groups of stakeholders ultimately supports or hinders the process remains to be seen.

Financing the Living Land

Sources of Finance

There are two challenges when it comes to finance – getting the money and then holding on to it long enough for it to add value to local economies and environments. This implies creatively activating a range of sources of finance from public sector to community contributions, and plugging the leaks in communities. Most communities leak pounds, francs, marks or dollars at an alarming rate. Money comes in through wages or social support, and then is spent on distantly sourced goods and services. In most circumstances, much more can be made of money before it leaks away.

There are four main sources of finance for encouraging the transition to more sustainable rural communities and agricultural and food systems:

- public sector or government support, both for agriculture and rural development;
- goods and services bought by the public, in particular food products and also recreational and aesthetic services bought by tourists and visitors;
- credit and private finance from banks, delivered locally by credit unions;
- community financing with local people's own contributions of time – local people contribute to social capital and cohesion with their time through cooperative efforts and local money systems.

Public Sector Support for Agriculture and Rural Development

All governments provide some sort of public support to their domestic agricultural and rural sectors. Through a wide range of direct and indirect monetary transfers from consumers and taxpayers to farmers, they have sought to ensure that agriculture provides the food and other products needed by the non-farming population. There are five types of agricultural policy measures (Pretty, 1995a; OECD, 1993):

1. market price support, in which producer and consumer prices are influenced by a range of policies that include levies or tariffs on goods entering the country, thereby raising the price of imports; guaranteed prices for domestic produce, usually at levels above world prices or those paid by domestic consumers; quotas on imports restricting penetration into domestic markets; and the subsidisation of exports to ensure their sales in markets elsewhere;
2. direct payments, in which money is transferred directly from taxpayers to farmers without raising prices to consumers; these include payments to encourage both production- and conservation-oriented technologies, and can be monetary, or in kind, such as support for training;
3. input cost variation, in which measures are taken to reduce the costs of inputs, such as pesticides, fertilisers, water, electricity and credit, encouraging greater use;
4. provision of general rural services, in which measures are taken to lower the long-term costs of research, extension, education and planning, ensuring that farms and other rural businesses have access to new technologies as well the capacity to adapt them to their own conditions;
5. indirect support, in which regions receive rural development support

for infrastructure, or tax concessions are granted to farmers who use particular activities or measures.

These policy measures have been widely used in the last half century to encourage the adoption of 'modern' methods of farming, and have led to substantial increases in food production in many parts of the world. Some of these types of support are set to fall under the trade agreements of the GATT and the establishment of the WTO. Signatories have agreed to remove taxes, levies, subsidies and quotas that differentiate prices according to place of production. The stated intention is that all prices will eventually approach world market prices.

Some argue that agriculture should receive no public support whatsoever, and all producers should be free to compete against one another. The Cairns group of Australasian and South American countries, for example, has substantially cut its subsidies to farmers, and now argues that other blocs, such as the EU and US, should do the same. However, one problem is that it is very difficult to assess what a subsidy is. Of the five categories above, the first and second are the easiest to measure, but it becomes more difficult when considering support for training, extension, community-based development or technology development.

A concept developed by the OECD in Paris for measuring the total amount of assistance to farmers is the producer subsidy equivalent (PSE). This measures the transfers to producers from consumers that result from agricultural policies (OECD, 1997). Two of the ways that the PSE can be expressed are the total PSE, which is the total value of transfers to farmers, and the percentage PSE, which is the total value of transfers as a percentage of the total value of production (valued at domestic prices), and adjusted to include direct payments and to exclude levies. By these measures, OECD countries provide substantial support to their farming sectors. The total support is about US$175 billion, which represents some $15,600 per full-time farmer-equivalent, or $175 per hectare of farmland. This comprises $80 billion in the EU, $34 billion in the US and $47 billion in Japan. There is huge variation between countries: from the high support in Japan of 74 per cent of the total value of agricultural production to just 3 per cent in New Zealand. The implication is that Japanese farmers get 74 per cent of their returns from subsidies. Although the GATT agreement and the WTO seek to reduce these levels of agricultural support in order to ensure some sort of 'level playing field', most countries have yet to embark on substantial reforms. Some have begun to reduce support prices and to replace them with systems of direct payments, such as in the CAP.

But the contrast between OECD and developing countries is enormous. Despite the desire to increase food production, many countries have pursued urban-biased policies that have strongly discriminated against agriculture (Pretty, 1995a). By selecting macro-economic policies that ensure high real-exchange rates and by protecting industry, many countries have distorted the domestic terms of trade against agriculture. In effect, many have imposed heavy direct taxes on agriculture and held farm gate prices below world prices.

Goods and Services Bought by the Public: Food and Tourism

Farming produces food and consumers buy food. Some of this money comes back to the farm, though as we have seen in earlier chapters, the proportion in industrialised countries has been falling all century. Now only about 10 per cent of the food pound, franc, mark or dollar gets to the farm and rural community. The choices that food consumers make play an important role in determining where the value goes. A box of vegetables bought from an organic box scheme supports a diverse farm, employing more than an average number of workers; a box bought from a supermarket costs the same but provides more of the value to the retailers' shareholders.

A much undervalued source of money is tourism. People value the countryside for its aesthetic value, its wildlife, its villages and communities. They like to visit, and when they do visit, they bring with them money to spend. Each year in the UK, urban people make some 660 million day visits to the countryside. They spend £4 to £12 each per visit (at 1996 prices), injecting £2.6 to £7.9 billion annually into the rural economy. In addition to day visitors, some 32 million Britons took holidays of four nights or more, spending a further £6.4 billion. Without counting the contribution made by the 23.8 million overseas visitors annually, this represents an injection of between £9 and £14.3 billion into rural areas each year. Tourist industry estimates suggest that about 70 per cent of this immediately leaks out of rural communities, putting the net contribution to communities and the countryside at £2.7 to £4.3 billion. A mid range figure of £3.5 billion represents about the same amount brought directly to farmers by the Common Agricultural Policy.

Tourism is vitally important in helping to sustain a living land. Again, though, people's choices are vital. Will they pay to visit a landscape devoid of hedges and birds; or would they rather visit a countryside rich in landscape features and vibrant with services and goods? Farmers already add value to their enterprises by offering bed-and-breakfast services. The longer people stay, the more they spend in local communities, and hopefully the more they enjoy themselves.

It is clear that there is already a substantial number of jobs depen-

Table 8.1 *The jobs dividend from some wildlife-rich sites in the UK*

Location	*Jobs created or safeguarded*
Abernathy Forest Reserve, Scotland	89 FTEs: 11 on the reserve and 78 in local communities generated by visitor spending
Shetland Islands wildlife	43 FTEs from visitor spending
Red Kite project, mid Wales	128.5 FTEs: 14 for project directly and 114.5 from visitor expenditure
Heathlands in Dorset	67 FTEs in total, of which 29 are in local communities

Source: Rayment M. 1997. *Working with Nature in Britain: Case Studies of Nature Conservation, Employment and Local Economies.* RSPB and Birdlife International, Sandy, Bedfordshire

dent on a well-managed rural environment. Matt Rayment and colleagues at the RSPB estimate that the nature conservation sector already provides more than 10,000 FTE jobs in Britain (Rayment, 1997). Conservation activities generate wildlife tourism, with visitors coming in and spending locally. The impacts of this can be greater diversification of more remote rural communities as well as generation of employment and income (see Table 8.1).

What visitors choose to buy is important. Staying a week in a farm cottage, and buying food from a multiple supermarket, simply brings money in and lets it leak out again. People are entitled to enjoy the 'free' public goods of the countryside – the views, the clean air, the tranquillity, the wildlife. This is not to suggest that they should pay for these. But while they are there enjoying themselves, they will spend on accommodation, food, drink, and handicrafts. Generally they want something distinctive in character to the locality – a local cheese or wine; a local ice-cream or handicraft. All these add value to the local community.

Rural communities need to develop more goods and services that will appeal to tourists if they are to capture more of this value. Cutting the leakage rate to 50 per cent would mean a net benefit to rural communities of £5.85 billion, rising to £9.36 billion if the leakage was cut to 20 per cent. As the Greek Prespa National Park and Welsh Red Kite projects show (see Chapter 7), some local initiatives have already been able to capture some of this value by building social capital, and then by finding ways of making the best of natural capital.

Credit Groups and Unions

One of the great revolutions to have occurred in the countries of Asia, Africa and Latin America in recent years has been the provision of affordable credit to poor families. People without access to credit are generally poor and lack assets. But people without assets cannot get hold of credit. Because they lack collateral, they are too high a risk for banks and have to turn to traditional money lenders – who, in turn, inevitably charge extortionate rates of interest. They are trapped in a vicious circle.

It has also long been assumed that poor people cannot save money and so are unable to help themselves. However, in the late 1970s, a Bangladeshi academic, Muhammad Yunus, realised that there was a way out of the trap. He recognised that when local groups are trusted to manage financial resources, they can be more efficient and effective than external bodies, such as banks. They are more likely to be able to make loans to poorer people. They also are able to recover a much greater proportion of loans. He helped women to get organised and raised money to lend to their groups. The groups themselves acted as the collateral, with the members vouchsafing for one another. This initiative became known as the Grameen Bank. It now has 1.6 million members who have been given the opportunity to escape the trap of indebtedness. Its principles are being widely applied elsewhere.

In the remote Northern Areas of Pakistan, the Aga Khan Rural Support Programme has established more than 2600 village or women's organisations which cater for some 53,000 households. Village groups first organised themselves to construct an irrigation channel, road or bridge, then helped members regularly to save small amounts of money, creating collateral for credit provision. Over time, and with local control and responsibility, groups have been able to save substantial sums. Other notable successes have emerged in southern India, where NGOs such as Myrada, SPEECH and Pradan have again shown the value of small groups (Fernandez, 1992). Years of relying on banks and local cooperative societies to supply credit rarely helped the poor. But when they started to work with small independent groups they noticed that: 'not only was the money managed more carefully, there was a far greater commitment and responsibility from the groups towards repaying the amount of money, something that had not unduly bothered them when they were part of the cooperative' (Ramaprasad and Ramachandran, 1989). What is particularly significant for the programmes is that some 95 to 98 per cent of loans are repaid in full. This contrasts with just 20 to 25 per cent for banks making loans under Integrated Rural Development programmes.

Another advantage is that locally lent money recycles quickly. It does not take long for the total advanced to exceed the total fund size. In the Myrada programme, some 108 million rupees (US$3.6 million) were loaned by more than 2000 local groups to their 48,000 members until the early 1990s. Yet the total common fund was 24 million rupees, implying that each rupee had been loaned and repaid five times. The number of loans made is much greater compared with those by banks. In four years, one group in the village of Talavadi advanced 26,454 loans. By contrast, one branch of a bank finds it difficult to handle 400 loans in a year. Most loans are small, often less than 100 rupees, and are more often for immediate needs, such as to pay for a funeral, marriage or food. As Aloysius Fernandez put it: 'while a farmer may be eligible for a loan package that can buy him 20 sheep, what he wants and can manage is only two sheep. A credit group understands such priorities much better than a bank can.'

Such micro-finance institutions are now receiving worldwide prominence, with thousands of groups engaged in savings and loans management in countries as diverse as China, Fiji, Kenya, Indonesia, and Vietnam. In 1996, the Bank-Poor meeting in Malaysia brought together most of the innovators in this movement to set out just how micro-credit should be making a difference for the poorest and most excluded groups. The 37 micro-finance institutions studied for Bank-Poor 1996 were found to have mobilised a total of US$132 million in savings from some 5.1 million savers. The Micro-Credit Summit has set an ambitious target of 100 million new savers by 2005 (Gibbons, 1996).

Credit unions in the UK are now beginning to provide similar services. A credit union is generally a cooperative, owned and run by its members who save regularly in a common fund. Members generally also share a common bond – they live in the same community or work in the same place. There are now some 600 credit unions throughout the UK, up from 50 ten years ago, with 80,000 members and assets of £30 million. In the US and Ireland, the extent is far greater since about one in four people use a credit union (NEF, 1997).

Community Financing with Time

An old aphorism states that time is money. But somehow this has become distorted. For those who work, there is often too little time; for those without jobs, time is so abundant that its value falls and it becomes unproductive. For those who work, the money they earn enters their community, and often leaks away without providing any local benefits. For those without work, they only have time.

Investing non-job time in local activities can, however, make a massive contribution to social capital. Indeed, we do this already. In 1997, there were some 22 million people in paid work in the UK. At the same time, 23 million people were active in the voluntary, or third sector. They contribute by giving their time to local schools, to parish councils, to youth groups, to looking after elderly or disabled relatives, to organising village fêtes or raising money. All of these contribute to social capital and help to keep value. A number of recent innovations linking people's skills and the value of time have shown how local investments can substitute for 'real' money, helping both to increase the cycling of money locally and to prevent it from leaking away. These innovations are centred on bartering and exchange, enabling people to obtain goods and services without resorting to cash, and to run up interest-free debts repayable in kind. Some involve the use of alternative money, such as 'time hours' in Ithaca, New York; 'farm notes' in Massachusetts; or 'barter bucks' in Kansas City. In other systems, such as in the Local Exchange Trading Systems (LETS) in Britain, the goods and services are bought with cheques and a central system records credits and debts for each member.

First established in the late 1980s, LETS now have some 20,000 members in 400 schemes across the UK. Each has its own currency, in which members gain credits when they sell goods or services and write cheques against their credit stock when they buy. Currencies are often chosen to give local flavour and include Ideals in Bristol, Readies in Reading and Groats in Stirling. Records are maintained at a central point, where there is also a directory of all the members and what they have to offer. As no interest is charged when an account is overdrawn, such debts represent a commitment to the rest of the community to provide goods or services in the future. LETS also make a very significant contribution to social capital. When LETS currency is used, it keeps its value within the community. LETS provide a mechanism for social interaction, increasing social cohesion amongst people of all types and ages.

Liz Shephard, coordinator of Letslink UK, an organisation set up to link individual groups and help new ones emerge, describes their effect: 'as popular grassroots initiatives, they can reach the parts that other currencies can't, mobilising all kinds of skills and resources to meet local needs...Community building and anti-poverty strategies feature as well as environmental considerations' (Liz Shephard, pers comm, 1997). Some LETS groups have a particular focus on food and there are now many experiments to add local value with small-scale production, community allotment schemes, box schemes, and local food distribution coops. In Stirling, Scotland, a LETS food coop combines a subscription system using real money with Groats from members to ensure the effective working of a local food system. Local activist Mark Rushell describes the effect:

> *The Stirling LETS is quite small, about 100 members, but has helped to develop the system in a number of ways: by increasing system turnover; by enabling members to meet each other at the monthly food coop delivery and trade fair; and most of all, giving them an ethical, cheap and environmentally responsible range of products (Rushell, 1996).*

In the US, the first effective system to value local people's time and to ensure that resources are recycled locally was set up by Paul Glover in Ithaca in the state of New York. It has now become the most spectacular success in local currency development. David Boyle of the New Economics Foundation describes how they came about:

> *Ithaca hours were introduced to keep local money circulating locally, to encourage local farmers and businesses and to provide income for these people. And it seems to be working. Hours are accepted in 300 local businesses, and have the enthusiastic backing of the chamber of commerce and the mayor (Boyle, 1996).*

Since 1991, some $51,000 of Ithaca hours have been issued to almost 1000 participants. Paul Glover estimates that about $500,000 of local trading has been added to the 'grassroots national product'. An Ithaca hour is set at $10, the average hourly wage in the region. These hours can be used to buy an extraordinarily wide range of goods and services – from plumbing and carpentry, to nursing and childcare, to mortgage and loan fees, to food in restaurants, to cinemas and bowling alleys. These are accepted by 40 farmers' market vendors.

Once again, the effect on social capital is remarkable. Says Glover (1996) 'we're making a community while making a living'. Another commentator, Michelle Silver (1993), said 'mostly what we saw in Ithaca were dozens of community members who addressed each other by first name, joked around, and negotiated for each other's goods and skills'. But social capital is not the only benefit of these schemes. They support local businesses and local producers of goods and services. Crucially, they also help to plug the leaks in the local economy, keeping money in the locality and contributing to employment and wealth creation.

Policies for Sustainable Agriculture and Rural Regeneration

The Integration of Policy?

Environment ministers of the OECD countries, meeting in January 1991, identified agriculture as one sector in which improved policy integration offered major returns. They noted that both environmental and agricultural goals could be pursued within the context of agricultural reform, with a view to moving towards sustainable agricultural practices. This was reinforced by the OECD Council at ministerial level, which later noted that: 'environmental policies should be integrated closely so that agriculture is carried out on an environmentally more sustainable basis' (OECD, 1993). This view still holds.

In recent years, there has clearly been an increasing number of policies which link agriculture with more environmentally sensitive management. But these are still highly fragmented. As yet there is little sign of integration. Sustainable agriculture can only be achieved by integrated action at farm and community level. If it is to succeed, this will require the better integration of policies. One problem is that these 'environmental' policies have tended to green the edges of farming. An essentially modernist agriculture remains much as it ever was, but is now tinged green. Non-crop habitats have been improved, including some hedges, woodlands and wetlands. But the food is largely produced in the conventional manner. The bigger challenge is to find ways of substantially greening the middle of farming – in the field rather than around the edges. A thriving and sustainable agricultural sector requires both integrated action by farmers and communities, and integrated action by policy-makers and planners. This implies both horizontal integration with better links between sectors, and vertical integration with better links from the micro to macro level.

The lack of integration in Europe was brought into sharp focus by Agriculture Commissioner Franz Fischler in his opening speech at the 1996 Cork Conference on integrated rural development: 'the current situation is also made more difficult by the enormous number of complicated administrative procedures at both Community and national level, including 62 Objective 1 programmes, 82 Objective 5b programmes, 101 LEADER programmes, 130 agri-environment schemes, 36 for Objective 5a measures, and others for reafforestation and early-retirement' (Fischler, 1996).

Most policy initiatives are still piecemeal. They affect a small part of an individual farmer's practices, but do not necessarily lead to substantial shifts towards sustainable agriculture. However, one of the first nations to convert the principles of sustainable development into a series of clear steps in a national strategy is The Netherlands. The National Environmental Policy Plan (NEPP) is probably the best example of an action-oriented plan, with 220 prescribed steps towards clear targets. It seeks to integrate land use, transport and energy plans with agricultural, industrial and economic planning. It declares that environmental problems are interconnected, and that society must end its practice of making others pay for the cost of degradation. Its governing principle is that polluters must pay for their action. This has not pleased conventional farmers, who feel coerced into action they would prefer not to take. Nonetheless, there remains widespread support for the basic principles of the NEPP (VROM, 1989; 1990; WRI, 1994). However, even without such a national strategy, policy-makers have found that the principle of integration can be furthered in agriculture by taking small steps to penalise polluters and to encourage resource conservers.

Tax the Polluters

There is a growing sense that green or ecotaxes are an efficient way to help meet environmental objectives, as well as to generate jobs and to raise government revenue (Tindale and Holtham, 1996; Jacobs, 1996; O'Riordan, 1996). These shift the burden of taxation away from economic goods, such as labour and sales, towards environmental bads, such as energy, transport, waste and pollution.

The market prices of food and other agricultural products do not reflect the full costs of the farming and food system as it currently operates. Some of these costs are not borne by the producer but dispersed through society. The producer does not need to pay for them and so has lower costs. Environmental taxes or pollution payments, however, attempt to internalise some of these costs in order to encourage individuals and businesses to use resources more efficiently. Green taxes, therefore, are a double dividend option. They cut environmental damage while promoting welfare.

Few European governments, however, recognise this potential dividend. They raise less than 10 per cent of their revenue from them (Tindale and Holtham, 1996). More than half of taxes are levied directly or indirectly on labour (through increases in tax and social security payments), a figure that has steadily grown from about 30 per cent in the 1960s. There is still a widespread view that eco-taxes are regressive, in that they stifle economic growth. Growing empirical evidence, however, suggests that this is not the case. There are already many green taxes and levies in place in European and North American countries (see Table 8.2).

The OECD indicates that the costs of complying with environmental regulations have had little or no impact on the overall competitiveness of countries. Environmental regulations and taxes appear to have made businesses more efficient and effective. There are also benefits in terms of employment. Stephen Tindale and Gerald Holtham's study for the IPPR suggests that 600,000 new jobs would be created in Britain by 2005 if a package of ecotaxes on commercial and industrial energy use, waste disposal, road fuel and quarrying were to be introduced (Tindale and Holtham, 1996).

Table 8.2 *Eco-taxes and charges in various European and North American countries (as of 1997)*

Measures	*Au*	*Ca*	*De*	*Fi*	*Fr*	*Ge*	*It*	*Ne*	*No*	*Po*	*Sp*	*Sw*	*UK*	*US*
Energy														
Carbon/ energy tax	✔		✔	✔				✔	✔	✔		✔		
Sulphur tax				✔	✔				✔	✔		✔		
NO_x charge					✔					✔		✔		
Agriculture														
Fertiliser tax									✔			✔		✔
Pesticide tax			✔	✔					✔			✔		✔
Manure charge								✔						
Water and wastes														
Waste disposal	✔		✔	✔	✔	✔	✔	✔	✔	✔	✔		✔	
Landfill tax								✔					✔	
Water effluent charge			✔		✔	✔		✔		✔				
Sewage charge			✔	✔		✔		✔	✔	✔	✔	✔	✔	✔
Other														
CFC and/or halon tax			✔							✔				✔

Note: Countries are Austria, Canada, Denmark, Finland, France, Germany, Italy, The Netherlands, Norway, Poland, Spain, Sweden, UK and US.
Sources: Sprenger, R-U. 1997. *Policy background paper on Environment and Employment.* Prepared for European Conference on Environment and Employment, 26–27th May, 1997, Brussels. OECD. 1997. *Environmental Policies and Employment*. OECD, Paris

Taxes on Pesticides and Fertilisers

Not many of these changes have directly affected the farming and food business, save for the introduction of taxes on fertilisers and pesticides by a small number of countries in Europe and most states in the US. In general the levels are low (OECD, 1995; 1994; Conway and Pretty, 1991; Center for Science in the Public Interest, 1995). For fertilisers they are currently of the order of US$0.1 to $0.4 per kilogramme of nitrogen, phosphorus and potassium in Austria, Norway and Sweden, though they are about a thousand times lower in most American states:

for example, $0.0006 to $0.00096 in Oregon, Wisconsin and Iowa, though 100 times higher in California at $0.02 per kg of nitrogen. (Finland's fertiliser taxes were phased out in 1994 prior to Finland's entry to the EU; subsidies are now used to encourage farmers to reduce fertiliser use.) For pesticides, they are of the order of 3 to 13 per cent of the retail price in Sweden, Finland, Norway and Denmark. Again, levels are much lower in the US, averaging about 0.7 per cent of sales prices (Center for Science in the Public Interest, 1995). In Iowa, for example, there are pesticide registration fees charged to manufacturers (0.2 per cent of gross sales of each product), to dealers (0.1 per cent), but no direct charges to farming. There are proposals to introduce similar taxes in Belgium, The Netherlands and Switzerland. (Levies have been used to limit pollution in the livestock sector. Levies are used in The Netherlands to penalise those farmers producing more livestock waste than their land can absorb.)

These taxes do raise reasonable sums of money, though it is not clear whether they have had any appreciable direct effect on encouraging a transition to sustainable agriculture. In the early 1990s, all states in the US raised about $59 million per year in pesticide fees and $20 million in fertiliser fees. California raises the most, about $24 million in total; Iowa just $2.5 million per year. Even though the tax levels are higher in European countries, the totals raised are generally smaller – about $3.5 million in Sweden and $20 million in Norway.

The Center for Science in the Public Interest in Washington has calculated that the totals raised could be very much higher. It calculates that a 5 per cent tax on pesticides (up from 0.7 per cent) and a 1.5 per cent fertiliser tax (up from 0.3 per cent) would result in a total revenue of $535 million per year (see Table 8.3). If this were to be invested in sustainable agriculture research and development, it would dwarf current USDA funding that amounts to just $8.7 million annually for sustainable agriculture. Even if all these costs were passed on to food consumers, it would only raise the annual food bill by $2.15 per person – equivalent to 16 cents per family of four per week.

A vital issue for sustainable agriculture is how governments use these funds. Most agree that they should be earmarked to support activities that further the transition towards sustainability. In Sweden they are used to support the input reduction programme; in Iowa and Wisconsin, to support research into sustainable agriculture; and in California to fund pesticide-related environmental programmes. But Finland used revenue to subsidise exports, and Norway to give income support to farmers (Pretty, 1995a; Governor's Office, 1994).

It is commonly felt that these levels have been set too low to affect consumption by farmers. Studies in the US, comparing alternative

Table 8.3 *Potential revenues from increased pesticide and fertiliser taxes in the US*

Tax level	*Pesticides ($ million/year)*	*Fertilisers ($ million/year)*
0.5%	42	37
1.0%	85	74
1.5%	127	111
3%	255	232
5%	424	370

Note: 1995 national average tax levels of 0.7% on pesticides and 0.3% on fertilisers.
Source: Center for Science in the Public Interest. 1995. *Funding Safer Farming.* Washington, DC

agriculture farms with conventional neighbours conclude that a 25 per cent tax on fertilisers and herbicides would not be sufficient alone to encourage farmers to convert from conventional to sustainable systems. Even though it would reduce net income by US$10 per hectare on conventional farms compared with just $1.25 on alternatives, taxes would have to be much higher to lead to big reductions in input use (Dobbs et al, 1991; Reichelderfer, 1990). In California, the low elasticity of demand for pesticide use suggests that a doubling of the pesticide tax would only reduce sales by 1 per cent. In The Netherlands, estimates show that the low elasticity of demand for pesticides means that a doubling of price would only produce a 12 per cent reduction in use (Robinson et al, 1996; OECD, 1994).

However, these concerns may be overstated. A well-designed package of taxes with regulations can increase price responsiveness. A nitrogen tax, for example, would be more effective when backed by minimum standards, alternative practices delivered by extension services, and codes of good practice. The impact will also grow over time. Demand is inelastic if there is an expectation that price rises will quickly be reversed. But if farmers come widely to accept that higher prices incorporating the ecotaxes are here to stay, then further behaviour changes will occur. Sweden's 65 per cent reduction in pesticide use may be the best example of a pesticide tax that has produced a significant effect. In 1990, the government, emboldened by success, mandated a further 50 per cent reduction by 1997.

In the UK, there are no pesticide or fertiliser taxes. In 1995, the Departments of Environment and Health jointly issued the UK Environmental Health Action Plan for consultation, which proposed limiting pesticide use; the DoE/MAFF 1995 White Paper *Rural England* proposed an Action Plan for the Responsible Use of Pesticides. Although these encourage farmers to use the minimum amount of pesticide consistent with protecting health and the environment and with producing food, they still lack clear goals and are poor at measuriing progress (DoE and DoH, 1995; Beaumont, 1995).

National Targets for Input Reduction

Several countries have set ambitious national targets in the mid 1990s for the reduced use of inputs. Sweden aimed to reduce nitrogen consumption by 20 per cent by the year 2000. The Netherlands also sought to cut pesticide use by 50 per cent by the year 2000 as part of its 'Multi-Year Plan for Crop Protection'. The cost of this reduction programme was estimated at US$1.3 billion, most of which was to be raised by levies on sales. Denmark aimed for a 50 per cent cut in its pesticide use by 1997, a plan which relied mostly on advice, research and training. Canada aimed for a 50 per cent reduction in pesticide use by 2000 in Quebec and by 2002 in Ontario. And in the US, the Clinton administration announced in 1993 a programme to reduce pesticide use while promoting sustainable agriculture. The aim was to see IPM programmes on 75 per cent of the total area of farmland by the year 2000.

Supplemented by other policy measures, such as new regulations, training programmes, provision of alternative control measures and reduced price support, there have been some substantial reductions in input use in recent years. In Sweden, pesticide consumption fell by 65 per cent between 1985 and 1993 (from 4.5 to 1.5 million kilogrammes of active ingredient); in Austria there was a decline in consumption of fertilisers, especially potassium; in Denmark, pesticide consumption fell by 40 per cent between 1985 and 1995 (from 7 to 4.3 million kilogrammes of active ingredient); and in The Netherlands, there was a 41 per cent fall between 1985 and 1995 (from 21.3 to 12.6 million kilogrammes of active ingredient) (Matteson, 1995; Beaumont, 1993; Jorgensen, 1997; Emmerman, 1997). However, the significance of these apparent sharp falls in use is disputed. The Netherlands was and still is one of the most intensive users of pesticides per hectare of farmland in the world. A 50 per cent cut in pesticide use still leaves farmers using very large amounts. In 1987, 20 kilogrammes of active ingredient per hectare of cropland were applied in the Netherlands, compared with 1.1 kilogrammes in Sweden, 1.8 kilogrammes in the US, and 2.6 kilogrammes in Denmark. Moreover, much of the decline is said to have been achieved through reducing soil sterilants.

In Sweden, half of the decline was due to lower dose applications; the rest was attributed to new lower dose compounds. For example, phenoxy-herbicides applied at one to two kilogrammes active ingredient per hectare were replaced with sulphonurea products applied only at

0.004 to 0.006 kilogrammes active ingredient per hectare. In Denmark, reduction has not been accompanied by a cut in the frequency of application, which remains at the 1981 level of 2.5 doses per hectare per year (Matteson, 1995). Success has been achieved without a diminished dependence on pesticides, which for many embodies the spirit of 'pesticide reduction'.

Although it is clear that there have been some reductions in pesticide use as a result of increasing taxes, combined with regulations, this has not happened yet for nitrogen. There have been recent designations of nitrate sensitive areas and nitrate vulnerable zones, in which farmers accept payments to change their practices in order to reduce the likelihood of leaching. But these represent only a very small proportion of the total farmed area. As a result, there is still debate over the best available mechanisms to limit or restrain nitrogen use and therefore nitrate leaching to groundwater. The nitrate quota is one mechanism that has been advanced in the past, though most now believe that the best option would be to create incentives for farmers to adopt resource-conserving and regenerative technologies, beginning with various forms of ICM (FoE, 1992; 1995). Since the early 1990s, more comprehensive evidence has emerged that farmers are able to maintain or increase gross margins while substantially cutting inputs (see Chapter 3). New technologies now exist that were simply not available five years ago. Encouraging the adoption of these technologies would be less costly and more efficient than a policy of nitrogen quotas.

Smaller-scale examples come from Germany, where water companies (like those throughout the EU) are having to respond to the EC directive on drinking water. This permits only 0.1 microgrammes per litre of any one active ingredient in drinking water, and 0.5 microgrammes per litre of all active ingredient. Water supply companies are responsible for supplying clean water to their customers, and they will have to pay to remove pesticide residues from water. What they have come to realise is that it is considerably cheaper to pay farmers in catchment areas to convert to organic farming than it is to wait until pesticides are applied and then try to remove them (Jewell, 1996).

According to environmental analyst Topsy Jewell, water distributors in Munich, Osnabrück and Leipzig are now paying farmers to convert their entire farms. They are also helping farmers to develop new marketing strategies, including supplying local markets and institutions such as the water companies' own canteens. There are some important components of the schemes – the incentives are high (550 Deutschmarks per hectare for three years), and success for the water companies only comes when all the farmers in a catchment make the transition to more sustainable practices. As of yet there are no plans by water companies or the

Environment Agency for similar schemes in Britain, though the cost of pesticide removal is some £121 million annually. One company, the Essex and Suffolk Water Company, aims to spend £15 million on pesticide removal to supply 250,000 people (£60 per person). But the DoE has concluded that curbs on pesticide use are a far cheaper way of controlling pesticides in water supplies than removing them in treatment (ENDS, 1995). Water protection zones farmed with low-input or organic farming are seen as the best option for both farmers and the general public.

Rewarding the Resource Conservers

The alternative to penalising farmers is to encourage them to adopt alternative low- or non-polluting or degrading technologies by acting on subsidies, grants, credit or low-interest loans. These could be in the form of direct subsidies for low-input systems or the removal of subsidies and other interventions that currently work against alternative systems. Acting on either would have the effect of removing distortions and making the sustainable options less unattractive.

Although some resource-conserving technologies and practices are currently being used, the total number of farmers employing them is still small. This is because adopting these technologies is not a costless process for farmers. They cannot simply cut their existing use of fertilisers or pesticides and hope to maintain outputs, making their operations more profitable. They will need to substitute something in return. They cannot simply introduce a new productive element into their farming systems and hope that it succeeds. They will need to invest labour, management skills and knowledge. They will need to experiment to innovate. But these costs do not necessarily go on forever, and much can be done to support the transition.

Regional Policies in the German Länder

In Germany, there are a wide range of agri-environmental schemes developed by the Länder (regional governments) (CPRE, 1995; Wilson, 1995). In 1997, some 200,000 farmers had joined the schemes, covering some 17 million hectares, about a tenth of total agricultural area. To date, the main uptake has been for extensive grassland management, about 80 per cent of the total.* However, some states have made important

* Altogether, some 4.57 million hectares are said to be registered under regulation 2078/92, but one scheme implemented on 3.25 million hectares in two states is highly criticised for being very weak extensification. Counting all schemes, 330,000 farmers have entered 4.57 million hectares. Not counting these two schemes, there are some 200,000 German farmers farming 1.322 million hectares under 2078/92.

steps towards positive environmental management.

The MEKA (*Marketentlastungs und Kulturlandschaftsausgleich*) scheme of Baden–Württemburg gives farmers an à la carte menu of technologies from which to choose, each one earning them ecopoints, and each point bringing them 20 Deutschmarks per hectare. For example, using no growth regulator attracts ten points; sowing a green manure crop in the autumn earns six points, applying no herbicides and using mechanical weeding gets five points; cutting back livestock to 1.2 to 1.8 adult units per hectare brings three points; and direct drilling on erosive soils earns six points. Direct environmental protection measures include up to 15 points for wetland conservation, habitat management and for keeping rare breeds.

The national cost of the scheme is split between the federal government and the regional government, with the CAP picking up the rest under the agri-environment regulation. 'By encouraging care of the environment and the traditional landscape with grant aid, the scheme is helping family farming businesses like ours to survive', said George Mayer, who farms a 60-hectare mixed farm. By 1997, 102,000 farmers had signed up, with the result that 220,000 hectares of grassland are now managed extensively, 225,000 hectares of arable are also managed extensively, with a considerable proportion no longer using pesticides or fertilisers: 97,000 hectares of protected vineyards and orchards. But only 2300 hectares of land have been entered for positive nature conservation, such as new hedge planting or riverine management. Some 14,000 hectares, however, have become organic.

In Hessen, some 82,000 hectares are farmed under the HEKUL programme, which extensifies farming and encourages the adoption of organic farming technologies. In Rheinland Pfalz, the FUL Programme has 1900 participants who farm 38,000 hectares; again, this includes payments for organic production, low-input integrated practices, and extensified grassland farming (HEKUL stands for *Hessiches Kulturlandschaftsprogramm*; FUL for *Förderprogram Umweltschonende Landbewirtschaftung*).

Policy Innovations in Switzerland

Switzerland's approach to increasing the sustainability of farming has now become one of the most forward looking in Europe (Roux and Blum, 1998; Michel Pimbert, pers comm, 1997). Article 31 of the Federal Agricultural Law now differentiates between three levels of support depending on the sustainability of agriculture. Tier one is support for specific biotypes, such as extensive grassland and meadows, high-stem fruit trees and hedges. Tier two supports integrated production with reduced inputs, meeting higher ecological standards than does conventional farming. Tier three is support for organic farming.

The most difficult policy issue was agreeing on standards for the reduced-input, integrated farming. Fortunately, since 1998 Switzerland has had a network of over 200 farms which test the economic and ecological viability of resource-conserving technologies and practices. This created a good empirical base and allowed for the sharing of both data and perceptions so that common standards could be agreed. There are six minimum conditions necessary if farmers are to receive payments for integrated production:

- At least 5 per cent of the land must be conserved as a nationally important biotype.
- Nitrogen and phosphorus nutrients must be in balance on the farm.
- Pesticides have to be reduced to established risk levels.
- Livestock husbandry must meet defined 'animal friendly' conditions.
- Records must be kept of all technical aspects of the farm.
- Participation in extension groups on integrated production is compulsory

Another vital difference between the Swiss style and most of those implemented under agri-environmental schemes in the EU is that responsibility to set, administer and monitor is delegated to farmers' unions and farm advisors, local bodies, and NGOs. Policy has always attempted to stop rural population decline in less favoured mountain areas, and so the importance of monitoring rural social capital is seen as a central part of agricultural policy. More than 20 per cent of all 75,000 Swiss farms are now participating in more sustainable agriculture, including 11,000 which manage 18,000 hectares of meadows and protect 1.5 million high-stem fruit trees; 9000 which meet the requirements of the integrated standards; and 1500 which are organic.

A New CAP for Rural Europe?

The main policy instrument for agriculture in the EU, the CAP, has undergone gradual greening in recent years. The 1992 MacSharry reforms introduced agri-environment support under regulation 2078/92, with the responsibility falling to individual member countries over the degree of implementation and support. As a result, uptake has been variable, from 100 per cent of farmland designated in Austria to just 0.2 per cent in Belgium (see Chapter 2).

At the time of writing, the CAP is once again under scrutiny, with a further round of reforms likely to be implemented by 1999. The pressures on the existing policy come from four sources. The first is budgetary: the cost of the CAP in 1997 was some 41.2 billion ECUs, which was over half of the EU's total budget and represents a cost of about 108 ECU per European citizen. This is no longer defensible. The CAP has encouraged farming that has damaged natural resources and contributed to rural community breakdown. The 18 million unemployed people in the EU have little or no assets, yet the EU's seven million farmers with assets receive more than half of the total budget. Birdlife International calculate that the CAP costs 108 times more than the EU's youth, culture and education budget; 312 times more than the EU's environment budget; and 2161 times more than the EU's consumer protection budget (Birdlife International, 1997).

The second pressure is over EU enlargement. Several east and central European countries are hoping to gain membership in the EU by 2002, with others to follow. The EU cannot afford to pay all these additional farmers similar support. MAFF calculated in 1996 that the cost of expanding the CAP to Poland, Hungary, the Czech Republic and Slovenia would be an additional 15 billion ECUs per year.

The third pressure comes from the WTO and indirectly from the US. The high levels of support for production are counter to the WTO's attempts to expose all producers across the world to similar prices. Payments to farmers will increasingly have to be for environmental and social goods if they are to continue. The US has already prepared itself for future negotiations by passing the Federal Agricultural Improvement and Reform Act in 1996, which decoupled support from production and abolished supply controls.

The fourth pressure comes from increasing consumer disquiet over the farming and food system. Consumers pay for farming several times over – through their taxes being used for direct support; through high food prices; and through the external costs that are imposed on natural and social capital. The OECD calculates that the CAP cost consumers 28 billion ECUs in 1990 by raising food prices (Birdlife International 1997). Consumers are unlikely to want to compensate farmers when they themselves bear all these costs. There is emerging consensus from a wide range of environmental, farming, consumer and community groups across Europe that the public's support through the CAP should be targeted much more at delivering public benefits. These would be in the form of improved natural capital, improved social capital in rural areas, and improved economic growth and job opportunities. In October 1996, the UK Agricultural Reform Group convened a seminar in Brussels where farming and environmental interests from all 15 member states were represented. The seminar was also attended by HRH The Prince of Wales, EU Commissioner for the Environment Ritt Bjerregaard, and the Chef du Cabinet of the Agriculture Commissioner, Franz Fischler.

Despite the great differences represented, this seminar agreed that all payments to farmers would have to be decoupled from production and recoupled to environmental and social goods. Payments would be in the form of a Basic Area Payment, and would be made conditional on farming protecting or regenerating the environment. Farming can deliver much, yet it has failed to do so. Most appreciate that the opportunity exists for farming to recapture a key role in rural regeneration. The advantage of such a Basic Area Payment is that it would be totally decoupled, and so WTO-friendly; and would be extremely simple and therefore compatible with the existing IACS system (ARG, 1997). This implies a policy framework that is much more integrated, putting rural development and support for farming together. Most agree that a reformed CAP would therefore decouple payments from productivity, create new jobs, protect and improve natural resources, and support social organisation and new partnerships. It would do this through environmental payments, support for rural development and some market stabilisation.

Environmental payments would be tied to a menu of options for farmers. These would have several tiers, giving farmers the choice to select their practices and the types of support they could expect.

- Tier 0: the basic regulatory floor, in which farmers receive no public support, farm at world prices, but follow codes of practice and regulations that seek to minimise environmental damage.
- Tier 1: public support for habitat protection and restoration, such as for woodland management, ponds, hedgerows and wetlands. This tier essentially supports greening of the edge but not the middle.
- Tier 2: payments for a transition towards sustainable agriculture, incorporating integrated crop and livestock management and the use of regenerative technologies; this level of support does much more than protect the environment since it seeks to rebuild natural capital.
- Tier 3: the highest level of support would be for organic farming and the redesign of food and farming systems.

The basic idea is that all farmers would have access to these payments. They would not be limited to designated parts of the EU territory. Set-aside would be eliminated. The second set of payments would be for rural development initiatives. These would be explicitly for supporting capital development, through local participatory processes and business development, in order to encourage the growth of job opportunities. Additional support in the system may also be necessary for both market and transitional assistance.

There are many other difficult and contentious features that need resolution. These include modulation – should there be an upper limit for payments according to the size of farm?; intervention – there may be a need for a limited intervention scheme for cereals, but should the rest go?; quotas – Agenda 2000 proposes to maintain dairy quotas to at least 2006, with no suggestion to get rid of the others affecting suckler cows, sheep and sugar beet. A basic area payment would render suckler and sheep quotas unnecessary; should milk and sugar beet both be continued?

In practice, however, it is unlikely that all this will happen in the next round of reforms. The EU organised a conference on Integrated Rural Development in Cork in November 1996. This set out new principles for more sustainable farming and rural community development. The later Agenda 2000 proposals issued in mid 1997 appeared, however, to reject entirely the agreements made in Cork (EC, 1997). Individual member countries have their own views, and many of these are in conflict. What will emerge is unlikely to be a result of rational and carefully analysed policy opportunities. It is more likely to be a function of personality differences within the European Commission and between ministries of agriculture in member states.

Others have emphasised the need to reform the EU Cohesion Policy (WWF, 1997). It accounts for about a third of EU funding. Through the Structural Funds and the Cohesion Funds, it targets regions where development is 'lagging behind', industrial regions, and regions in decline. Many argue that funds have not been best targeted at local needs (see Chapter 6). The opportunity to integrate CAP and cohesion policy funds into one single coherent and integrated 'Sustainable Regional Development Policy' has been put forward by WWF. But, once again, defence of territory and funds within the EC make this an unlikely outcome, despite the rationality of the proposal.

It is likely, nonetheless, that some progress will be made, with more payments shifted to the agri-environment envelope. A move from the 1997 level of 4.1 per cent of CAP budget for agri-environment schemes to, say, 25 per cent would represent substantial progress. But opportunities to make huge strides towards rebuilding natural and social capital for the benefit of rural people throughout Europe will probably be missed in this round. However, these pressures are unlikely to go away. It is possible that all this can still be achieved during the first decade of the 20th century as consensus continues to grow.

An Emerging Policy Focus on Rural Jobs?

In addition to the changes in the CAP, there is a growing policy focus on the need to create jobs. The EC and member states are increasingly concerned about the impact of their policies on rural employment. As we have seen, the CAP and associated rural development policies have only succeeded in cutting jobs in rural areas (by nearly two million in the agriculture sector during the 1980s alone). In a speech in June 1996, President of the EC Jacques Santer (EC, 1996), said: 'the Commission intends to highlight the priority attached to employment'. The structural funds have a budget of 170 billion ECUs for 1994–1999 and 'it is essential that they be managed in such a way as to ensure an optimum impact on employment'. The Commission has suggested that member states select towns or regions for an 'exceptional mobilisation campaign' to promote local employment on a 'territorial pact'. This type of geographic focus is essential if meaningful partnerships of different interests are to emerge.

Another policy shift which should increasingly help community groups and farmers is the emerging focus on small- and medium-sized enterprises (SMEs). Small businesses in Britain have created 2.5 million jobs in recent years, at the same time as big companies have been shedding jobs. The EC authorises aid to SMEs, particularly where they develop new technology, save natural resources, and develop new products. As the commission itself put it: 'state aid comes to billions of ECUs a year, most of it going to large companies. It is time a very close look was taken at the effect of current policies on employment' (EC, 1996). Some countries already do so. In France and Italy, domestic policies to support rural communities have always been stronger. Since 1971, France has taken specific measures to aid investments by farmers, entitled *indemnité spéciale montagne* (special mountain grants), as remuneration for public services provided by farmers (Lerot, 1996). In Italy, the 1994 act for mountain development provides for soft loans and tax breaks to allow families to buy land and prevent fragmentation; free technical support to farmers; a high ceiling for small businesses before they have to register for VAT; support to young entrepreneurs; payments for land stewardship; and territorial pacts to promote partnerships between commerce, government and banks for employment projects.

In the Bavarian Alps of Germany, the state government is helping to support farmers in Hindelang, a major resort dependent on tourism. The area is famous for its rich ecological diversity, which is closely linked to traditional farming. Yet the number of farms has fallen from 200 to 90 in recent years, and meadows are reverting to scrubland. Realising this could mean the disappearance of the area's outstanding beauty, which would threaten the 80 per cent of jobs in the region dependent on tourism, authorities have encouraged new partnerships, developed labelling for agricultural products, and are paying farmers to convert to ecological farming (Haug, 1996).

Policy Processes

Green Conditionality or Public Participation?

It is increasingly clear that there are public policy mechanisms available to support the move towards more sustainable rural communities and the countryside. But whether these will work in the long run depends also on the processes by which policies are established. Broadly speaking, a coercive policy intended to achieve a particular outcome tends not to work in the long run. A democratically developed policy, arising out of consultation and participation, and reflecting local people's and farmers' needs and specific circumstances, is much more likely to receive widespread support (Pretty, 1995a).

It is important to be clear about just how policies should try to address sustainability. Sustainable agriculture is not a set of practices to be fixed in time and space. It requires the capacity amongst farmers and other actors to adapt and change as external and internal conditions are altered. Yet there is a danger that policy, in the name of sustainable development, will prescribe acceptable practices for farmers, preventing the emergence of locally generated and adapted technologies. During the course of this century, environmental policy worldwide has taken the view that rural people are mismanagers of natural resources. The history of soil and water conservation, of rangeland management, and of national parks and protected areas shows a common pattern. Technical prescriptions derived from controlled and uniform conditions are applied widely with little or no regard for diverse local needs and conditions. Differences in each environment then often make the technologies unworkable. When they are rejected locally, policies seek success through manipulating social, economic and ecological environments, and eventually through outright enforcement (Pretty 1995a).

For sustainable agriculture to work, policy formulation must not repeat these mistakes. Policies will have to emerge in a new way. They must be more enabling, creating the conditions for sustainable development based on locally available resources, skills and knowledge. Achieving this will be difficult. In practice, policy is the net result of the actions of different interest groups which pull in complementary and opposing directions. It is not just the normative expression of governments. To be effective in the long term, policy will have to recognise this and invest in processes that bring together different actors and institutions.

But good ideas and intentions do not alone suffice. Research into the implementation of the recent Netherlands environmental legislation (both the Multi-year Crop Protection Plan and the Nature Conservation Plan) is showing that the old approach to policy development often ends up in impasse. Niels Röling, Professor of Agricultural Knowledge Systems at Wageningen, stated recently: 'implementing cut and dried policies decided at the top does not work. What is required is policy for local interactive processes among stakeholders' (Röling, 1995).

Centrally Planned Environmentally Sensitive Areas (ESAs)

A similar problem is widely reported by farmers in Britain who are part of, or could be if they chose, an ESA. Farmers in ESAs are able to claim extra payments if they agree to farm according to a number of conditions set by government (see Chapter 2). The basic objective is to encourage farmers to protect nationally valued habitats and landscapes. ESAs now cover 10 per cent of the agricultural land of the UK. But many farmers chose not to join ESAs – the 1997 take-up was just 28 per cent of all the designated land, about 483,000 hectares. Furthermore, many farmers who have accepted support are unhappy with the process. They feel that the prescriptions offer them little flexibility, even if they wish to farm sustainably.

One study of the Welsh Cambrian Mountains ESA by Geoff Wilson and colleagues at King's College found that, although the uptake at 48 per cent was high relative to other ESAs, it was the larger farmers who were benefiting most (Wilson, 1997). The ESA, by targeting specific habitats such as semi-natural rough grazing or woodlands, tends to favour larger farmers as they are more likely to have farms with these habitats. As a result, larger farmers get more income from the ESA scheme, and some have now bought smaller family farms. This has brought new divisions to the close-knit rural communities. Geoff Wilson put it this way: 'some participants... were getting increasingly disillusioned with farming since the ESA scheme started. They felt insufficiently rewarded for their environmental management practices as stewards of the land.' Even more importantly, participation in the ESA appears not to have affected the attitudes of farmers: 'only a small fraction seem to be influenced in both their attitudes and behaviour by the scheme'.

The Blackdown Hills are a remote and traditional farming area of Devon. More than 80 per cent of the farms are less than 50 hectares in size, and farm numbers have fallen by a third between 1986 and 1996 (David Dixon, pers comm, 1996). Although it was designated an ESA in 1994, the first two years saw only 17 per cent of the land entered into the scheme. The reasons for the poor take-up are contested. At a meeting in the Exmoor National Park in 1996, the local project officer, David Dixon, indicated that 'the top-down approach was badly received in an area known for its independent spirit'. But an ADAS officer responsible for the centralised design and implementation of the ESA put it differently: 'I fully refute that it [the scheme] was top-down; the sole reason why farmers' take-up was so poor was that payments weren't pitched high enough'. However, offers of a 24 per cent rise in payments during 1996 attracted no more entrants to the scheme.

To some people, participation is simply a matter of paying enough and getting farmers to do what they want. But this type of 'bought' participation simply does not work. Payments buy short-term acquiescence but not long-term changes in attitudes and values. Two farmers at the same meeting were able to be clear about what was needed. One, Humphrey Temperley, said: 'I am a victim of an ESA.' Another, Bill Geen, put it this way: 'ESAs should be FSAs – farmer sensitive areas. The schemes must be flexible and farmer friendly. The most important thing is the project staff: they must be sympathetic, knowledgeable, flexible and consistent.'

But it does not have to be like this. When plans are developed jointly and openly, then farmers are much more willing to give their support. The nearby Exmoor National Park is a good example. When the Exmoor National Park Authority (NPA) piloted a moorland protection scheme with local farmers, it was designed jointly by the authority and the farmers. As a result, farmers were presented with a range of alternative practices from which they could chose. Whole farm plans are drawn up around the kitchen table; each reflects the individual farm conditions; and 20-year index-linked agreements are signed. The plans have joint ownership, and Exmoor NPA are flexible. Says David Lloyd, of the park authority, 'if things don't work, we change them'. The pilot scheme was more expensive in management terms; this type of interactive participation costs more than imposition. But the long-term returns are likely to be much greater. Similar principles have been adopted by the Tir Cymen project in Wales and the North Yorks Moor Farm Scheme. Both seek to develop whole farm plans with the full involvement of farmers, but with clear principles and objectives about what is a desired outcome for natural capital.

Working Policies and Local Partnerships

Many innovative programmes have emerged in recent years which make good use of public money for sustainable agriculture and community regeneration. Some have the support of national policies, such as the Plans de Développement Durable in France; others are regional, such as those of the German Länder. Yet others are a lower level in designated areas, such as national parks or areas under particular threat.

The Plan de Développement Durable, France

In 1992, three French government ministries, the Ministère de l'Agriculture de la Pêche et de l'Alimentation, the Ministère de l'Amenagement du Territoire de la Ville et de l'Integration and the Ministère de l'Environment, launched an initiative to bring together a wide range of farmers to develop new sustainable approaches (Viaux and Rieu, 1995). This was the Plan de Développement Durable and involves some 1200 farmers spread over 59 regions. These are part of a nation-wide experiment to find ways to improve farm revenue, local working conditions for rural people, and the state of natural capital. The aim is to develop policies, rules and financial support that will initiate sustainable agriculture but that will accommodate local needs and conditions.

Many partnerships have been developed. In the Gironde on the Atlantic coast, wet meadows in the Arcachon Basin were bought by local authorities to stop them from being abandoned; as a result, extensive grazing has been reintroduced, bird diversity maintained and farm income sustained. In the Marne-et-Loire, the president of the Group of Beef Producers described the effect: 'for four years, our group has chosen to manage beef under our own label (*Label Rouge*). There is a balance between production and land management, with beef fattened on natural meadows. The PDD farmers are showing us the way.' In the Seine–Maritime, the mayor described the value of local involvement in the programme for watershed and erosion management: 'the community must engage with farmers in these preventative actions rather than slipping into curative control when it is always too late and costs too much'.

Philippe Viaux and his colleagues at the Institut Technique des Céréales et Fourages in Biogneville have been working with a group of farmers in a watershed that provides drinking water for Paris. La Ferté–Vidame has seen nitrate concentrations rise for 30 years, and the water company is increasingly concerned that it will not be able to supply wholesome water to Parisians. The water company had first offered to buy the farmland located in the sensitive areas. But this simply served to antagonise farmers and they rejected it out of hand.

Following earlier experimental work on farms, it was found that integrated farming with substantially reduced inputs, including a reduction by half for fertilisers, could still mean equal or better economic returns (Viaux and Rieu, 1995; Ansay and Viaux, 1996). These technologies were spread to 16 more pilot farms, and extensionists and researchers were brought together on a technical support committee. This provides training, crop walks and other technical support for farmers. Farmers were initially anxious, as the agronomic and economic solutions were not 'on the shelf' for them to use. However, over the four years to 1996, variable costs fell steadily from about 1900 to 1500 French francs per hectare (a fall of 21 per cent), while gross margins rose year on year from 5800 to 6300 French francs per hectare (a rise of 9 per cent).

As farmers have learned new technologies and practices, they have become better at management. The challenge for the partnership, however, is to expand its operations from 10 per cent of the watershed to at least 80 per cent, if any permanent impact on water quality is to be achieved.

The North Yorks Moors (NYM) National Park Farm Scheme, North Yorkshire

The NYM National Park receives some ten to 12 million day visitors each year. In recent years, there have been major environmental changes, with the amalgamation of farms and a sharp decline in employment opportunities. The National Parks Committee had been working, on a small scale, with farmers during the 1980s on projects such as tree planting, drystone wall repair, bracken control and woodland management, but realised it needed to take a more coordinated approach. The NYM Farm Scheme was launched in 1990, with support from the National Park budget (NYM, 1994; Peter Barfoot, pers comm, 1996; Anya McCracken, pers comm; NYM National Park reports and documents).

The NYM Farm Scheme encourages sensitive land management while maintaining farm viability. At present it works with farmers within selected dales. Management agreements are developed with each farmer. These aim to guarantee the conservation of landscape, wildlife and historic features in the National Park, while creating new environmental benefits; they maintain the rural fabric of the dales; they restore and conserve vernacular farm buildings; they stimulate the rural economy and create jobs; they ensure that farmers still have flexibility in their land management decisions. The scheme is offered to all farmers in a designated area, and by 1996 there was a 90 per cent uptake amongst eligible farmers – far better than for ESAs. The personal contacts between staff of the scheme and local farmers is seen as crucial to long-term success, and peer pressure amongst local people helps to keep support high. The management agreements are for five years, are individually tailored, and make provisions for maintenance and management payments. The NYM Farm Scheme has resulted in a marked improvement in the landscape (particularly farm buildings and walls) as well as in farm incomes and has raised farmers' awareness and knowledge of conservation. A recent survey of visitors who regularly walk the landscape showed that they recognised the changes and valued them. Altogether, 16 new jobs have been created by the 110 farm agreements, equivalent to one for every seven farms.

National Park Partnerships in France

There are a number of important initiatives in the Parcs Naturels Régionaux in France that seek to reconcile local economic development with nature conservation (Club de Bruxelles, 1994). The common features of success are that action occurs at a platform level, allowing an integrated approach; there is good negotiation and communication between the various local actors; and the contracts for farmers are voluntary. As the schemes were not imposed, but rather developed through an open and participatory process, farmers do not feel coerced.

Haute Chaumes and Haute Vallées Vosgiennes

The high pastures of the Vosges Mountains, long sustained by farming, are under two pressures. Land use has intensified in some areas, leading to declining biodiversity; and at the same time land is being abandoned in other areas, leading to tree encroachment and again loss of conservation value. Technical committees were established by the park authorities to draw up a common land management policy with all the main actors. Surveys were carried out with farmers. It was particularly important to get everyone involved as the area falls between two regions and is under the control of many local authorities. Farmers have now entered into contracts to follow more sustainable farming practices, such as controlling livestock numbers, not using fertilisers, and not burning.

Marais du Cotentin et du Bessin

The marshes on the Cherbourg Peninsula are of international ecological importance, but deterioration of the habitat is now widespread. Neglected wet grasslands turn to peat bogs and become waterlogged; the soils become acidic; plant biodiversity falls; and the area becomes less attractive to agriculture. This encourages greater intensification on the remaining grasslands, the productivity of which is increasingly maintained with high fertiliser inputs. A working group was established with local and central government, farmers, conservationists, hunters and anglers. The group conducted surveys and experiments with alternative practices. As a result, farmers now enter into contracts to use extensive farming practices that preserve the unique landscape. This partnership approach has also meant that some 3500 hectares of communally owned land have also been entered into the scheme.

La Albufera National Park, Spain

The irrigated rice-producing area to the south of Valencia is a 20,000-hectare site of international importance to birds and fish (Barrés, 1996). Rice has been cultivated traditionally for several hundred years, though recently there has been an increasing reliance on machinery, fertilisers and pesticides. These have increased yields but have led to environmental and habitat damage. In 1995, a new scheme under the 2078/92 agri-environment regulation was established with 75 per cent funding from the EC. Farmers are subsidised to reintroduce traditional measures, such as mechanical weeding and premature flooding of rice fields in November to March, and reduce their use of inputs. In two years, 2750 agreements were signed for 9500 hectares of land. The scheme has already protected one fish species that lives in the rice fields, and populations of teal and shovelduck are rising.

Tir Cymen Farm Scheme, Wales

One of the best rural partnership schemes in the UK is Tir Cymen, run by the Countryside Council of Wales in the areas of Swansea, Dinefwr and Meirionydd between 1992 and 1997, and now to be expanded to the whole of Wales (CCW, 1992; 1996; ADAS, 1996). It seeks to reward farmers for using their skills and resources to look after the landscape and wildlife, as well as to improve their farming practices. Tir Cymen translates roughly as 'a well-crafted landscape'.

The Welsh countryside was created, and much of it is still managed, by farming practices which have become increasingly difficult to sustain in the face of declining agricultural prices. Tir Cymen has shown how environmental management can be integrated with agricultural production on ordinary farms. It is entirely voluntary, and offers an annual payment in return for farmers agreeing to follow sustainable management guidelines as part of a whole farm plan. Priority is given to activities which offer the most public benefit in environmental terms. Farmers are obliged to follow the Tir Cymen code on the whole farm for ten years, making improvements to arable and livestock components, woodlands, archaeological features, stone and slate walls, and buildings. The management guidelines encourage environmental improvements throughout farms, including the transition to more sustainable infield farming practices.

To mid 1996, 718 farmers had signed up to the scheme, with a total area of nearly 74,000 hectares. This included the safeguarding of 22,200 hectares of heather moorland, 10,700 hectares of upland grassland, 3260 hectares of woodlands, 3170 hectares of marshy grassland, and 1070 hectares of flower-rich meadows, together with smaller amounts of highly targeted areas, such as 65 hectares of winter grazing for swans and 57 hectares of cliff-top grazing. By 1997, there were some 930 farmers participating, and the scheme is now being made available to all farmers in Wales.

The recent ADAS evaluation of the scheme has shown just how beneficial this scheme is to farming, wildlife, the environment and local economies (see Box 8.1). Tir Cymen was found to maintain and enhance farm incomes while farming became more environmentally sensitive. It generated and maintained employment on farms through the need to carry out environmental maintenance work; it generated and maintained jobs in local economies through the purchase of local materials and services by farmers in the scheme; and it generated additional benefits for local economies over and above the payments to farmers – the expenditure on the scheme was found to lever up a further 123 per cent of spending by local farmers, particularly on capital works.

The Darby Watershed Project, Ohio, US

The Darby Project in Ohio is a collaborative effort funded by the WK Kellogg Foundation to preserve, enhance and maintain some 163,000 hectares of watershed and its diverse ecological and farming systems. A wide range of government, non-government, private and community groups are working together to change the attitudes and practices of both farmers and local people (Wes Beery, pers comm, 1997; Darby Project reports; Fisk et al, 1996).

The watershed is primarily farmed – some 80 per cent is farmed intensively to produce maize and soybeans, with little grassland, legumes or small grains as ground cover. Some 10 per cent of the watershed is poor quality pasture. Urbanisation is rapidly consuming large portions of agricultural land. Both modern farming and urban sprawl are threatening a biologically rich landscape: there are 86 species of fish and 40

Box 8.1 The impacts of the Tir Cymen scheme to the end of 1994*

A Farming Performance

Average farm income	up £1616
Arable costs	
pesticide costs	down 18 per cent
fertiliser costs	down 10 per cent (on non-Tir Cymen farms, fertiliser costs were up 10 per cent)
organic fertiliser costs	up 47 per cent
Livestock and grassland	
fertiliser costs	down 4 per cent
contract labour costs	up 9 per cent
grass keep costs	up 95 per cent
sheep numbers	down 26 per farm (total fall of 3466)
ewe quality & lambing %	both up
dairy cow numbers	up 3.5 per farm
milk yields	up 125 litres per cow

B Environmental Performance

Substantial increase in environmental works, including walling, woodland and moorland management

C Labour

Family employment on farm	no change
Permanent jobs created on farm	no change
Casual labour jobs created	204 created (rising to 263 at end 1995–96, and 511 by end 1996–97)
Casual labour jobs safeguarded**	49 existing jobs
Environmental work generated	62 person years (rising to 154 after year 5)

D Local Economy

Businesses showing increased demand for services***	83 per cent
Jobs created and retained in 35 local businesses	16
Ratio of jobs created to existing jobs	1 to 7.6
Extra value of capital payments accruing to local economy through farmers' increased spending	up 23 per cent
Expenditure by sample farmers on wages and materials for environmental work	£448,246

* The study by ADAS compared 131 farmers in the scheme with their performance before the scheme started, and compared current performance with 30 non-Tir Cymen farmers.

** Calculated by comparing job losses on non-Tir Cymen farms.

*** Demand was for timber, fencing, gates, stiles, bird boxes, contract fencing, hedging and walling.

Sources: Countryside Council of Wales. 1992. *Tir Cymen.* Bangor, Wales; ADAS. 1996. *Socio-Economic Assessments of Tir Cymen.* Report by ADAS for Countryside Council of Wales, ADAS, Ceredigion. CCW. 1996. *Tir Cymen: 4 Years of Achievement.* CCW, Bangor, Gwynedd

species of freshwater mussels in the rivers and creeks, including 104 species of birds, 35 of mammals, and 33 of reptiles and amphibians. More than 25 endangered species of plants are also found.

Conservation activities are not new to the area. Authorities have been trying to promote soil and water conservation in farming since the 1940s. What is new, though, is the way that many agencies are now collaborating to solve conservation problems on a whole watershed basis. This means involving not just farmers, but also leisure users and environmental and community groups. Nearly 5000 children are involved in stream-quality monitoring programmes; others are engaged in labelling storm drains to discourage household waste dumping; group walks are organised for keen naturalists; households are encouraged to manage their lawns and gardens sustainably; families are urged to get involved in local planning and decision-making.

There has been progress on the farming front with the formation of a group called the Operation Future Association. This began with 12 farmers and has now grown to 170. Many felt stuck in the old type of farming, and needed help to begin to take small steps towards sustainability. As Wes Beery put it: 'the most common thing I hear now is "what methods can I use to farm in an environmentally benign way and what can I do to reduce chemical inputs"'. The hope is that as these farmers make progress, more farmers in the watershed will be exposed to their approaches and technologies, increasing the impact on the watershed as a whole. Until 1996, one third of farms in the watershed had implemented land conservation plans, 18 new wetlands had been created, and a 35,000-tonne reduction in sediment load per year was recorded.

The Pang Valley Countryside Project, Berkshire

The PVCP was launched after the River Pang dried up in the 1980s, caused by overextraction and a series of droughts. Eight parish councils came together to form the Pang Valley Conservation Trust and with the National Rivers Authority (now part of the Environment Agency), put pressure on Thames Water to change their extraction policy. Their success in restoring water flow made local communities realise that the countryside was vulnerable and that they could make improvements. It was also felt that there was a deep rift between farmers and the community, and that something should be done to bridge that gap.

The project seeks to conserve and enhance the existing landscape and wildlife habitats of the Pang Valley; to promote environmentally responsible farming throughout the project; to improve and promote enjoyment of the valley; to promote an understanding of the need to integrate agriculture, conservation, forestry and access to the countryside; and to further environmental education (Ed Cooper, pers comm, 1997; PVCP annual reports, 1994–1996; Pretty and Raven, 1994). It is a proactive project, working in a defined geographic area. Farmers are visited and encouraged to help in the production of a farm report, which acts as the basis for a whole farm agreement. There are now between 40 and 50 farmers in the project. The project officer for the first four years, Ed Cooper, played a key role as facilitator, linking different institutions together. He also helped farmers to complete forms, ensuring they had access to all possible sources of money. Some £332,500 of grants have been brought into the area over four years. Much of this has been used for capital works and represents an important input to the local economy. A new charcoal burning enterprise employing four people has emerged. This is helping to restore SSSI woodlands back to working coppice.

The most important changes for the long term are probably the attitudes and values of local people. Farmers now trust the project and are thinking much more about what they can do for conservation and the environment. As a local councillor, Roger Carter, put it in 1994, 'this has helped to create a sense of community... attitudes have changed dramatically'. The project works closely with schools and local voluntary groups. Between 1994 and 1996, there were some 140 to 160 school visits annually to farms, with 5000 children exposed, including many other visits by youth and community groups. A similar project is being replicated in nearby Kennet Valley.

The Next Steps

New Policy Partnerships and Alliances

To make the land a living land again, it is clear that a wide range of different stakeholders will have to become involved in policy processes. They will have to overcome their differences and build trust and confidence in each other. In time, more complex and sophisticated views on the nature of the challenge and the opportunities for change will emerge.

In some places, this is beginning to occur. In Sweden, there is formal cooperation between ecological and conventional farmers for joint research and lobbying activities. Both the Ecological Farmers' Association (EFA) and the conventional Farmers' Federation acknowledge their differences, but also recognise the benefits to each other of cooperation. It is not easy, but it is certainly working. The president of the EFA, Inger Källander, put it like this:

> *When my ecological farmer members ask me why we work together, I think the best way to justify this is that we have one agriculture... Even if we have differences, we are actively finding ways to talk to each other. In some regions, farmers discuss and cooperate well; in others there are still too big barriers and a lot of prejudice remaining.*

One outcome of this collaboration was the Swedish parliament's 1994 decision to work towards 10 per cent of the land as organic by the year 2000; another was the launch, by the Federation of Swedish Farmers, of a campaign called Towards the Cleanest Agriculture in the World. These are helping to improve the environmental practices of all farmers, as well as to win over Swedish consumers. Eva Tjelke Eckborn, board member of the Federation of Swedish Farmers, says: 'the most important point is, all farmers in Sweden, not only organic producers, are working together to make improvements for a better environment and better animal welfare' (pers comm, 1997). This working together is manifested in the 20 or so 'catchment-based groups' that have emerged to bring together farmers, rural residents and public officials to help reduce plant nutrient leakage to rivers and seas.

The Agricultural Reform Group is an example of an alliance created by both farming and environmental interests. Established in 1993, it has helped to map out common ground both within the UK and across various other EU countries. Again, these alliances are not easy to sustain. They need individuals to invest heavily in creating social capital and in developing trust. This does not mean lowest common-denominator consensus since individuals retain their particular interest. But it does mean creating sufficient common ground upon which serious change in policies and practices can be established (ARG-Europe Policy Notes, 1997).

In the long run, it is clear that fundamental values and principles will have to change. All of this will be difficult. New policies and policy processes can play a crucial lead in establishing clear objectives for a more sustainable agricultural and food system, and can provide appropriate support to encourage widespread transition. Individual citizens can also play a vital role by taking action of their own, whether in the food they chose, the places they visit, or the relationships they make with others.

Delivering the Dividend

As has been shown throughout this book, there is a significant sustainability dividend to be derived from a system-wide transition towards sustainable agriculture, localised food systems and rural community partnerships. The benefits extend to natural and social capital, and would help to transform the European countries into job-rich economies.

The total sustainability dividend in Britain's rural communities from just a 50 per cent shift towards sustainable practices is calculated to be between £9 and £16 billion in cash terms annually, and 320 to 590,000 jobs. These financial resources are locked up in the current modernist and exogenous systems of farming and rural development. Yet, they are freely available to us all. They come from reducing the external costs of farming on the environment, from shifting inputs to regenerative technologies and labour, from direct marketing of foods, and from enhanced rural tourism. The sustainability dividend brings net benefits to rural communities and farming. It does not, however, represent the full benefit to the economy at large. Not all stakeholders in the food and farming system will be winners. Some will be losers. What is important is that the losers are the groups already doing well – and it can be argued on equity grounds that they can afford to capture less of the value from the system.

Farmers as a whole are winners. They are better off as gross margins improve with sustainable agriculture. Their environments are resourceful and more healthy. They benefit from increased business opportunities through increased jobs and spending in rural communities. They also benefit through their repaired relationships in society as stewards of the countryside. Rural communities are major winners. There are more jobs and wealth created in the countryside. There is more social capital developed through better participation in development. There is more natural capital delivered by a more sustainable farming industry.

Food consumers are also winner. Better quality food is available for all. Closer links to farming encourages people to eat more varied and better diets, reducing ill-health. Wildlife and other natural resources are winners, since these are valued and protected through sustainable systems of land use. Tourists stay longer and spend more in the landscapes and environments that give them pleasure. Governments, both local and national, are also winners. Fewer taxes are spent on cleaning up the environmental damage caused by modern farming. The health costs brought about by poor food and diets are reduced. A more robust and resilient natural capital delivers goods and services to many sectors of the economy. And rural communities with reinforced social capital are more self-regulating and more pleasant places in which to live.

The major loser is clearly the input companies, currently supplying fertilisers, pesticides, feedstuffs and seeds. They lose out in the shift towards sustainable agriculture and localised food systems unless they change the nature of their business. Losers too, but not significantly, are the food manufacturers, processors and retailers. As greater value from food is captured by rural communities, so some market share is lost by several of these larger players unless they, too, change the nature of their business.

National governments can do much to help deliver this dividend. No industrialised country has a national framework for sustainable agriculture and rural regeneration. This is the first step. A national policy for sustainable agriculture sets out a vision and embodies different values. It reveals to all stakeholders what is expected to occur, and how we are to achieve it. It clarifies policies and policy processes that will support the transition. A national policy for sustainable agriculture would contain a mix of approaches and instruments that will either penalise the polluters or reward the resource conservers. Other key goals include reform of the Common Agricultural Policy in order to switch all payments away from production to environmental and social goods; national schemes for rural partnerships and rural development support; national strategies for integrated pest management, integrated nutrient conservation and soil regeneration; a national scheme to guarantee the quality of food through approved standards and certification schemes; and support for education and extension in agricultural colleges, universities and training organisations for sustainable agriculture and rural development.

Will it happen? I have argued that there will be many more winners than losers. But, of course, vested interests in maintaining the status quo will clearly resist any change. It will be up to all of the citizens and politicians of Europe to seize these opportunities and to set in motion changes that will fundamentally reform the lives of millions of people. In doing so, they will be giving life to the land again.

References

ACORA. 1990. *Faith in the Countryside.* The Archbishops' Commission on Rural Areas, London.
ACRE. 1995. *Rural England: The White Paper.* Volume II. Minutes of Evidence and Appendices. House of Commons Select Committee on Environment. HMSO, London
ACRE. 1996. *Rural England: The White Paper.* Volume II. Minutes of Evidence and Appendices. House of Commons Select Committee on Environment. HMSO, London, pp 58–71
ADAS. 1996. *Socio-Economic Assessment of Tir Cymen.* Research Report for Countryside Council of Wales, Bangor, Gwynedd
Adnan S, Barrett A, Nurul Alam S M, and Brustinow A. 1992. *People's Participation: NGOs and the Flood Action Plan.* Research and Advisory Services, Dhaka.
Advisory Commission on Food Policy. 1996. *Annual Report.* City of Hartford Court of Common Council and Hartford Food System, Hartford, Conn.
Agne S and Waibel H. 1997. 'Pesticide policy in Costa Rica'. *Pesticides News* 36, pp 8–10
Agrow. 15 March, 1996, p 16
Allies P and Derounian J G. 1997. 'Parish appraisals'. In *Participation Works.* Centre for Community Visions, New Economics Foundation, London
Alternative Agriculture News. 1993. 'Census bureau drops survey of farm residents'. Vol 11 (11), p 2
Amr M. 1995. Reported in *Pesticides News* 30, p 10
Anderson A. 1986. 'It's raining pesticides in Hokkaido'. *Nature* 370, p 478
Ansay F and Viaux P. 1996. 'Technology transfer study of integrated farming systems in a small arable crops region with environmental constraints'. *Second European Symposium on Rural and Farming Systems Research.* 27–29 March, Granada, Spain
ARG. 1997. *Beyond Agenda 2000: An Alternative Proposal.* Edited by Simon Gourlay. ARG, London
ARG-Europe Policy Notes. 1997. Number 1, December, University of Essex, Colchester and The Wildlife Trusts, London
Argyris, C, Putnam, R and Smith, D M 1985. *Action Science.* Jossey-Bass Publishers, San Francisco and London
Assouline, G 1997. *Conditions et Obstacles à la Difference de Modeles d'Agriculture Durable en Europe.* Report for Ministere de l'Environment, Paris. QAP Decision, Theys
Avery D and Avery A. 1996. *Farming to Sustain the Environment.* Hudson Briefing Paper 190. Hudson Institute, Indianapolis
Avery D. 1995. *Saving the Planet with Pesticides and Plastic.* The Hudson Institute, Indianapolis
Avery N, Drake M and Lang T. 1993. *Cracking the Codex.* National Food Alliance, London
Baldock D and Bishop K. 1996. *Growing Greener: Sustainable Agriculture in the UK.* CPRE and WWF-UK, London
Balfour E B. 1943. *The Living Soil.* Faber and Faber, London
Barber D. 1991. *State of Agriculture in the UK.* Report to RASE prepared by a study group under the chairmanship of Sir Derek Barber, RASE, Stoneleigh
Barr C, Bunce R C H, and Clarke RT. 1993. *Countryside Survey 1990.* DoE, London
Barrés T. 1996. La Albufera National Park scheme. Presented at Agri-Environment seminar organised by Federation of Nature and National Parks of Europe, Dulverton, November 1996, Devon
Beaumont P. 1993. *Pesticides, Policies and People.* The Pesticides Trust, London;
Beaumont P. 1995. 'Pesticide minimization'. *Pesticides News* 30, p 14
Bell A. 1996. *Saltmarsh Lamb Study.* Taw Torridge Estuary Project, Devon
Bianchi M. 1996. 'National perceptions and key challenges and steps towards solutions for sustainable mountain development'. In Proc. of Scottish Session of European Intergovernmental Consultation on Sustainable Mountain Development, The Cairngorms, 22–27 April 1996. The

Scottish Office
Bignall E M and McCracken D I. 1996. 'Low intensity farming systems in the conservation of the countryside'. *Journal of Applied Ecology* 33, pp 416–424
Birdlife International. 1997. *A Future for Europe's Rural Environment: Reforming the CAP.* Sandy, Beds.
Blunden J and Curry N. 1988. *A Future for Our Countryside.* Basil Blackwell, Oxford
Boardman J and Evans R. 1991. *Flooding at Steepdown.* A report to Adur District Council, West Sussex.
Boardman J. 1990. 'Soil erosion on the South Downs: a review'. In Boardman J, Foster I D L and Dearing J A (eds). *Soil Erosion on Agricultural Land.* John Wiley and Sons, Chichester.
Body, Sir R. 1996. Opening Address. in Carruthers SP and Miller FA (eds). *Crisis on the Family Farm: Ethics or Economics?* CAS Paper 28, CAS, Reading
Böge S. 1993. *Road transport of goods and effects on the spatial environment.* Wupperthal Institute, Wupperthal
Bollman R A and Bryden J M (eds). 1997. *Rural Employment: An International Perspective.* CAB International, London
Bontron J-C and Lasnier N. 1997. 'Tourism: a potential source of rural employment'. in Bollman and Bryden (eds), op cit
Booth, E. 1996. 'Local food links'. Paper at The Vegetable Challenge conference, 21 May. The Guild of Food Writers.
Borlaug, N. 1992. 'Small-scale agriculture in Africa: the myths and realities'. *Feeding the Future* (Newsletter of the Sasakawa Africa Association) 4:2.
Borlaug, N. 1994a. 'Agricultural research for sustainable development'. Testimony before US House of Representatives Committee on Agriculture, 1 March, 1994.
Borlaug, N. 1994b. 'Chemical fertiliser "essential"'. Letter to *International Agricultural Development* (Nov–Dec), p 23.
Bowers J K and Cheshire P. 1983. *Agriculture, the Countryside and Land Use.* Methuen and Co, London
Bowler I. 1979. *Government and Agriculture. A Spatial Perspective.* Longman, London
Boyle D. 1996. 'The transatlantic money revolution'. *New Economics Magazine* 40, pp 4–7 (New Economics Foundation, London)
Briggs M. 1994. 'Status, problems and solutions for a sustainable shrimp industry'. Report to ODA. Development of Strategies for Sustainable Shrimp Farming, ODA, London
Brown L R and Kane H. 1994. *Full House: Reassessing the Earth's Population Carrying Capacity.* W W Norton and Co, New York
Brown L R. 1994. 'The world food prospect: entering a new era'. In *Assisting Sustainable Food Production: Apathy or Action?* Winrock International, Arlington, VA
Brueninghaus G. 1996. Agriculture in Finland. Presented at ARG conference Farmers and Environmentalists: The Land We Share. Brussels, 29–31 October
Bunch, R. 1983. *Two Ears of Corn.* World Neighbors, Oklahoma City.
Bunch, R. 1990. *Low Input Soil Restoration in Honduras: The Cantarranas Farmer-to-Farmer Extension Programme.* Gatekeeper Series SA 23. Sustainable Agriculture Programme, IIED, London
Bunch, R and López, G. 1996. *Soil Recuperation in Central America: Sustaining Innovation after Invention.* Gatekeeper Series SA 55. Sustainable Agriculture Programme, IIED, London
CADISPA. 1996. *Reaching people in Aragón and Catalunya.* University of Strathclyde, Jordanhill Campus, Glasgow
California Department of Food and Agriculture. *Summary of Illnesses and Injuries Reported by Californian Physicians as Potentially Related to Pesticides* 1972–current. Sacramento, CA
Campbell A. 1994. *Landcare: Communities Shaping the Land and the Future.* Allen and Unwin, St Leonards, NSW
Campbell L H and Cooke A S (eds). 1995. *The Indirect Effects of Pesticides on Birds.* JNCC, Peterborough
Cannon G. 1992. *Food and Health: The Experts Agree.* Consumers Association, London
Carruthers, I. 1993. 'Going, going, gone! Tropical agriculture as we knew it'. *Tropical Agriculture Association Newsletter,* 13 (3): pp 1–5
CCW. 1992. *Tir Cymen.* Countryside Council of Wales, Bangor, Gwynedd
CCW. 1996. *Tir Cymen: 4 Years of Achievement.* Countryside Council of Wales, Bangor, Gwynedd
Center for Rural Affairs. 1996. *The Beginning Farmer.* Issue 19, June. Hartington, Nebraska

Center for Science in the Public Interest. 1995. *Funding Safer Farming: Taxing Pesticides and Fertilizers.* Washington, DC
Central Statistical Office. 1994. *Family Planning – Report on the 1993 Family Expenditure Survey.* HMSO, London
Centre for Community Visions. 1997. *Chattanooga, Case Study No 4.* CCV, London
CGIAR. 1994. *Sustainable Agriculture for a Food Secure World: A Vision for International Agricultural Research.* Expert Panel of the CGIAR, Washington, DC, and SAREC, Stockholm
Chamberlain D, Fuller R and Brooks J. 1996. 'The effects of organic farming on birds'. *Elm Farm Research Centre Bulletin* 21, pp 5–9, Elm Farm Research Centre, Newbury
Chambers R, Pacey A and Thrupp L A. (eds). 1989. *Farmer First. Farmer Innovation and Agricultural Research.* IT Publications, London
City of Bradford Council. 1996. *Springfield – People Growing Together.* Bradford
CLA. 1995. *Towards a Rural Policy. A Vision for the 21st Century.* Country Landowners Association, London
Clifford S. 1997. 'Barefoot through the learning locality'. In *Particpation Works.* Centre for Community Visions, New Economics Foundation, London
CLM. 1994. *Integrating the Environment with the EU Common Agricultural Policy.* Centre for Agriculture and Environment, Utrecht, The Netherlands
Clout H. 1984. *A Rural Policy for the EEC.* Methuen, London and New York
Club de Bruxelles. 1994. *Rural Environment and Sustainable Development.* Brussels, Belgium
Clunies-Ross T and Hildyard N. 1992. *The Politics of Industrial Agriculture.* SAFE Alliance and Earthscan Publications Ltd, London
Coleman J. 1990. *Foundations of Social Theory.* Harvard University Press, Mass.
Common Ground. 1991. *Apple Day and Community Orchard leaflets.* Common Ground, London
Common Ground. 1996. *Avon Parish Maps Project.* London
Community Council for Wiltshire. 1996. In *Rural England: The White Paper.* Volume II. Minutes of Evidence and Appendices. House of Commons Select Committee on Environment. HMSO, London, pp 130–132
Community Food Security News, published by CFS Coalition, Austin, Hartford and Los Angeles
Community Food Security News. 1996. Winter/Spring. Venice, California
Concise Oxford English Dictionary. 1976. 6th edition. OUP, Oxford
Conway G R and Pretty J N. 1991. *Unwelcome Harvest: Agriculture and Pollution.* Earthscan Publications Ltd, London
Cook C D and Rodgers J. 1996. 'Community food security: a growing movement'. *Global Pesticide Campaigner* Vol 6(3) 1, pp 8–11
Coote A and Lenaghan J. 1997. *Citizens' Juries: Theory into Practice.* IPPR, London
Costanza R, d'Arge R, de Groot R, Farber S, Grasso M, Hannon B, Limburg K, Naeem S, O'Neil R V, Parvelo J, Raskin R G, Sutton P and van den Belt M. 1997. 'The value of the world's ecosystem services and natural capital'. *Nature* 387, pp 253–260
Council for the Protection of Rural England and Countryside Commission. 1995. *Tranquil Areas.* CPRE, London
Council for the Protection of Rural England. 1992. *The Lost Land.* CPRE, London
Council for the Protection of Rural England. 1993. *Regional Lost Land.* CPRE, London
Council for the Protection of Rural England. 1994. *Down to Earth: Environmental Problems Associated with Degradation in the English Landscape.* CPRE, London
Council for the Protection of Rural England. 1995. *Local Influence.* CPRE, London
Council for the Protection of Rural England. 1996. *Lost Lanes.* CPRE, London
Council for the Protection of Rural England. 1997. *Drystone Walls.* CPRE, London
Countryside Commission. 1994. 'People vote with their feet'. *Countryside*, No 67
Cranbrook C. 1997. *The Rural Economy and Supermarkets.* Great Glemham, Suffolk
Crosson, P and Anderson, J. 1995. *Achieving a Sustainable Agricultural System in Sub-Saharan Africa.* Building Block for Africa Paper No 2, AFTES, The World Bank, Washington, DC
Cuff J and Rayment M. 1997. *Working with Nature: Economies, Employment and Conservation in Europe.* RSPB and Birdlife International, Sandy, Beds.
Curtis J, Profeta T and Mott L. 1993. *After 'Silent Spring': The Unsolved Problems of Pesticide Use in the US.* Natural Resources Defense Council, Washington, DC

Dabbert S. 1990. 'Der Begriff des Betriebsorganisms'. *Lebendige Erde.* 90 (5), pp 333–337
Daily Telegraph. 1994. 'Set Aside? Its a Waste of Space'. 2 July, cited in CPRE, 1994. *Down to Earth: Environmental Problems Associated with Degradation in the Engish Landscape.* CPRE, London
de Vries J. 1996. 'Sustainable farming in Holland – green labels and environmental yardsticks'. *Pesticides News* 33, p 6
Derounian J G. 1993. *Another Country: Life Beyond the Rose Cottage.* NCVO Publications, London
Derounian J G. 1998. *Effective Working with Rural Communities.* Packard Publishing, London
Devon County Council. 1996. *Local Food Links: A Proposal.* Environment Dept, DCC, Exeter
DeVore B. 1997. 'Sustainable farmers tap grocery chain market. Wisconsin CSAs become first source'. News Release, Land Stewardship Project, 24.4.97
Dinham B. 1993. *The Pesticides Hazard.* The Pesticides Trust, London; *Pesticides News* 30 (1995), pp 10–11
Dinham B. 1996. *Growing Food Security.* Pesticides Trust and PAN, London
Dobbs T L, Becker D L and Taylor D. 1991. 'Sustainable agriculture policy analyses: South Dakota on-farm case studies'. *Journal of Farming Systems Research-Extension* 2(2), pp 109–124.
DoE. 1991. *Taking the Neighbourhood Initiative: A Facilitators Guide.* Department of the Environment, London
DoE. 1993. *Critique of CPRE Report The Lost Land.* Department of the Environment, London
DoE. 1996. *Greening the City.* Department of the Environment, London
DoE. 1996. *Indicators of Sustainable Development for the United Kingdom.* Department of the Environment, London
DoE. 1997. *Sustainable Use of Soil: Government Response to 19th Report of RCEP.* HMSO, London
DoE/MAFF. 1995. *Rural England: A Nation Committed to a Living Countryside.* The Rural White Paper. HMSO, London
Donázar J A, Navesi M A, Tella J L, Campión D. 1997. 'Extensive grazing and raptors in Spain'. In Pain D J and Pienkowski M W (eds). *Farming and Birds in Europe.* Academic Press Ltd, London
Douthwaite R. 1996. *Short Circuit: Strengthening Local Economies for Security in an Unstable World.* Green Books, Dartington, Devon
DowElanco. 1994. 'What makes agriculture sustainable'. *The Bottom Line*, Indianapolis, US
DTI. 1994. *Short Rotation Coppice Production and the Environment.* Dept of Trade and Industry, London
Dubois D, Fried P M, Malitius O and Tschachtli R. 1995. 'Burgrain: direktvergleich dreier anbausysteme'. *Agrarforschung* 2(10) pp 457–460
Duke of Westminster (DoW). 1992. *The Problems in Rural Areas.* A report of recommendations arising from an inquiry chaired by His Grace the Duke of Westminster DL. Brecon, Powys
Duncan C M and Lamborghini N. 1994. 'Poverty and social context in remote rural communities'. *Rural Sociology* 59 (3), pp 437–461
The Economist. 1982. 'Scottish Highlands: life in the outback'. 24 July. Quoted in Clout, 1984, op cit
Ehrlich P. 1968. *The Population Bomb.* Ballantine, New York
El Titi A. 1992. 'Integrated farming: an ecological farming approach in European agriculture'. *Outlook in Agriculture* 21(1), pp 33–39
El Titi A. 1996. 'Veränderung der Unkrautzusammensetzung nach 16 jahren integrieter Bewirtschaftung auf dem Lautenbach Hof'. *Z. Pflkrankh. PflSchutz.*, Sonderh. XV, pp 201–209
El Titi A and Landes H. 1990. 'Integrated farming system of Lautenbach: a practical contribution toward sustainable agriculture in Europe'. In Edwards C et al (eds). *Sustainable Agricultural Systems.* Soil and Water Conservation Society, Ankeny, Iowa
El Titi A, Boller E F and Gendrier J P(eds). 1993. *Integrated Production.* IOBC/WPRS Bulletin XVI/I
Emmerman A. 1997. 'Sweden's reduced risk pesticide policy'. *Pesticides News* 34, p 6
ENDS. 1995. *DoE pesticides study supports Water Protection Zones.* Vol 242, p 7
English Nature, quoted in HC. 1996. *Rural England: Rural White Paper.* Ev Vol II, para 329, p 99. House of Commons Select Committee on the Environment, Westminster; English Nature reported in FW 2.5.97
Etzioni A. 1995. *The Spirit of Community: Rights, Responsibilities and the Communitarian Agenda.* Fontana Press, London

European Commission. 1995. *Progress report on implementation of EU programme of policy and action 'Towards Sustainability'.* EC, Brussels
European Commission. 1996. *Action for Employment in Europe.* EC, Brussels
European Commission. 1997. *Agenda 2000. Volume 1 – For a Stronger and Wider Vision.* COM (97) 2000, Brussels
Eurostat. 1995. *Agriculture Statistical Yearbook.* EC, Brussels
Evans R. 1990. 'Soils at risk of accelerated erosion in England and Wales'. *Soil Use and Management* 6(3), pp 125–131.
Evans R. 1990. 'Water erosion in British farmers' fields: some causes, impacts, predictions'. *Progress in Physical Geog.* 14(2), pp 199–219
Evans, R. 1996. *Soil Erosion and its Impact in England and Wales.* Report to Friends of the Earth, London
Faeth, P (ed). 1993. *Agricultural Policy and Sustainability: Case Studies from India, Chile, the Philippines and the United States.* World Resources Institute, Washington, DC
FAO. 1993a. *Harvesting Nature's Diversity.* FAO, Rome
FAO. 1993b. *Strategies for Sustainable Agriculture and Rural Development (SARD): The Role of Agriculture, Forestry and Fisheries.* United Nations Food and Agriculture Organization, Rome
Farmer P. 1995. 'Debate about LIFE and low input farming'. *Pesticides News* 30, p 12
Farmers' Link. 1995. *Producer–Consumer Linkages. A Discussion Paper.* Norwich
Farmers' Weekly. 1997a. 'Ten signs that could be caused by OP exposure'. 26 September
Farmers' Weekly. 1997b. 'Cereal farm assurance schemes totally mad'. 4 July.
Feber R. 1996. *The effects of organic and conventional farming systems on the abundance of butterflies.* Report to WWF (UK). Dept. Of Zoology, South Parks Road, Oxford
Federation of Swedish Farmers (LRF). 1997. *Agriculture and Environment in Sweden*
Fernandez, A. 1992. *The MYRADA Experience: Alternate Management Systems for Savings and Credit of the Rural Poor.* MYRADA, Bangalore
Festing H. 1994. *Should Farmers Market Direct to Consumers? America Says Yes.* Wye College Food Industry Perspectives, Wye College, Ashford
Festing H. 1995. *Direct Marketing of Fresh Produce.* MPhil thesis, Wye College, University of London
Festing H. 1997. 'The potential for direct marketing by small farms in the UK'. *Farm Management* 9(8), pp 409–421
Festing H and Hamir A. 1997. 'Community Supported Agriculture and Vegetable Box Schemes'. Presented at conference Agricultural Production and Nutrition, School of Nutrition Sciences and Policy, Tufts University, Mass. 19–21 March, 1997
Fischler F. 1996. Reported in Birdlife International. 1997. *A Future for Europe's Rural Environment: Reforming the CAP.* Sandy, Beds
Fisher A. 1996. 'Food security unites community advocates'. *Community Greening Record,* pp 12–13
Fisk J W, Hesterman O B and Thorburn T L. 1996. Integrated farming systems: a sustainable agriculture learning community in the US. In Roling N and Wagemakers M (eds). *Social Learning for Sustainable Agriculture.* Cambridge University Press, Cambridge
Food Share. 1996. *Field to Table Program Evaluation Results.* Toronto, Ontario.
Fowler, C. and Mooney, P. 1990. *The Threatened Gene: Food, Policies and the Loss of Genetic Diversity.* Lutterworth Press, Cambridge
Friends of the Earth. 1992. *Off the Treadmill.* FoE, London
Friends of the Earth. 1995. *Nitrogen restraint as a policy tool.* FoE, London
Fukuyama F. 1992. *The End of History and the Last Man.* Penguin Books, London
Fuller R J, Gregory R D and Gibbons D W. 1994. *Recent Trends in the Numbers and Distribution of Birds on Farmland in Britain.* BTO, Norfolk
Furusawa K. 1994. 'Cooperative alternatives in Japan'. In Conford P (ed). *A Future for the Land: Organic Practice from a Global Perspective.* Resurgence Books, Bideford
G W Stokes, pers comm 16.5.97. National Society of Allotment and Leisure Gardeners
Garnett T. 1996. *Growing Food in Cities: A report to highlight and promote the benefits of urban agriculture in the UK.* SAFE Alliance and National Food Alliance, London
Garreau J. 1992. *Edge City. Life on the New Frontier.* Anchor Books, New York
Gedde-Dahl T. 1992. 'Identifying development needs and decisions with local experimentation: experimental groups in Norwegian agriculture'. *J. International Farm Management* 1(3), pp 74–80

Geier B. 1996. 'Organic agriculture: part of the food security solution'. In Dinham B (ed). 1996. *Growing Food Security: Challenging the Link between Pesticides and Food Security.* PAN and Pesticides Trust, London

Georg-August-Universität Göttingen. 1996. *Entwicklung Integrieter Anbausysteme am Beispiel Einer Rapsfruchfolge.* Forschungs-und Studienzentrum Lanwirtschaft und Unwelt

Georghiou G P. 1986. 'The magnitude of the problem'. In National Research Council. *Pesticide Resistance, Strategies and Tactics.* National Academy Press, Washington, DC

Ghimire K and Pimbert M. 1997. *Social Change and Conservation.* Earthscan Publications Ltd, London

Gibbons D S. 1996. 'Resource mobilisation for maximising MFI outreach and financial self-sufficiency'. Issues Paper No. 3 for Bank-Poor 96, 10–12 December, Kuala Lumpur

Gibson T. 1993. *Danger: Opportunity.* Meadowell Community Development. Neighbourhood Initiatives Foundation, Telford

Gibson T. 1997. *The Power in Our Hands.* Jon Carpenter Publishing, Oxford

Glotfelty D E, Seiber J N and Lidjedahl C A. 1987. 'Pesticides in fog'. *Nature* 325, pp 602–605

Glover P. 1996. 'Creating community economics with local currency'. *Forest, Trees and People Newsletter.* No 30, p 51 (Uppsala, Sweden)

Goldschmidt W. 1978 (1946). *As You Sow: Three Studies in the Social Consequences of Agri-Business.* Allanheld, Monclair, NJ

Gómez-Pompa A and Kaus A. 1992. 'Taming the wilderness myth'. *Bioscience* 42 (4), pp 271–279

Governor's Office. 1994. *Governor's Budget 1994–95.* Sacramento, California

Greenpeace. 1992. *Green Fields, Green Future.* Greenpeace, London

Greig-Smith P, Frampton G and Hardy T (eds). 1993. *Pesticides, Cereal Farming and the Environment: The Boxworth Experiments.* HMSO, London

Griffiths K. 1996. *The Economic Impact of the Kite Country Project in Mid Wales.* MSc thesis, quoted in Cuff and Rayment, 1997, op cit

Guo Y L et al. 1996. 'Prevalence of dermatoses and skin sensitisation associated with the use of pesticides in fruit farmers of southern Taiwan'. *Occupational and Environmental Medicine* 53, pp 427–431.

H M Government. 1992. *Health of the Nation.* HMSO, London

Habermas J. 1987. *Theory of Communicative Action: Critique of Functionalist Reason.* Vol. II. Polity Press, Oxford.

Hajesz D and Dawe S P. 1997. 'De-mythologising rural youth exodus'. In Bollman and Bryden (eds), op cit

Hamilton G. 1995. *Learning to Learn with Farmers.* PhD Thesis, Wageningen Agricultural University, The Netherlands

Hampden-Turner C. 1996. 'The enterprising stakeholder'. *The Independent.* 5 February, London

Hart A, Boddy P, Shequist K, Huber G and Exner D. 1996. 'Iowa, US: An effective partnership between the Practical Farmers of Iowa and Iowa State University'. In Thrupp L A (ed). *New Partnerships for Sustainable Agriculture.* WRI, Washington, DC

Hart R A 1992. *Children's Participation: From Tokenism to Citizenship.* UNICEF Innocenti Essays No 4. Florence, UNICEF

Harvey D. 1989. *The Condition of Postmodernity.* Basil Blackwell Ltd, Oxford

Haug R. 1996. 'Hindelang – an alliance of extensive mountain farming with tourism'. *La Cañada* 6 pp 7–8

Hazell P and Lutz E. 1998. 'Agriculture, development and the environment'. In Lutz E, Binswanger H, Hazell P and McCalla A (eds). *Agriculture, Development and the Environment: Policy, Institution and Technical Perspectives.* The World Bank, Washington, DC (in press)

HC. 1996. *Rural England: Rural White Paper.* Ev Vol II, para 329, p 99. House of Commons Select Committee on the Environment, Westminster, London

Hewitt T I and Smith K R. 1995. *Intensive Agriculture and Environmental Quality: Examining the Newest Agricultural Myth.* Henry Wallace Institute for Alternative Agriculture, Greenbelt, MD

Hinchcliffe F, Thompson J and Pretty J N. 1996. *Sustainable Agriculture and Food Security in East and Southern Africa.* A report for Swedish International Development Cooperation Agency, Stockholm. IIED, London

Hird, V. 1997. *Double Yield: Jobs and Sustainable Food Production.* SAFE Alliance, London
HL. 1990. *The Future of Rural Society.* House of Lords Select Committee on the European Communities. HMSO, London.
Hoyt R, O'Donnell D and Mack K Y. 1995. 'Psychological distress and size of place: the epidemiology of rural economic stress'. *Rural Sociology* 60 (4), pp 707–720
HSE. 1996. *Pesticide Incidents Report 1995–1996.* HSE, Nottingham
Hughes D. 1996. 'Dancing with an elephant: building partnerships with multiples'. Paper presented at The Vegetable Challenge conference, London, 21 May. The Guild of Food Writers
Hutton W. 1995. *The State We're In.* Vintage Books, London
IEEP and WWF. 1994. *The Nature of Farming. Low Intensity Systems in Nine European Countries.* Institute for European Environmental Policy, London, and World Wide Fund for Nature, Geneva
IFAD. 1992. *Soil and Water Conservation in Sub-Saharan Africa.* IFAD, Rome
IFPRI. 1995. *A 2020 Vision for Food, Agriculture and the Environment.* International Food Policy Research Institute, Washington, DC
IGER. 1996. *Conversion to Organic Milk Production.* Institute of Grassland and Environmental Research, Aberystwyth
The Independent. 1996. 'Leominster's traders play the loyalty card in attempt to trump out-of-town supermarket'. 2 March
The Independent. 1997. 'Eat up your greens and grow up like ancient Greeks'. March 17
The Independent. 1997. 'Parents lose battle with the greens'. 21 January
Independent on Sunday. 1996. 'Asian tigers lured by low pay in the valleys'. p 10
Jackson T and Marks N. 1994. *Measuring sustainable economic welfare: a pilot index 1950–1990.* Stockholm Environment Institute and New Economics Foundation, London
Jacobs M. 1996. *The Politics of the Real World.* The Real World Coalition. Earthscan Publications Ltd, London
James P. 1996. 'The nation's diet: can it be changed?' Paper presented at The Vegetable Challenge conference, London, 21 May. The Guild of Food Writers
Jewell T. 1996. *Opportunities for campaigning on agriculture and food policy.* Report for Friends of the Earth, London
Jishi M and Hirschern N. 1995. 'Relationships of pesticide spraying to signs and symptoms in Indonesian farmers'. *Scandinavian J. of Work and Environmental Health* 21, pp 124–133
Jordan V W L and Hutcheon J A. 1994. 'Economic viability of less-intensive farming systems designed to meet current and future policy requirements: 5 year summary of the LIFE project'. *Aspects of Applied Biology* 40, pp 61–68.
Jordan V W L and Hutcheon J A. 1995. 'Less intensive farming and the environment (LIFE): an integrated farming systems approach for UK arable crop production'. In Glen D M, Greaves M P and Anderson H M (eds). *Ecology and Integrated Farming Systems: Proceedings of the 13th Long Ashton International Symposium.* Long Ashton, Bristol
Jordan V W L and Hutcheon J A. 1996. 'Alternative production systems for efficient arable crop production and environmental protection'. *Proceedings of 1996 National Tillage Conference.* Ireland.TEAGASC
Jordan V W L, Hutcheon J A and Glen D M. 1993. *Studies in Technology Transfer of Integrated Farming Systems. Considerations and Principles for Development.* AFRC Institute of Arable Crops Research, Long Ashton Research Station, Bristol
Jorgensen L N. 1997. 'Measures to cut Danish fungicide use'. *Pesticides News* 34, p 7
Kalburtji K L and Kostopoulov S. 1996. 'Converting to organic farming at the watershed of Lake Kerkini'. In Isart J and Herene J S (eds). *Steps in the Conversion and Development of Organic Farming.* Proc. of Second ENOF Workshop, Barcelona 3–4 October, 1996, pp 45–55
Kelly J. 1997. 'Community audit in North Kesteven'. In *Participation Works.* Centre for Community Visions, New Economics Foundation, London
Keys A. 1980. *Seven Countries: A Multivariate Analysis of Death and Coronary Heart Disease.* Harvard University Press, London
King T. 1996. 'Precision agriculture'. *CornerPost MISA.* Spring 1996, No. 13, at www.centers.agric.umn.edu/misa/
Kings Fund. 1997. *Citizens' Juries: Current Work and Development.* At http://www.kingsfund.org.uk/grantguide/juries.htm

Klinkenborg V. 1995. 'A farming revolution: sustainable agriculture'. *National Geographic* December, pp 62–89
Kloppenberg J. 1991. 'Social theory and the de/reconstruction of agricultural science: a new agenda for rural sociology'. *Sociologia Ruralis* 32(1), pp 519–548
Knutson, R D, Taylor, J B, Penson, J B and Smith, E G 1990. *Economic Impacts of Reduced Chemical Use.* Texas, A&M University
Kothari A, Pande P, Singh S and Dilnavaz R. 1989. *Management of National Parks and Sanctuaries in India.* Indian Institute of Public Administration, New Delhi
Kroese R and Butler Flora C. 1992. 'Stewards of the land'. *ILEIA Newsletter* 2/92, pp 5–6
Kurokawa K. 1991. *Intercultural Architecture.The Philosophy of Symbiosis.* Academy Editions, London.
La Fondation pour le Progrès de l'Homme and Solagral. 1995. *The CAP and International Relations.* Preparatory document, 10–12 May, Chantilly, France
Lampkin N and Padel S (eds). 1994. *The Economics of Organic Farming: An International Perspective.* CAB International
Lampkin N. 1996. Impact of EC Regulation 2078/92 on the development of organic farming in the European Union. Welsh Institute of Rural Affairs, University of Wales, Aberystwyth, Dyfed
Lang T. 1995a. 'Local sustainability in a sea of globalisation? The case for food policy'. Paper for the Planning Sustainability conference organised by the Political Economy Research Centre, University of Sheffield, 8–10 September,
Lang T. 1995b. 'The contradictions of food labelling policy'. *Information Design Journal* 8/1, pp 3–16
Lang T. 1997a. 'Dividing the cake: food as social exclusion'. In Walker A and Walker C (eds). *Britain Divided. The Growth of Social Exclusion in the 1980s and 1990s.* CPAG, London
Lang T. 1997b. Food policy for the 21st century: can it be both raised and reasonable? The Paul Hamlyn Lecture series, 22 April, TVU Centre for Food Policy, Thames Valley University, London
Lang T and Hines C. 1993. *The New Protectionism.* Earthscan Publications Ltd, London
Leach, G. 1995. 'Global land and food in the 21st Century'. *Polestar Series Report*, No 5. Stockholm Environment Institute, Stockholm.
LEAF, 1996. Results of ICM and conventional farm comparisons. Linking Environment and Farming, National Agriculture Centre, Stoneleigh
Leake A. 1996. 'Setting new standards'. *Pesticides News* 31, p 17
Lean G. 1997. 'Designer gene fails farmers'. *The Independent on Sunday*, December
LeQuesne C. 1996. *Reforming the World Trade. The Social and Environmental Priorities.* Oxfam, Oxford
Lerot F. 1996. Ministère de l'Environment, in European Intergovernmental Conference on Sustainable Mountain Development, Aviemore, Scotland
Liberal Democrats. 1994. *Reclaiming the Countryside.* Proposals for Rural Communities and Agriculture in England and Wales. Liberal Democrats, London
Lobao, L. 1990. *Locality and Inequality: Farm and Industry Structure and Socio-Economic Conditions.* State University of New York Press
Lobao L M, Schulman M D and Swanson L E. 1993. 'Still going: recent debate on the Goldschmidt hypothesis'. *Rural Sociology* 58 (2), pp 277–288
Lobley, M. 1996. 'Small farms and agricultural policy: a conservationist perspective'. In Carruthers and Miller (eds), op cit
Luick R. 1996. 'The demise of cattle farming in the Black Forest'. *La Cañada* No 5 pp 4–5 (JNCC, Peterborough)
MacRae R. 1994. 'So why is the City of Toronto concerned about food and agriculture policy? A short history of the Toronto Food Policy Council'. *C and A Bulletin* 50, pp 15–18
MacRae R. 1997. *Eco-labelling: too great a threat to the food industry.* Mimeo, Toronto, Canada
MacRae R J, Henning J and Hill S B. 1993. 'Strategies to overcome barriers to the development of sustainable agriculture in Canada: the role of agribusiness'. *Journal of Agricultural and Environmental Ethics* 6, pp 21–51
Madden P. 1992. *Raw Deal.* Christian Aid, London
MAFF. 1970. *Modern Farming and the Soil.* Report of the Agricultural Advisory Council on Soil Structure and Soil Fertility: The Strutt Commission. HMSO, London

MAFF. 1997. *Departmental Report. The Government's Expenditure Plans 1997–98 to 1999–2000.* MAFF, London
MAFF. *Agriculture in the UK 1996.* HMSO, London
MAFF. 1997. *Agriculture in the United Kingdom.* Agricultural Statistics. HMSO, London
MAFF CAP Review Group. 1995. *European Agriculture: the case for radical reform – working papers.* MAFF, London
Matteson, P. 1995. 'The 50% pesticide cuts in Europe: a glimpse of our future?' *American Entomologist* 41(4), pp 210–220
Maughan J. 1995. 'Willapa Watershed'. *Ford Foundation magazine.* Ford Foundation, New York
McCalla, A. 1994. *Agriculture and Food Needs to 2025: Why We Should be Concerned.* Sir John Crawford Memorial Lecture, October 27. CGIAR Secretariat, The World Bank, Washington, DC
McLaren D. 1994. *Soils and sustainability: a view from FOE.* Friends of the Earth, London
McLaren D. 1995. *Counter-urbanisation: how to achieve a trend-breach.* FOE Trust, London
Media Villa V, Meister E, Walther U, Fried P, Malitus O and Sidler A. 1995. 'Vileseitige fruchtfolge: integriert und intensiv bewirtschaftet'. *Agrarforschung* 2(6) pp 231–234
MF. 1995. *Precision Farming.* Massey Ferguson, Coventry
Milo N. 1990. *Nutrition Policy for Food-Rich Countries.* Johns Hopkins University Press, Baltimore MD
Ministère de l'Agriculture de la Pêche et de l'Alimentation. 1996. *Projet pilote et de démonstration relatif à de Développement Durable en France.* Paris
Mintel. 1992. *Regional Lifestyles.* Mintel International Group, London
Mitchell, D O and Ingco M D. 1993. *The World Food Outlook.* International Economics Department. World Bank, Washington, DC
Mndeme K C H. 1992. 'Combating soil erosion in Tanzania: the HADO experience'. In Tato K and Hurni H (eds). *Soil Conservation for Survival.* SCS, Ankeny
Moseley M, Derounian J G and Allies P J. 1996. 'Parish appraisals – a spur to local action?' *TPR* 67 (3), pp 309–329
Moss, J. 1996. 'Pluriactivity and survival? A study of family farms in Northern Ireland'. In Carruthers and Miller (eds), op cit
Mulgan G and Wilkinson H. 1995. *Freedom's Children. Work, Relationships and Politics for the 18–30 Year Olds in Britain Today.* Demos, London
Munton R and Marsden T. 1991. 'Occupancy change and the farmed landscape: an analysis of farm-level trends'. *Environment and Planning A* 23, pp 499–510
Musema-Uwimana A. 1983. 'La conservation des terraces au Rwanda'. *Recherche Agricole* 16, pp 86–93
N W Food Alliance. 1995. *Green marketing in Massachusetts.* Volume I (1). Portland, Oregon
N W Food Alliance. 1995–96. *Strategic plans and research plans.* Briefing Papers, NWFA, Olympia, Washington State
NAF. 1994. *A Better Row to Hoe. The Economic, Environmental and Social Impact of Sustainable Agriculture.* Northwest Area Foundation, Minnesota, US
Narayan D. 1993. *Focus on Participation: Evidence from 121 Rural Water Supply Projects.* UNDP-World Bank Water Supply and Sanitation Program, World Bank, Washington, DC
National Association of Citizens Advice Bureaux. 1995. *Rural Benefits.* NACAB, London
National Federation of Women's Institutes. 1996. *Rural England: The White Paper.* Volume II. Minutes of Evidence and Appendices. House of Commons Select Committee on Environment. HMSO, London, p 164
National Food Alliance. 1995. *Easy to Swallow, Hard to Stomach: the results of a survey of food advertising on television.* NFA, London.
National Food Alliance. 1997. *If They Don't Eat a Healthy Diet, It's Their Own Fault.* NFA, London
National Gardening Association. 1986. *National Gardening Survey Fact Sheet.* NGA, Burlington, Vermont
NEF. 1997. *Community Works.* New Economics Foundation, London
Newby H. 1980. *Green and Pleasant Land? Social Change in Rural England.* Hutchinson, England
Newman M. 1997. 'In search of food security'. In Community Food Security at http://www.marketreport.com/security.htm

Nugent R. 1997. *The sustainability of urban agriculture: a case study in Hartford, Connecticut.* Draft report of work in progress. Published at City Farmer, http://www.cityfarmer.org

NYM National Park. 1994. *Farm Scheme Briefing Document.* North Yorks Moors National Park, Helmsley, York

O'Brien D J, Hassinger E W and Dershem L. 1994. 'Community attachment and depression among residents in two rural midwestern communities'. *Rural Sociology* 59 (2), pp 255–265

OECD. 1992. *Investing in Food*, quoted in Tansey and Worsley, 1995, op cit

OECD. 1993. *Agricultural and Environmental Policy Integration. Recent Progress and New Directions.* OECD, Paris

OECD. 1994. *Economic instruments for achieving environmental goals in the agricultural sector.* COM/AGR/CA/ENV/EPOC (94) 102. OECD, Paris

OECD. 1995. *Sustainable agriculture: a survey of policy issues and OECD country experience.* Committee for Agriculture AGR/CA (95) 3. OECD, Paris

OECD. 1997. at http://www.oecd.org

Office of Science and Technology. 1995. *Progress through Partnership Report No 11: Agriculture, natural resources and environment.* HMSO, London

Ogilvy S, Turley D B, Cook S K, Fisher N M, Holland J, Prew R D and Spink J. 1995. *LINK integrated farming systems: a considered approach to crop protection.* BCPC/SFS Symposium Proceedings No 63, 11–14 September

Oma R. 1995. *Rural Partnerships for a Better Farming Future.* SAFE Alliance, London

Organophosphate Information Network (OPIN). 1996. Briefing to House of Commons, 30 April. OPIN, Cornwall

O'Riordan T. 1996. *Environmental Taxation.* Earthscan Publications Ltd, London

Ostrom E. 1990. *Governing the Commons: The Evolution of Institutions for Collective Action.* Cambridge University Press, New York

OTA. 1988. *Enhancing Agriculture in Africa: A Role for US Development Assistance.* US Office of Technology Assessment. US Government Printing Office, Washington, DC

Pang Valley Countryside Project Annual Reports, 1994–96, Theale, Berkshire

Paxton A. 1994. *The Food Miles Report: The Dangers of Long Distance Food Transport.* SAFE Alliance, London

Pearce D. 1995. *Blueprint 5: The True Cost of Road Transport.* Earthscan Publications Ltd, London

Pearce D and Tinch R. 1998. 'The true price of pesticides'. In Vorley W and Keeney D (eds), op cit

Pearce J. 1996. 'The village that refused to die'. *New Economics Magazine* 40, pp 8–9 (New Economics Foundation, London)

Perelman, M. 1976. 'Efficiency in agriculture: the economics of energy'. In Merril R (ed) *Radical Agriculture.* Harper and Row, New York

Philippe A. 1974. *Mortalité par suicide en France.* INSERM, quoted in Fairburn D. 1996. *Stress in the Farming Community: Its Effects and Mitigation.* Nuffield Farming Scholarship Trust, Uckfield, East Sussex

Pimbert M and Pretty J N. 1995. *Parks, People and Professionals: Putting 'Participation' into Protected Area Management.* UNRISD Discussion Paper DP 57, United Nations Research Institute for Social Development and WWF International, Geneva

Pimentel D, Acguay H, Biltonen M, Rice P, Silva M, Nelson J, Lipner V, Giordano S, Harowitz A and D'Amore M. 1992. 'Environmental and economic cost of pesticide use'. *Bioscience*, 42 (10), pp 750–60

Pingali P L and Roger P A. 1995. *Impact of Pesticides on Farmers' Health and the Rice Environment.* Kluwer Academic Press

Post J and Terluin I. 1997. 'The changing role of agriculture in rural employment'. In Bollman and Bryden (eds), op cit

Practical Farmers of Iowa. 1995. *Shared Visions.* Report to the W K Kellogg Foundation. By Gary Huber, PFI, Iowa

Prest and Bowen. 1996. 'Vegetable magic'. *Independent on Sunday*, Business. 7 July, pp 1–2

Pretty J N. 1991. 'Farmers' extension practice and technology adaptation: agricultural revolution in 17th–19th century Britain'. *Agriculture and Human Values*, VIII (1 & 2), pp 132–148

Pretty J N. 1995a. *Regenerating Agriculture: Policies and Practice for Sustainability and Self-Reliance.* Earthscan Publications Ltd, London
Pretty J N. 1995b. 'Participatory learning for sustainable agriculture'. *World Development* 23(8), pp 1247–1263
Pretty J N. 1996a. 'A three step framework for agricultural change'. *Pesticides News* 32 (June), pp 6–8.
Pretty J N. 1996b. 'Can sustainable agriculture feed the world?' *Biologist* 43 (3) pp 130–133
Pretty J. 1997. *The Association for Better Land Husbandry in Kenya: The Challenge of Scaling Up and Spreading Success for Project Extension (Years 3–5).* Report to ABLH and DFID, Kenya
Pretty, J N. 1998. *Of Pandora, Progress and Pollution. A Review of Biotechnology and Genetic Engineering.* IIED, London
Pretty J N and Chambers, R. 1993. *Towards a learning paradigm: new professionalism and institutions for sustainable agriculture.* IDS Discussion Paper DP 334. IDS, Brighton
Pretty J N and Howes R. 1993. *Sustainable Agriculture in Britain: Recent Achievements and New Policy Challenges.* Research Series Vol 2, No 1. IIED, London
Pretty J N and Raven H. 1994. *Partnerships for farming and rural land management.* Summary of an ARG national seminar, Royal Geographical Society, 8 June, 1994. IIED and SAFE, London
Pretty J N and Shah P. 1997. 'Making soil and water conservation sustainable: from coercion and control to partnerships and participation'. *Land Degradation and Development*, 8, pp 39–58
Pretty J N and Thompson J. 1996. *Sustainable Agriculture and the Overseas Development Administration.* Report for Natural Resources Policy Advisory Department, DFID (formerly ODA), London
Pretty J N, Thompson J and Hinchcliffe F. 1996. *Sustainable Agriculture: Impacts on Food Production and Challenges for Food Security.* Gatekeeper Series SA 60, International Institute for Environment and Development (IIED), London
Proost J and Matteson P. 1997a. 'Integrated farming in the Netherlands: flirtation or solid change?' *Outlook on Agriculture* 26 (2), pp 87–94
Proost J and Matteson P. 1997b. 'Reducing pesticides use in the Netherlands with stick and carrot'. *Journal of Pesticide Reform* 17 (3), pp 2–8
Pugh J. 1996. 'Crisis on the family farm: a matter of life or death'. In Carruthers and Miller (eds), op cit
Putnam R. 1995. 'Bowling alone: America's declining social capital'. *Journal of Democracy* 6(1), pp 65–78
Putnam R D, with Leonardi R and Nanetti R Y. 1993. *Making Democracy Work: Civic Traditions in Modern Italy.* Princeton University Press, Princeton, New Jersey
Rahnema, M. 1992. 'Participation'. In Sachs, W (ed). *The Development Dictionary.* Zed Books Ltd, London
Ramaprasad V and Ramachandran V. 1989. *Celebrating Awareness.* MYRADA, Bangalore and Foster Parents Plan International, New Delhi
Raven H. 1996. *Supermarkets: powers and responsibilities.* Paper presented at The Vegetable Challenge conference, London, 21 May. The Guild of Food Writers
Raven H and Brownbridge M. 1996. 'Why small farmers?' In Carruthers and Miller (eds), op cit
Raven H and Lang T. 1995. *Off Our Trolleys? Food Retailing and the Hypermarket Economy.* IPPR, London
Raver A. 1997. 'Houses before gardens, the city decides'. *The New York Times*, 8 January
Rawles K and Holland A. 1996. 'Small farm agriculture: ethics and economics'. In Carruthers and Miller (eds), op cit
Rayment M. 1997. *Working with Nature in Britain: Case Studies of Nature Conservation, Employment and Local Economies.* RSPB and Birdlife International, Sandy, Beds
RCEP. 1996. *Sustainable Use of Soil.* 19th report of the Royal Commission on Environmental Pollution. Cmnd 3165, HMSO, London
RDC. 1994. *Lifestyles in Rural England.* Rural Development Commission, Salisbury
RDC. 1996. *Research Finding: the employment impact of changing agricultural policy.* Rural Development Commission, Salisbury
Redman M. 1995. 'A response on integrated crop management'. *Pesticides News* No 30, p 13
Reichelderfer K. 1990. 'Environmental protection and agricultural support: are trade-offs

necessary?' In Allen K (ed). *Agricultural Policies in a New Decade.* Resources for the Future, Washington, DC

Remmers G. 1996. *Hitting a moving target: endogenous development in marginal European areas.* Gatekeeper Series No 63, IIED, London

Rennie F. 1996. 'Sustainable rural development'. In Mitchell K (ed). *The Common Agricultural Policy and Environmental Practices.* Proc. of a seminar organised by European Forum on Nature Conservation and Pastoralism, 24 Jan 1996, Brussels. EFNCP and WWF, Brussels

Repetto R and Baliga S S. 1996. *Pesticides and the Immune System: The Public Health Risks.* WRI, Washington, DC

Rew L J and Cussans G W. 1995. 'Patch ecology and dynamics – how much do we know'. In *Proceedings of the Brighton Crop Protection Conference*, Vol 3, pp 1059–1068

Rhoades, R. 1987. *Farmers and Experimentation.* Agric. Admin (R and E) Network Paper 21, ODI, London

Rhône-Poulenc. 1991–1997. *Annual Reports of Boarded Barns and Bundish Hall Farms.* Ongar, Essex

Robins N and Roberts S. 1997. *Unlocking the Trade Opportunities.* IIED, London and UN Dept of Policy Coordination and Sustainable Development, NY

Robinson D A and Blackman J D. 1990. 'Soil erosion and flooding'. *Land Use Policy* 7, pp 41–52

Robinson J C, Tuden D J and Pease W S. 1996. *Taxing pesticides to fund environmental protection.* Center for Occupational Health, University of California, Berkeley

Rola A and Pingali P. 1993. *Pesticides, Rice Productivity and Farmers – An Economic Assessment.* IRRI, Manila and WRI, Washington, DC

Röling N. 1995. *Towards Interactive Agricultural Science.* Inaugural address at occasion of appointment to Extra-Ordinary Chair in Agricultural Knowledge Systems, 21 September, 1995, Wageningen Agricultural University, The Netherlands

Rorty, R. 1989. *Contingency, Irony and Solidarity.* Cambridge University Press, Cambridge

Rose, Sir J. 1996. 'The farmer and the market'. In Carruthers and Miller (eds), op cit

Rosegrant M W and Agcaolli M. 1994. *Global and regional food demand, supply and trade prospects to 2010.* IFPRI, Washington, DC

Rosenblum G. 1994. 'On the Way to Market: Roadblocks to Reducing Pesticide Use on Produce'. *Public Voice for Food and Health Policy*, Washington, DC

Rosvold M, Ebbesvik M and Kerner K N. 1996. 'Organic agriculture in Norway'. *Ecology and Farming (IFOAM)*, No. 12, pp 17–18

Roux M and Blum A. 1998. 'Developing standards for sustainable farming within the Swiss Agricultural Knowledge system'. In Röling N and Wagemakers M (eds). *Social Learning for Sustainable Agriculture.* Cambridge University Press, Cambridge

Ruddell E. 1997. 'Empowering farmers to conduct experiments'. In van Veldhuizen L, Waters-Bayer A, Ramírez R, Johnson D A and Thompson J (eds). *Farmers' Research in Practice.* IT Publications, London

Rushell M. 1996. *Food first.* From Letslink UK materials, Warminster, Wiltshire

SAFE Alliance. 1997. *How Green are our Strawberries?* SAFE, London

Sasakawa Global 2000. 1993–1995. *Annual Reports.* Sasakawa Africa Association, Tokyo

Scenario 2010 Working Group. 1996. *Vision 2020: Scenarios for a Sustainable Europe.* Scenario 2020 Working Group of the General Consultative Forum to DG XI, European Commission, Brussels

Schaller N. 1993. 'Sustainable agriculture: where do we go now?' *Forum for Applied Research and Public Policy* (Fall), pp 78–82

Schwarz D and Schwarz W. 1998. *Living Lightly: Travels in Post-Consumer Society.* Green Books, Bideford

Senge P. 1990. *The Fifth Discipline: The Art and Practice of the Learning Organisation.* Doubleday-Currency, NY

Silver M. 1993. 'The ultimate barter'. *Mother Earth News*, Aug–Sept, pp 32–33

Singh G S. 1996. 'Pesticides: a growing menace'. Society for Participatory Research in Asia. New Delhi – reported in *Pesticides News* No 35, p 10; *Pesticides News.* 1997. Pesticide facts in Thailand. No 35, p 8

Skauge O. 1996. 'Norway: sustainable development of mountain areas'. *Sustainable Mountain Development. Proceedings of the Scottish Session of the European Intergovernmental Consultation.* The Cairngorms, 22–27 April, The Scottish Office, Edinburgh

Small Farm Viability Project. 1977. *The Family Farm in California*. Governor's Office of Planning and Research, Sacramento, California

Smit J and Ratta A. 1995. *Urban Agriculture: Neglected Resource for Food, Jobs and Sustainable Cities*. UNDP, New York

Social Justice Report, Chapter 7, quoted in Conaty P. 1997. *Rebuilding Working Neighbourhoods – A New Agenda*. NTT, London

Soil Association, 1997. *Counting the Cost of Industrialised Agriculture*. Interim Research Findings, November 1997. Soil Association, Bristol

Somers B. 1998. 'Learning about sustainable agriculture: the case of Dutch arable farmers'. In Röling N G and Wagemakers M A (eds). *Sustainable Agriculture: Participatory Learning and Action*. Cambridge University Press, Cambridge (in press).

Steenland K. 1996. 'Chronic neurological effects of OP pesticides: subclinical damage does occur, but longer follow up needed'. *British Medical Journal*, 25 May

Steiner R, McLaughlin L, Faeth P and Janke R. 1995. 'Incorporating externality costs in productivity measures: a case study using US agriculture'. In Barbett V, Payne R and Steiner R (eds). *Agricultural Sustainability: Environmental and Statistical Considerations*. John Wiley, New York, pp 209–230

Stewart J, Kendall E and Cooke A. 1994. *Citizens' Juries*. IPPR, London

Strathclyde University Study, 1997, quoted in *The Independent*, 21 January 1997

Suárez F, Naveso M and de Juana E. 1997. 'Farming in the drylands of Spain: birds of the pseudosteppes'. In Pain D J and Pienkowski M W (eds). *Farming and Birds in Europe*. Academic Press Ltd, London

Tansey G and Worsley T. 1995. *The Food System*. Earthscan Publications Ltd, London

Taylor M. 1982. *Community, Anarchy and Liberty*. Cambridge University Press, Cambridge

Temperley H. 1996. In *Rural England: The White Paper*. Volume II. Minutes of Evidence and Appendices. House of Commons Select Committee on Environment. HMSO, London, p 18

TEST. 1988. *Trouble in Store: retail locational policy in Britain and Germany*. Transport and Environmental Studies, London

TFPC. 1994. *Reducing urban hunger in Ontario: policy responses to support the transition from food charity to local food security*. TFPC Discussion Paper Series No 1, Toronto, Ontario

ThumbPrint, *passim*. GreenThumb News, Parks and Recreation, City of New York, New York

Tindale S and Holtham G. 1996. *Green Taxes: pollution payments and labour tax cuts*. IPPR Briefing, London

Trapp R. 1995. 'Off with their overheads'. *The Independent on Sunday*, Business, 10 December, London

Trebay G. 1996. 'Uprooted: expulsion from the community gardens'. *The Village Voice*, 10 December 1996

Tucker G M and Heath M F (eds). 1994. *Birds in Europe: Their Conservation Status*. Birdlife Conservation Series No 3. Birdlife International, Sandy, Beds

Tye R. 1997. 'Are water companies bearing an unfair burden?' *Pesticides News* 35, p 18

UK Environmental Health Action Plan. 1995. *Public Consultation document*. DoE and DOH, August 1995

UNICEF. 1993. *Food, Hunger and Care: the UNICEF Vision for a World Free from Hunger*. UNICEF, New York

United Nations Food and Agriculture Organization, Rome. FAO. 1995. *World Agriculture: Toward 2010*. Edited by N. Alexandratos. United Nations Food and Agriculture Organization, Rome

Unnevehr L J. 1993. 'Suburban consumers and exurban farmers: the changing political economy of food policy'. *American Journal of Agricultural Economics* 75, pp 1140–44

USDA. 1997. *Passim*. US Census of Agriculture. Washington, DC at http://www.usda.gov

van Weperen W and Röling N. 1995. 'Integrated arable farming in the Netherlands'. In *Hetveranderingsproces* (The Change Process), Wageningen, The Netherlands

Vereijken P, Wijnands F, Stol W and Visser R. 1995. *Progress Reports 1 and 2*. Research Network on Integrated and Ecological Arable Farming Systems for EU and Associated Countries. DLO Research Institute for Agrobiology and Soil Fertility, Wageningen, The Netherlands

Viaux P. 1995. 'Les systemes intégrés, approche agronomique du développement agricole durable'. *Aménagement et Nature* 117, pp 31–44. ANDA

Viaux P and Rieu C. 1995. *Integrated farming systems and sustainable agriculture in France.* BCPC Symposium Proceedings No 63: *Integrated Crop Protection: Towards Sustainability.* Brighton
Viera M and Eden P. 1995. Portuguese montados. *La Cañada*, No 3 p 5, JNCC, Argyll
Vision 21. 1997. *Sustainable Gloucestershire. A Vision 21 Handbook for Creating a Brighter Future.* Vision 21, Cheltenham
von Meyer H. 1997. 'Rural employment in OECD countries: structure and dynamics of regional labour markets'. In Bollman and Bryden (eds), op cit
Vorley W and Keeney D (eds). 1998. *Bugs in the System: Redesigning the Pesticide Industry for Sustainable Agriculture.* Earthscan Publications Ltd, London
VROM (Ministry of Housing, Physical Planning and Environment). 1989. *To Choose or To Lose: National Environmental Policy Plan.* VROM, The Hague, The Netherlands.
VROM (Ministry of Housing, Physical Planning and Environment). 1990. *National Environmental Policy Plan Plus.* VROM, The Hague, The Netherlands.
Waldegrave W. House of Commons Questions, 18 April 1995
Ward H. 1997. *Citizens' juries and valuing the environment.* Paper presented to Green Circle, University of Essex, 12 November, 1997
Ward, N. 1996. 'Environmental concern and the decline of dynastic family farming'. In Carruthers and Miller (eds), op cit
Wates N. 1996. *Action Planning.* The Prince of Wales's Institute of Architecture, London
Weisbord M and Janoff S. 1995. *Future Search.* Philadelphia
Weissman J (ed). 1995. *City Farmers: Tales from the Field.* GreenThumb, New York
Weissman J (ed). 1996. *Tales from the Field. Stories by GreenThumb Gardeners.* GT, NY
Whitehead A and Smyth J. 1996. 'Creating Social Capital'. *Demos Quarterly* 9, pp 36–17
Wijnands F G. 1992. 'Evaluation and introduction of integrated farming in practice'. *Netherlands Journal of Agricultural Science* 40, pp 239–249
Wijnands F G, van Asperen P, van Dongen G J M, Janssens S R M, Schröder J J and van Bon K B. 1995. *Pilot Farms and Integrated Arable Farming: Technical and Economical Results.* PAGV verslag No. 196. Leystad, The Netherlands
Wilson G. 1995. 'German agri-environmental schemes II. The MEKA programme in Baden-Württemburg'. *Journal of Rural Studies* 11(2), pp 149–159
Wilson, G. 1997. 'Assessing the environmental impact of the Environmentally Sensitive Areas scheme: a case for using farmers' environmental knowledge?' *Landscape Research* 22 (3), pp 303–326
Wilson J. 1993. 'The BTO Birds and Organic Farming Project'. *New Farmer and Grower*, Spring pp 51–53
Wise C. 1996. 'Defending ICM – NFU response'. *Pesticides News* 31, pp 16–17
World Bank. 1993. *Agricultural Sector Review.* Agriculture and Natural Resources Department, Washington, DC
World Bank. 1994. *The World Bank and Participation.* Report of the Learning Group on Participatory Development. April 1994. World Bank, Washington, DC
World Commission on Environment and Development. 1997. *Our Common Future.* OUP, Oxford
World Neighbors, reported in Hinchcliffe et al, 1996, op cit
WRI. 1994. *World Resources 1994–95.* World Resources Institute, Washington. Oxford University Press, Oxford
Wright S. 1992. 'Rural community development'. *Journal of Rural Studies* 8(1), pp 15–28
WWF. 1997. *At Cross Purposes: How EU Policy Conflicts Undermine the Environment.* WWF, Godalming
Yellachich N. 1993. *Towards Sustainable Systems: Critical Analysis of the Agri-Environment Measures in Alentejo, Portugal.* MSc thesis. Centre for Environmental Technology, Imperial College, London
Young A. 1804. *A General Review of the Agriculture of Norfolk.* Board of Agriculture, London
Zilberman D, Schmiyz A, Casterline G, Lichtenben E and Siebert J. 1991. 'The economics of pesticide use and regulation'. *Science*, 253 (5019), pp 519–523

Index

UNIVERSITY OF GLAMORGAN
PRIFYSGOL MORGANNWG
Learning Resources Centre